Pests of Ornamental Trees, Shrubs and Flowers

A Colour Handbook

Second Edition

David V Alford

BSc, PhD
Formerly Regional Entomologist and
Head of the Entomology Department
Ministry of Agriculture, Fisheries & Food
Cambridge, UK

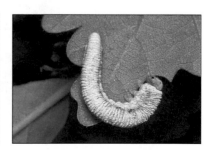

CRC Press
Taylor & Francis Group
Boca Raton London New York

CRC Press is an imprint of the
Taylor & Francis Group, an **informa** business

CRC Press
Taylor & Francis Group
6000 Broken Sound Parkway NW, Suite 300
Boca Raton, FL 33487-2742

First issued in paperback 2017

© 2012 by Taylor & Francis Group, LLC
CRC Press is an imprint of Taylor & Francis Group, an Informa business

No claim to original U.S. Government works

ISBN 13: 978-1-138-03406-8 (pbk)
ISBN 13: 978-1-84076-162-7 (hbk)

**Visit the Taylor & Francis Web site at
http://www.taylorandfrancis.com**

**and the CRC Press Web site at
http://www.crcpress.com**

A CIP catalogue record for this book is available from the British Library.

Commissioning editor: Jill Northcott
Project manager: Paul Bennett
Layout: DiacriTech, Chennai, India
Colour reproduction: Tenon & Polert Colour Scanning Ltd, Hong Kong

Plant Protection Handbooks Series
Alford: *Pests of Fruit Crops – A Colour Handbook*
Alford: *Pests of Ornamental Trees, Shrubs and Flowers, Second Edition – A Colour Handbook*
Biddle & Cattlin: *Pests and Diseases of Peas and Beans – A Colour Handbook*
Blancard: *Cucurbit Diseases – A Colour Atlas*
Blancard: *Tomato Diseases – A Colour Atlas*
Blancard: *Diseases of Lettuce and Related Salad Crops – A Colour Atlas*
Fletcher & Gaze: *Mushroom Pest and Disease Control – A Colour Handbook*
Helyer *et al*: *Biological Control in Plant Protection – A Colour Handbook*
Koike *et al*: *Vegetable Diseases – A Colour Handbook*
Murray *et al*: *Diseases of Small Grain Cereal Crops – A Colour Handbook*
Wale *et al*: *Pests & Diseases of Potatoes – A Colour Handbook*

Contents

Contents

To Iona and Xakiera

Preface

Ornamental trees, shrubs and flowers are important components of modern life, lending great attraction to our domestic, leisure and working environments. The market for ornamentals is, therefore, very wide, and includes a demand for alpines, bedding plants, cacti, cut-flowers, house plants and pot plants, as well as herbaceous plants, ornamental grasses, shrubs and trees for gardens, parks and other amenity areas.

The care and cultivation of ornamental plants often leads to the discovery of pests or pest damage which, unless correctly diagnosed, controlled or managed, may prove disastrous. This book provides a means of recognizing the various pests associated with ornamental plants in the British Isles and much of mainland Europe (particularly from the Alps northwards). Biological details and information on the importance of such pests are also given. Details of pest lifecycles (including the relative abundance and significance of particular pests, and the number of generations completed in a season) often vary according to local conditions, not least due to geographical differences in climate and other factors. Therefore, where cited, dates (months) for the occurrence of the various stages of pests must be taken merely as a general guide.

Many of the pests included here as members of the European fauna also occur in other parts of the world, including North America, Asia and Australasia, several as accidental introductions from Europe. The European fauna too is constantly being 'enriched' by alien species arriving from abroad. This is particularly so since international trade in ornamental plants has become a major industry, with an ever-increasing demand for novel and exotic products to be imported. The likelihood of non-indigenous pests gaining a foothold in a new country is lessened by stringent plant health regulations and inspections, but the risk can never be entirely eliminated or total success claimed. Attempts to eradicate alien pests have also not always proved successful – tobacco whitefly and western flower thrips are relatively recent examples of aliens that have beaten European pest eradication programmes. Plants or plant products despatched to Europe from subtropical or tropical countries pose a particular threat, many consignments having to be rejected or destroyed because of the presence of alien pests, particularly aphids, leaf miners, mites, scale insects, thrips and whiteflies. Global warming is also a factor, increasing the likelihood of non-indigenous and invasive pests becoming established in new areas, either following their accidental introduction or as a result of an expansion in their natural range.

It would be impractical to include details of every pest likely to occur on ornamental trees, shrubs and flowers. However, information is provided for those most commonly causing damage and those that, although of little or no economic importance, are often noticed upon such plants (perhaps because of their large size, or colourful or unusual appearance). Details are also given of various non-indigenous pests, with emphasis on those which have become temporarily if not permanently established in Europe. Several forestry pests that cause damage to ornamental trees and shrubs are also mentioned. However, limitations of space have largely precluded consideration of wood-boring insects that, although sometimes attacking ornamental trees and shrubs, are primarily of significance to the timber industry.

Largely in response to differences in regional if not national guidelines, and moves within the industry towards integrated pest management (IPM) strategies, specific details of pest control measures have been excluded. Readers requiring such information are encouraged to consult up-to-date, often annually revised, literature available from national or local pest management services.

Various other changes have been made since the first edition. In particular, for ease of reference, genera within families are now listed alphabetically, rather than in their recognized systematic sequences. Subfamilies are then ignored, except in the case of aphids (family Aphididae). Names of authorities for species are retained in full, but those for species embedded within the text (i.e. those for species not delineated as headings) have been moved to the General Index.

David V Alford
Cambridge

Acknowledgements

In compiling this and the earlier account, I am grateful to many people for their help in various ways. Particular thanks are due to Dr D. J. L. Agassiz, R. W. Brown, Dr J. H. Buxton, C. I. Carter, D. J. Carter, R. Coutin, J. V. Cross, G. R. Ellis, B. J. Emmett, A. Fraval, the late C. Furk, Miss M. Gratwick, A. J. Halstead, N. J. Hurford, A. W. Jackson, M. J. Lole, D. MacFarlane, H. Riedel, the late Dr G. Rimpel, Dr M Saynor, P. R. Seymour, S. J. Tones, R. A. Umpelby, F. Wellnitz and Dr K. B. Wildey. Finally, I acknowledge the invaluable and continued help and encouragement of my wife and family.

Chapter 1

Introduction

Ornamental plants are attacked by a wide range of pests, most of which are arthropods (phylum Arthropoda). Arthropods are a major group of invertebrate animals, characterized by their hard exoskeleton or body shell, segmented bodies and jointed limbs. Insects and, to a lesser extent, mites are of greatest significance as pests of cultivated plants.

Insects

Insects differ from other arthropods in possessing just three pairs of legs, usually one or two pairs of wings (all winged invertebrates are insects), and by having the body divided into three distinct regions: head, thorax and abdomen.

The outer skin or integument of an insect is known as the cuticle. It forms a non-cellular, waterproof layer over the body, and is composed of chitin and protein, the precise chemical composition and thickness determining its hardness and rigidity. The cuticle has three layers (epicuticle, exocuticle and endocuticle) and is secreted by an inner lining of cells which forms the hypodermis or basement membrane. When first produced the cuticle is elastic and flexible, but soon after deposition it usually undergoes a period of hardening or sclerotization and becomes darkened by the addition of a chemical called melanin. The adult cuticle is not replaceable, except in certain primitive insects. However, at intervals during the growth of the immature stages, the 'old' hardened cuticle becomes too tight and is replaced by a new, initially expandable one secreted from below. Certain insecticides have been developed that are capable of disrupting chitin deposition. Although ineffective against adults, they kill insects undergoing ecdysis (i.e. those moulting from one growth stage to the next).

The insect cuticle is often thrown into ridges and depressions, is frequently sculptured or distinctly coloured, and may bear a variety of spines and hairs. In larvae, body hairs often arise from hardened, spot-like pinacula (often called tubercles) or larger, wart-like verrucae. In some groups, features of the adult cuticle (such as colour, sculpturing and texture) are of considerable value for distinguishing between species.

The basic body segment of an insect is divided into four sectors (a dorsal tergum, a ventral sternum and two lateral pleurons) which often form horny, chitinized plates called sclerites. These may give the body an armour-like appearance, and are either fused rigidly together or joined by soft, flexible, chitinized membranes to allow for body movement. Segmental appendages, such as legs, are developed as outgrowths from the pleurons.

The head is composed of six fused body segments, and carries a pair of sensory antennae, eyes and mouthparts. The form of an insect antenna varies considerably, the number of antennal segments (strictly speaking these are not segments) ranging from one to more than a hundred. The basal segment is called the scape and is often distinctly longer than the rest; the second segment is the pedicel and from this arises the many-segmented flagellum. In geniculate antennae, the pedicel acts as the articulating joint between the elongated scape and the flagellum; such antennae are characteristic of certain weevils, bees and wasps. Many insects possess two compound eyes, each composed of several thousand facets, and three simple eyes called ocelli, the latter usually forming a triangle at the top of the head. Compound eyes are large, and particularly well developed in predatory insects, where good vision is important. The compound eye provides a mosaic, rather than a clear picture, but is well able to detect movement. The ocelli are optically simple and lack a focusing mechanism. They are concerned mainly with registering light intensity, enabling the insect to distinguish between light and shade. Unlike insect nymphs, insect larvae lack compound eyes but they often possess several ocelli, arranged in clusters on each side of the head. Insect mouthparts are derived from several modified, paired segmental appendages; they range from simple biting jaws (mandibles) to complex structures for piercing, sucking or lapping. Among phytophagous insects, biting mouthparts are found in adult and immature grasshoppers, locusts, earwigs, beetles etc., but may be restricted to the larval stages, as in butterflies, moths and sawflies. Some insects (e.g. various dipterous larvae) have rasping mouthparts which are used to tear plant tissue, the food material then being ingested in a semi-liquid state. Style-like, suctorial mouthparts are characteristic of aphids, mirids, psyllids and other bugs; such insects may introduce toxic saliva into plants and cause distortion or galling of tissue. Certain insects (notably some aphids) carry and transmit virus diseases to host plants.

The thorax has three segments – prothorax, mesothorax and metathorax – whose relative sizes vary from one insect group to the next. In crickets, cockroaches and beetles, for example, the prothorax is the largest section and is covered on its upper surface by an expanded dorsal sclerite called the pronotum; in flies, the mesothorax is greatly enlarged, and the prothoracic and metathoracic segments are much reduced. Typically, each thoracic segment bears a pair of jointed legs. Their form varies considerably but all legs have the same basic structure. Wings, when present, arise from the mesothorax and metathorax as a pair of fore wings and hind wings, respectively. In many insects the base of each fore wing is covered by a scale-like lobe, known as the tegula. Basically, each wing is an expanded membrane-like structure supported by a series of hardened veins, but considerable modification has taken place in the various insect groups. In cockroaches, earwigs and beetles, for instance, the fore wings have become hardened and thickened protective flaps, called elytra, and only the hind wings are used for flying; in true flies, the fore wings retain their propulsive function but the hind wings have become greatly reduced in size and are modified into drumstick-like balancing organs known as halteres. Wing structure is of importance in the classification of insects, and the names of many insect orders are based upon it. Wing venation is also of considerable significance.

The abdomen is normally formed from 10 or 11 segments, but fusion and apparent reduction of the most anterior or posterior segments are common. Although present in many larvae, abdominal appendages are wanting on most segments of adults, their ambulatory function, as found in various other arthropods, having been lost. However, appendages on the eighth and ninth segments remain to form the genitalia, including the male claspers and female ovipositor. In many groups,

cerci are formed from a pair of appendages on the last body segment. These are particularly long and noticeable in primitive insects, but are absent in the most advanced groups. Abdominal sclerites are limited to a series of dorsal tergites and a set of ventral sternites; these give the abdomen a distinctly segmented appearance.

The body cavity of an insect extends into the appendages and is filled with a more or less colourless, blood-like fluid called haemolymph. This bathes all the internal organs and tissues, and is circulated by muscular action of the body and by a primitive, tube-like heart which extends mid-dorsally from the head to near the tip of the abdomen.

The brain is the main co-ordinating centre of the body. It fills much of the head and is intimately linked to the antennae, the compound eyes and the mouthparts. The brain gives rise to a central nerve cord which extends back mid-ventrally through the various thoracic and abdominal segments. The nerve cord is swollen at intervals into a series of ganglia, from which arise numerous lateral nerves. These ganglia control many nervous functions (such as movement of the body appendages) independent of the brain.

The gut or alimentary tract is a long, much modified tube stretching from the mouth to the anus. It is subdivided into three sections: a fore gut, with a long oesophagus and a bulbous crop; a mid-gut, where digestion of food and absorption of nutriment occurs; and a hind gut, concerned with water absorption, excretion and the storage of waste matter prior to its disposal. A large number of blind-ending, much convoluted Malpighian tubules arise from the junction between the mid-gut and the hind gut. These tubules collect waste products from within the body and pass them into the gut.

The respiratory system comprises a complex series of branching tubes (tracheae) and microscopic tubules (tracheoles) which ramify throughout the body in contact with the internal organs and tissues. This tracheal system opens to the outside through segmentally arranged valve-like breathing holes or spiracles, present along each side of the body. Air is forced through the spiracles by contraction and relaxation of the abdominal body muscles. Spiracles also occur in nymphs and larvae (they are often very obvious in butterfly and moth caterpillars) but they are often much reduced in number. In some groups (e.g. various fly larvae) the tracheae open via a pair of anterior spiracles, commonly located on the first body segment (prothorax), and a pair of posterior spiracles, usually located on the anal segment; these spiracles are often borne on raised processes. Morphological details of the spiracular openings and processes are often used to distinguish between species (as in agromyzid leaf miners).

Female insects possess a pair of ovaries, subdivided into several egg-forming filaments called ovarioles. The ovaries enter a median oviduct and this opens to the outside on the ninth abdominal segment. Many insects have a protrusible egg-laying tube, called an ovipositor. The male reproductive system includes a pair of testes and associated ducts which lead to a seminal vesicle in which sperm is stored prior to mating. The male genitalia may include chitinized structures, such as claspers which help to grasp the female during copulation. Examination of the male or female genitalia is often essential for distinguishing between closely related species.

Sexual reproduction is common in insects, but in certain groups fertilized eggs produce only female offspring and males are reared only from unfertilized ones. In other cases, male production may be wanting or extremely rare and parthenogenesis (reproduction without a sexual phase) is the rule.

Although some insects are viviparous (giving birth to active young), most lay eggs. A few, such as aphids, reproduce viviparously by parthenogenesis in spring and summer but may produce eggs in the autumn (after a sexual phase). Insect eggs have a waterproof shell. Many are capable of withstanding severe winter conditions on tree bark or shoots, and are the means whereby many insects survive from one year to the next.

Insects normally grow only during the period of pre-adult development, as nymphs or larvae, their outer cuticular skin being moulted and replaced between each successive growth stage or instar. The most primitive insects (subclass Apterygota) have wingless adults, and their eggs hatch into nymphs that are essentially similar to adults but smaller and sexually immature. The more advanced, winged or secondarily wingless, insects (subclass Pterygota) develop in one of two ways. In some, there is a succession of nymphal stages in which wings (when present in the adult) typically develop as external wing buds that become fully formed and functional once the adult stage is reached. In such insects, nymphs and adults are frequently of similar appearance (apart from wing buds or wings), and often share the same feeding habits. This type of development, in which metamorphosis is incomplete, is termed hemimetabolous. The most advanced insects show complete metamorphosis, development (termed holometabolous) including several larval instars of quite different structure and habit from the adult. Here, wings develop internally and the transformation from larval to adult form occurs during a quiescent, non-feeding pupal stage. Insect larvae are of various kinds. Some, commonly called caterpillars, have three pairs of jointed thoracic legs (true legs) and a number of fleshy, false legs (prolegs) on the abdomen. Many butterfly, moth and sawfly larvae are of this type. Unlike sawfly larvae, those of butterflies and moths are usually provided with small chitinous hooks known as crotchets. Some larvae, including many beetle grubs, possess well-developed thoracic legs but lack abdominal prolegs. In other insect larvae, legs are totally absent; fly and various wasp larvae are examples.

Classification of insects

Class **INSECTA**	
Subclass APTERYGOTA	
Order Thysanura	Bristle-tails, silverfish
Order Diplura	Diplurans
Order Protura	Proturans
Order Collembola	Springtails
Subclass PTERYGOTA	
Order Ephemeroptera	Mayflies
Order Odonata	Dragonflies
Order Plecoptera	Stoneflies
Order Grylloblattodea	Grylloblattodeans
Order Orthoptera	Crickets, grasshoppers
Order Phasmida	Stick-insects, leaf-insects
Order Dermaptera	Earwigs
Order Embioptera	Web-spinners
Order Dictyoptera	Cockroaches, mantids
Order Isoptera	Termites
Order Zoraptera	Zorapterans
Order Psocoptera	Psocids or booklice
Order Mallophaga	Biting lice
Order Anoplura	Sucking lice
Order Hemiptera	True bugs
Order Thysanoptera	Thrips
Order Neuroptera	Alder flies, lacewings, etc.
Order Coleoptera	Beetles
Order Strepsiptera	Stylopids
Order Mecoptera	Scorpion flies
Order Siphonaptera	Fleas
Order Diptera	True flies
Order Lepidoptera	Butterflies, moths
Order Trichoptera	Caddis flies
Order Hymenoptera	Ants, bees, wasps, etc.

Pests of ornamental plants are found in many different orders, the main groups being characterized as follows:

Collembola: small, wingless insects, often with a forked springing organ ventrally on the fourth abdominal segment; biting mouthparts; antennae usually with four segments; metamorphosis slight: *family Sminthuridae* (p. 20); *family Onychiuridae* (p. 20).

Orthoptera: medium-sized to large, stout-bodied insects with a large head, large pronotum and usually two pairs of wings, the thickened fore wings termed tegmina; fore wings or hind wings reduced or absent; femur of hind leg often modified for jumping; tarsi usually 3- or 4-segmented; chewing mouthparts; cerci usually short and unsegmented; development hemimetabolous, including egg and several nymphal stages: *family Gryllotalpidae* (p. 21); *family Gryllidae* (p. 21).

Dermaptera: elongate, omnivorous insects with biting mouthparts; fore wings modified into very short, leathery elytra; hind wings semicircular and membranous, with radial venation; anal cerci modified into pincers; development hemimetabolous, including egg and several nymphal stages: *family Forficulidae* (p. 22).

Dictyoptera: small to large, stout-bodied but rather flattened insects with a large pronotum and two pairs of wings, the thickened fore wings called tegmina; hind wings folded longitudinally like a fan; chewing mouthparts; antennae very long and thread-like; legs robust and spinose, and modified for running; tarsi usually 3- or 4-segmented; cerci many-segmented; development hemimetabolous, including egg and several nymphal stages: *family Blattidae* (p. 23).

Hemiptera: minute to large insects, usually with two pairs of wings and piercing, suctorial mouthparts; fore wings frequently partly or entirely hardened; development hemimetabolous, including egg and several nymphal stages (the egg stage often omitted).

   Suborder Heteroptera: usually with two pairs of wings, the fore wings (termed hemelytra) with a horny basal area and a membranous tip; hind wings membranous; wings held flat over the abdomen when in repose; the beak-like mouthparts arise from the front of the head and are flexibly attached; prothorax large; some species are phytophagous but many are predacious: *family Tingidae* (p. 24); *family Miridae* (p. 26).

   Suborder Auchenorrhyncha: wings (when present) typically held over the body in a sloping, roof like posture; fore wings (termed elytra) uniform throughout and horny; hind wings membranous; mouthparts arising from the base of the head and the point of attachment rigid; entirely phytophagous. *Superfamily Cercopoidea* – tegulae absent; hind legs modified for jumping, with long tibiae bearing one or two long spines: *family Cercopidae* (p. 28). *Superfamily Fulgoroidea* – elytra with anal vein Y-shaped; antennae 3-segmented: *family Flatidae* (p. 30). *Superfamily Cicadelloidea* – tegulae absent; hind legs modified for jumping, with long tibiae bearing longitudinal rows of short spines: *family Cicadellidae* (p. 31).

   Suborder Sternorrhyncha: wings (when present) typically held over the body in a sloping, roof like posture; fore wings and hind wings membranous and uniform throughout; mouthparts arising from a rearward position relative to the head and the point of attachment rigid; entirely phytophagous. *Superfamily Psylloidea* – antennae usually 10-segmented; tarsi 2-segmented and with a pair of claws: *family Psyllidae* (p. 36); *family Triozidae* (p. 41); *family Carsidaridae* (p. 43); *family Spondyliaspidae* (p. 44). *Superfamily Aleyrodoidea* – antennae 7-segmented; wings opaque and coated in whitish wax: *family Aleyrodidae* (p. 45). *Superfamily Aphidoidea* – females winged or wingless; wings, when present, usually large and transparent, with few veins; abdomen often with a pair of siphunculi: *family Aphididae* (p. 49); *family Adelgidae* (p. 95); *family Phylloxeridae* (p. 100). *Superfamily Coccoidea* – females always wingless; males usually with a single pair of wings and vestigial mouthparts, and developing through a pupal stage; tarsi, if present, 1-segmented and with a single claw: *family Diaspididae* (p. 101); *family Coccidae* (p. 107); *family Eriococcidae* (p. 115); *family Pseudococcidae* (p. 116); *family Margarodidae* (p. 118).

Thysanoptera: small or minute, slender-bodied insects with short antennae and asymmetrical, piercing and sucking mouthparts; a protrusible bladder at the tip of each tarsus; wings, when present, very narrow with hair-like fringes and greatly reduced venation. Nymphs are similar in appearance to adults but are wingless.

Development includes an egg, two nymphal and two or three inactive stages (termed propupae and pupae), and is intermediate between that of hemimetabolous and holometabolous insects: *family Thripidae* (p. 119); *family Phlaeothripidae* (p. 124).

Coleoptera: minute to large insects with biting mouthparts; fore wings modified into horny elytra which usually meet in a straight line along the back; hind wings membranous and folded beneath the elytra when in repose, but often reduced or absent; prothorax normally large and mobile. Development holometabolous, including egg, larval and pupal stages. The largest insect order, with more than a quarter of a million species worldwide.

Superfamily Scarabaeoidea – a large group of often very large, brightly coloured insects, some of which possess enlarged horns on the head and thorax: *family Scarabaeidae* (p. 125). *Superfamily Buprestoidea* – minute to medium-sized, shiny, metallic beetles, with the head sunk into the thorax, eyes very large and antennae short and toothed: *family Buprestidae* (p. 128). *Superfamily Elateroidea* – elongate beetles with a hard exoskeleton, the head sunk into the prothorax, antennae toothed or comb-like, and hind angles of the prothorax sharply pointed and often extended: *family Elateridae* (p. 129). *Superfamily Cucujoidea* – beetles usually with five visible abdominal segments, and antennae often clubbed: *family Nitidulidae* (p. 130); *family Byturidae* (p. 130). *Superfamily Chrysomeloidea* – mostly phytophagous beetles with 4-segmented tarsi (the fourth segment very small), and larvae usually with well-developed thoracic legs: *family Cerambycidae* (p. 130); *family Chrysomelidae* (p. 132). *Superfamily Curculionoidea* – a very large group, including weevils and bark beetles. Antennae are typically clubbed and usually geniculate, with a long basal segment (scape); however, in some all antennal segments are of a similar length. Larvae are usually apodous: *family Rhynchitidae* (p. 148); *family Attelabidae* (p. 148); *family Apionidae* (p. 151); *family Curculionidae* (p. 152).

Diptera: minute to large insects with a single pair of membranous wings; hind wings reduced to small, drumstick-like balancing organs (halteres); mouthparts suctorial and sometimes adapted for piercing. Larvae apodous, and usually with a reduced retractile head.

Development holometabolous, including egg, larval and pupal stages.

Suborder Nematocera: antennae of adults with a scape, pedicel and flagellum, the flagellum comprising numerous similar-looking segments, each bearing a whorl of hairs. Larvae usually (not in the Cecidomyiidae) with a well-defined head and horizontally opposed mandibles: *family Tipulidae* (p. 170); *family Bibionidae* (p. 172); *family Chironomidae* (p. 173); *family Sciaridae* (p. 173); *family Cecidomyiidae* (p. 174).

Suborder Cyclorrapha: antennae of adults with a scape, pedicel and flagellum, the flagellum usually forming an enlarged, compound segment tipped by a short, bristle-like arista. Larvae are maggot-like, often tapering anteriorly; they possess distinctive, rasping 'mouth-hooks', but the head is small and inconspicuous; pupation occurs within the last larval skin, which then forms a protective barrel-like puparium from which the adult eventually escapes by forcing off a circular cap (the operculum): *family Syrphidae* (p. 189); *family Tephritidae* (p. 192); *family Psilidae* (p. 193); *family Ephydridae* (p. 194); *family Drosophilidae* (p. 194); *family Agromyzidae* (p. 195); *family Anthomyiidae* (p. 206).

Lepidoptera: minute to large insects with two pairs of membranous wings; cross-veins few in number; body, wings and appendages scale-covered; adult mouthparts suctorial but those of larvae adapted for biting; the larvae are mainly caterpillar-like and phytophagous. Development holometabolous, including egg, larval and pupal stages.

Superfamily Eriocranioidea – adults with a short proboscis; females with a piercing ovipositor; pupae with functional mandibles: *family Eriocraniidae* (p. 208). *Superfamily Hepialoidea* – adults with non-functional, vestigial mouthparts and short antennae: *family Hepialidae* (p. 210); *Superfamily Nepticuloidea* – adults with wing venation reduced; ovipositor soft: *family Nepticulidae* (p. 212); *family Tischeriidae* (p. 214). *Superfamily Incurvarioidea* – small, day-flying moths, with antennae of males often very long: *family Incurvariidae* (p. 215). *Superfamily Cossoidea* – heavy-bodied moths with a primitive wing venation: *family Cossidae* (p. 216); *family Castniidae* (p. 217). *Superfamily Zygaenoidea* –

a group of moths with complete venation but some rudimentary features: *family Zygaenidae* (p. 218). **Superfamily Tineoidea** – primitive moths with narrow or very narrow wings: *family Lyonetiidae* (p. 219); *family Hieroxestidae* (p. 222); *family Gracillariidae* (p. 223); *family Phyllocnistidae* (p. 234). **Superfamily Yponomeutoidea** – an indistinct and rather diverse group: *family Sesiidae* (p. 235); *family Choreutidae* (p. 236); *family Yponomeutidae* (p. 237). **Superfamily Gelechioidea** – a large group of moderately small moths: *family Coleophoridae* (p. 246); *family Oecophoridae* (p. 248); *family Gelechiidae* (p. 251); *family Blastobasidae* (p. 253). **Superfamily Tortricoidea** – a major group of moderately small moths with mainly rectangular fore wings, and mainly leaf-folding or leaf-rolling larvae: *family Tortricidae* (p. 254). **Superfamily Pyraloidea** – a very large group of mainly slender-bodied, long-legged moths, often with narrow, elongate fore wings: *family Pyralidae* (p. 286). **Superfamily Papilionoidea** – adults diurnal, and with clubbed but terminally unhooked antennae: *family Pieridae* (p. 290); *family Lycaenidae* (p. 291). **Superfamily Bombycoidea** – often large to very large moths, with non-functional mouthparts; male antennae strongly bipectinate: *family Lasiocampidae* (p. 291). **Superfamily Geometroidea** – mainly slender-bodied moths with broad wings; larvae with a reduced number of abdominal prolegs: *family Thyatiridae* (p. 294); *family Geometridae* (p. 294). **Superfamily Sphingoidea** – large-bodied, strong-flying moths, often with a large proboscis; larvae usually possess a characteristic dorsal horn on the eighth abdominal segment: *family Sphingidae* (p. 311). **Superfamily Notodontoidea** – a small group of heavily bodied, mainly dull-coloured moths with elongated wings, sometimes included within the Noctuoidea: *family Notodontidae* (p. 315); *family Dilobidae* (p. 318); *family Thaumetopoeidae* (p. 318). **Superfamily Noctuoidea** – the largest group of lepidopterous insects, with a wide variety of forms: *family Lymantriidae* (p. 320); *family Arctiidae* (p. 325); *family Noctuidae* (p. 330).

Trichoptera: small, medium to large insects, with two pairs of wings which are held in a roof-like position when in repose; wings with few cross-veins and coated with small, inconspicuous hairs. Larvae have biting mouthparts and are omnivorous; they live submerged in water, most species occupying characteristic cases constructed from silk and pieces of vegetation or grains of sand. Development holometabolous, including egg, larval and pupal stages: *family Limnephilidae* (p. 354).

Hymenoptera: minute to large insects with, usually, two pairs of membranous wings, the hind wings smaller and interlocked with the fore wings by small hooks; mouthparts adapted for biting but often also for lapping and sucking; females possess an ovipositor, modified for sawing, piercing or stinging. Development holometabolous, including egg, larval and pupal stages.

Suborder Symphyta: includes sawflies, insects with a well-developed ovipositor, and the abdomen and thorax joined without a constriction or 'waist'; larvae are mainly caterpillar-like and phytophagous. **Superfamily Megalodontoidea** – a small group of primitive sawflies, with a flattened abdomen: *family Pamphiliidae* (p. 355). **Superfamily Tenthredinoidea** – the main group of sawflies, adults with a saw-like ovipositor: *family Argidae* (p. 356); *family Cimbicidae* (p. 359); *family Diprionidae* (p. 360); *family Tenthredinidae* (p. 361).

Suborder Apocrita: the main group of hymenopterous insects, the first abdominal segment being fused to the thorax and separated from the rest of the abdomen (known as the gaster) by a wasp-like 'waist'. The suborder is composed of two groups: the *Parasitica* (most of which are parasitic and have the ovipositor adapted for piercing their hosts) and the *Aculeata* (in which the ovipositor is modified into a sting). **Superfamily Cynipoidea** – minute or very small, mainly black insects, including gall wasps: *family Cynipidae* (p. 393). **Superfamily Chalcidoidea** – minute or very small, often metallic-looking insects, most of which are parasitoids or hyperparasitoids: *family Eulophidae* (p. 402); *family Eurytomidae* (p. 402). **Superfamily Scolioidea** – a large, primitive group of Aculeates, including ants: *family Formicidae* (p. 403). **Superfamily Vespoidea** – social wasps, the pronotum extending back to the tegulae; larvae are fed on meat: *family Vespidae* (p. 403). **Superfamily Apoidea** – generally hairy insects (solitary bees or social bees), with broad hind tarsi and the pronotum not extending back to the tegulae; larvae are fed on nectar and pollen: *family Andrenidae* (p. 403); *family Megachilidae* (p. 404).

Mites

Mites and ticks (the subclass Acari) form part of the Arachnida, a major class of arthropods. Unlike insects, they have no antennae, wings or compound eyes, are usually 8-legged and possess chelicerate mouthparts adapted for biting or piercing. The body is composed of a gnathosoma, which bears a pair of sensory palps (pedipalps) and paired chelicerae, and a sac-like idiosoma with no obvious segmentation; they are thus readily distinguished from other arachnids that have the body divided into a distinct cephalothorax and a usually (but not spiders) clearly segmented opisthosoma. The respiratory system in the Acari often includes a pair of breathing pores, also known as stigmata; their position on the body, or their absence, forms a basic character for naming the various acarine orders.

Unlike members of other arachnid groups, the Acari includes many phytophagous species, mainly in the order Prostigmata. The chelicerae of most prostigmatid mites are needle-like and are used to penetrate plant cells; also, the idiosoma is subdivided by a subjugal furrow into the propodosoma and the hysterosoma, each region bearing two pairs of legs. The body and limbs of a mite are adorned by various setae, the arrangement and characteristics of which are of considerable value in the classification and identification of species.

Development from egg to adult usually includes a six-legged larva and two or three eight-legged nymphal stages: proto-, deuto- and tritonymphs. Larvae and nymphs are generally similar in appearance and habit to adults but are smaller and sexually immature.

Many phytophagous mites are free-living but others (notably members of the Eriophyoidea) inhabit distinctive galls formed in response to toxic saliva injected into host plants during feeding. A few species are important vectors of plant virus diseases.

Classification of mites

Subclass **ACARI**
Superorder OPILIOACARIFORMES
    Order **Notostigmata** — weakly sclerotized, brightly coloured, harvestman-like mites, with four pairs of hysterosomal stigmata
Superorder PARASITIFORMES
    Order **Tetrastigmata** — a small group of large, heavily sclerotized mites
    Order **Mesostigmata** — a large, diverse group of mites, including many predacious species
    Order **Metastigmata** — ticks
Superorder ACARIFORMES
    Order **Prostigmata** — an extremely varied group of minute to large, usually lightly sclerotized mites, including many phytophagous species
    Order **Astigmata** — soft-bodied, lightly sclerotized mites without stigmata
    Order **Cryptostigmata** — dark, strongly sclerotized mites, often known as oribatids or 'beetle mites'

Brief details of the main groups containing pests of ornamental plants are given below:

Prostigmata: mites with the stigmata placed between the chelicerae, and often with one or two pairs of sensory hairs (trichobothria) on the propodosoma.

    Superfamily Eriophyoidea: minute, sausage-shaped or pear-shaped mites with two pairs of legs, each leg terminating in a branched feather-claw; body with a distinct prodorsal shield; hysterosoma annulated with a dorsal series of tergites and a ventral series of sternites.

Family Phytoptidae (p. 405); prodorsal shield bearing three or four setae; feather-claw simple: genus *Phytoptus*.

Family Eriophyidae (p. 406); similar to the Phytoptidae but prodorsal shield bearing two or no setae; feather-claw either simple or divided. *Subfamily Cecidophyinae* – elongate mites without prodorsal shield setae: genus *Cecidophyopsis*. *Subfamily Eriophyinae* –

elongate, worm-like mites with a pair of setae on the prodorsal shield; hysterosoma subdivided into numerous tergites and sternites, typically subequal anteriorly: genera *Acalitus, Aceria, Artacris, Eriophyes*. **Subfamily Phyllocoptinae** – cigar-shaped to pear-shaped mites with a pair of setae on the prodorsal shield; hysterosoma subdivided by relatively few, broad tergites and several narrow sternites: genera *Acaricalus, Aculus, Epitrimerus, Phyllocoptes, Tegonotus, Vasates*.

Superfamily Tarsonemoidea: mites with short, needle-like mouthparts: *family Tarsonemidae* (p. 423).

Superfamily Tetranychoidea: spider-like mites with long, needle-like chelicerae: *family Tetranychidae* (p. 426); *family Tenuipalpidae* (p. 432).

Superfamily Eupodoidea: includes tarsonemid-like species with claws on each pair of tarsi: *family Siteroptidae* (p. 432).

Astigmata: soft-bodied, semitransparent mites; chelicerae forceps-like.

Superfamily Acaroidea: features as for order: *family Acaridae* (p. 433).

Woodlice

Woodlice (phylum Arthropoda: class Crustacea) are terrestrial, 14-legged crustaceans, forming a distinct order, the Isopoda. They feed on decaying vegetation, but also sometimes attack the roots, stems and leaves of healthy plants; animal matter is also included in the diet, individuals commonly feeding on dried-blood fertilizer: *family Armadillidiidae* (p. 434); *family Oniscidae* (p. 435); *family Porcellionidae* (p. 435).

Millepedes

Millepedes (phylum Arthropoda: class Diploda) usually have elongate bodies composed of a variable number of double abdominal segments, most of which bear two pairs of legs. The head is armed with biting mouthparts, and bears simple eyes and a pair of short, clubbed antennae. Millepedes are secretive, light-shy creatures, and usually inhabit moist, sheltered situations. Although some species are scavengers, and a few are phytophagous or predacious, most feed on decaying vegetation. Millepedes sometimes damage germinating seeds and seedlings, but are of little or no importance on older plants.

Symphylids

Symphylids (phylum Arthropoda: class Symphyla) are small, soft-bodied creatures with three pairs of mouthparts, 12 pairs of legs and a posterior pair of cerci. Most species inhabit the soil and feed on decaying vegetation; a few attack the underground parts of plants.

Nematodes

Nematodes (phylum Nematoda) are unsegmented worm-like invertebrates that lack circulatory and respiratory systems, and are devoid of cilia (both external and internal); also, they possess a stiff yet flexible external cuticle which, unlike that of insects, lacks chitin. Nematodes are often abundant in soils, feeding on various micro-organisms, but many species are parasitic. Those attacking vertebrates (including man) are commonly known as 'roundworms', whereas those associated with plants are often known as 'eelworms'. Plant-parasitic nematodes are microscopic, commonly no more than 0.1–0.5 mm long; they are unique in possessing a distinctive spear-shaped structure in the oesophagus, with which they pierce the walls of plant cells. The detailed form of this spear is often useful for distinguishing between genera. To be active, a nematode is dependent upon the presence of moisture; individuals usually travel through the soil or over plant tissue in films of water, progressing with serpentine movements of the body.

Slugs and snails

Slugs and snails (phylum Mollusca: class Gastropoda) are soft-bodied, non-segmented invertebrates with the body composed of three regions: head, foot and visceral mass. The last-mentioned is covered by a layer of epithelial cells, called the mantle; this secretes a shell of calcium carbonate and encloses a mantle cavity. The mouth usually contains a rasping tongue or radula, armed with thousands of minute chitinous teeth. Most molluscs, such as clams, cuttlefish, octopuses, oysters, sea-slugs and squids, are marine animals; several species live in freshwater habitats but only certain slugs and snails in the order Pulmonata are able to survive on land.

Terrestrial slugs and snails are hermaphroditic creatures with a slimy, asymmetrical body. Their mantle cavity is vascularized and functions as a lung, with an aperture on one side of the body. The visceral mass is often contained within a hard, helical shell. The head is well developed and bears a long and a short pair of retractile tentacles, each longer tentacle having a simple eye at its tip. The foot is muscular, broad and flattened, and functions as the propulsive organ, the animal gliding along on a bed of slime. Slugs and snails are most active in warm, humid conditions. They feed on plant material of various kinds, and some species are regarded as important pests.

Earthworms

Earthworms (phylum Annelida: class Oligochaeta) are well-known hermaphroditic creatures with long, thin and distinctly segmented bodies. They burrow in the soil and feed mainly on decaying vegetative matter, thereby contributing to soil fertility, aeration and drainage. Although primarily beneficial, a few species can be a nuisance in lawns and sports turf.

Birds and mammals

Birds (class Aves) and mammals (class Mammalia) are of only minor importance as pests of ornamental plants. Birds are of particular significance as pests of flower buds or open blossoms. Mammals are mainly damaging to young plants, new shoots, bulbs, corms and seeds; they may also strip bark from trees.

Pest damage

The kind of damage inflicted upon plants by pests varies according to feeding habit and methods (e.g. whether the pest's mouthparts are adapted for biting, piercing, rasping or sucking). Some pests attack the roots or other underground parts, but most affect the leaves, stems, shoots, buds or flowers. Damage also varies from minor, often imperceptible blemishes, colour changes or loss of vigour, to complete death of plants. Leaves, for example, may become blistered, discoloured, disfigured, distorted, dwarfed, galled, malformed, mined, punctured, ragged, skeletonized, speckled, thickened, webbed, wilted or withered, and they may fall off prematurely. Symptoms are sometimes in themselves sufficiently characteristic to enable (at least with experience) the causal organism to be identified; leaf mines formed by certain insect larvae are good examples. However, in many cases the cause of plant damage cannot be determined with confidence unless the pest itself is found and identified.

Some pests (termed polyphagous) are indiscriminate feeders, and attack a wide range of plants. Others are more specific, and often feed on only a restricted group of hosts – perhaps those from a single group (family or genus) of plants, or even a single species; such pests are termed oligophagous or monophagous, respectively. Plant susceptibility to pests may also vary at the specific level, and sometimes differs markedly from cultivar to cultivar. Unlike native plants, exotics introduced from abroad may prove largely if not entirely immune to pest attack. However, they may be damaged by pests that arrived along with them (or subsequently) from the country of origin; in Europe, Australasian pests on plants such as *Eucalyptus* and *Pittosporum* are examples.

In some instances (e.g. aphids, leaf beetles, mites and slugs), all active feeding stages (adults and juveniles) of a pest cause similar damage. In others the type of damage caused by phytophagous adults and juveniles may be different: chafer adults, for example, attack the leaves and other aerial parts of plants whereas their grubs are root-feeders. In many cases (e.g. true flies, butterflies and moths), damage is usually caused only by the larvae.

Several pests (e.g. certain aphids, midges, wasps, mites and nematodes) produce characteristic galls on host plants. Such gall-formers inject a toxin into the plant cells, thereby stimulating abnormal development of the plant tissue. In other cases injection of toxins merely causes distortion of the plant tissue, affected shoots, leaves or flowers becoming malformed and often discoloured.

Many pests of ornamental plants have little or no direct effect on growth but their depredations might be disfiguring; such damage is often of little or no consequence on established plants but on young ones (particularly in commercial nurseries) may have a significant impact on plant quality. The mere presence of certain pests (e.g. wax-secreting or honeydew-excreting aphids, scale insects and other bugs) can be unacceptable, even when infestations are slight.

Control of pests

Good husbandry will reduce the likelihood of pest problems developing on ornamentals but, in some instances, specific control measures may be necessary to protect plants from attack or to keep pests and their damage within acceptable bounds. Pest attacks can be lessened by using traps or physical barriers (e.g. grease tree-bands for pests such as winter moth, and netting for birds or mammals) but such methods are not always practical and are certainly not available for combating the majority of pests.

Attention to hygiene is important for lessening the impact of pests, especially in greenhouses – plant debris should be cleared as soon as cropping is completed, and buildings, pots and other equipment disinfected before new plants are introduced. Efficient weed control, both within greenhouses and outdoors, will reduce the range of places where pests can find shelter and will also limit the number of possible alternative host plants upon which certain pests might survive or breed. Regular cultivation of soils will help to control weeds and will also keep soil pests in check, either destroying them directly or exposing them to desiccation or to the attention of birds and other predators.

Wherever practical, plants should be examined regularly for signs of pests, so that appropriate action can be taken at the earliest possible stage. Newly acquired plants, including the roots and adhering soil or compost, should always be inspected to prevent the accidental introduction of pests into clean sites; this is of particular importance for combating insidious pests such a nematodes.

On a small scale, some pests may be controlled by hand, any egg clumps, larvae or other stages found on plants being squashed or picked off and destroyed; in some cases affected parts of plants, such a shoots containing galls or webbed by caterpillars, may be removed and destroyed. Prunings and other plant debris, whether thought to be harbouring pests or not, should never be left lying around but should be gathered up immediately and burnt.

Various pesticides are available for use against pests of ornamental plants. Some are broad-spectrum materials (capable of killing various kinds of pest) but others are more selective and some may be highly specific; modes of action also vary. Choice of product will depend on many different factors. Systemic materials (which are absorbed through the leaves or taken up by the roots, and then translocated through the plant in the sap) are particularly effective against sap-feeding pests, especially aphids and leaf miners. Contact materials have a variety of uses, and stomach poisons are useful for killing pests such as caterpillars, leaf beetles and weevils. In some situations, fogs and fumigants may be useful; in others, granules, pellets or sprays will be more appropriate. Some pests (e.g. spider mites, and certain aphids and thrips) have developed resistance to pesticides, and this has limited the effectiveness of many products. Whichever pesticide is selected, the directions on the manufacturer's label should be followed, and care taken to ensure that treatments are applied appropriately, effectively and efficiently.

Some ornamental plants, such as *Begonia*, *Calceolaria* and *Hydrangea*, may be intolerant of pesticides, and susceptibility sometimes varies from cultivar to cultivar. In some cases, growth is checked, perhaps imperceptibly; in others, tissue becomes discoloured or distorted and, in extreme cases, affected plants might be killed. Where information regarding the safety of a pesticide to any particular plant species or cultivar is lacking, or if doubt exists, a few plants should be treated first and these later checked for signs of phytotoxicity before larger-scale treatment is undertaken. Young, tender plants are particularly susceptible to chemicals; also, certain sprays otherwise considered safe may have an adverse effect on open blooms, causing a range of undesirable symptoms such as speckling or overall discoloration of tissue; spray damage of this type is well-known on greenhouse-grown chrysanthemums. As a general rule, spraying of open flowers should be avoided, not only because of the risk of phytotoxicity but also to safeguard pollinating insects which might be foraging upon them. Further, spraying with pesticides should not be undertaken during bright sunlight, the risk of damage from excessive temperatures being particularly serious in greenhouses and when plants are under stress. In general, problems of phytotoxicity are more likely to occur on protected plants than on those grown outdoors.

Most pesticides recommended for use on ornamental plants are available only to commercial growers. However, some products are specifically formulated and recommended for amateur use in private gardens. These products will protect plants against the majority of important pests. However, non-chemical methods (see below) are often preferable and frequently just as effective.

The application of pesticides in amenity areas poses particular problems and is often impractical. Not only are there increased risks of killing non-target species, but potential hazards to the public must also be considered. A few insecticides are specifically formulated and recommended for control of pests in amenity areas; such treatments are, of course, 'safe' when used as recommended by the pesticide manufacturers. Even so, on environmental grounds, their use should be kept to the absolute minimum.

In some situations, both in amenity areas and elsewhere, it is possible to use a biological control agent rather than a chemical pesticide, and this has obvious attractions. Examples include the application of the bacterium *Bacillus thuringiensis* to kill caterpillars, the use of predatory mites to combat spider mites, and the release of parasitoid wasps to attack leaf miners on greenhouse crops. Nowadays, various biological control agents are available for use by both amateur and commercial growers.

In nature, phytophagous insects, mites and so forth are subject to attack by a wide range of natural enemies, including a vast array of parasitoids and predators. Pests also succumb to other naturally occurring controlling agents, such as bacterial, fungal and viral diseases. Pests of ornamental plants are no exception, and in many situations their populations will remain below economically important levels unless the balance of this natural control is overturned. Although some pesticides are intrinsically safe to beneficial insects and mites, many have adverse effects upon them. It is prudent, therefore, to restrict the use of chemicals and to ensure that, when treatment with a pesticide is required, the one chosen from the list of those available will have the least deleterious impact on non-target organisms. Recommendations relating to pest management often vary according to local circumstances. They may also differ in detail from country to country, if not from region to region. Readers seeking information on pest control or pest management, therefore, should refer to information relevant to their regional or local circumstances. On occasions, it might also be prudent to seek expert advice.

Finally, an ability to identify pests correctly and to recognize the symptoms of pest damage is an essential starting point for good pest management. Knowledge of the habits and biology of the various pests, and of the risks they pose, is also required if correct decisions concerning their possible control are to be made.

Chapter 2
Insects

Order COLLEMBOLA (springtails)

Family SMINTHURIDAE

Globular-bodied springtails with thorax and abdomen fused; antennae long; ocelli usually present.

Bourletiella hortensis (Fitch)
syn. *B. signatus* (Nicolet)
Garden springtail

An often common pest, particularly in wet, acid soils. Damage is caused to seedling plants, including various ornamentals; conifer seedlings in forest nurseries, especially beach pine (*Pinus contorta*), are seriously affected. Widely distributed in Europe; also present in North America.

DESCRIPTION

Adult: 1.5 mm long; black to dark green, often spotted with white; head large, with long antennae and prominent, black, yellowish-bordered eyes; abdomen globular, with a small ventral tube-like sucker and a forked springing organ.

LIFE HISTORY

Eggs are laid in the soil, usually in small groups, each female depositing up to 100 in about three weeks. The eggs swell rapidly after laying and hatch shortly afterwards. Under favourable conditions maturity is reached in 2–3 months but development can take much longer, individuals continuing to moult even after the adult stage is attained. Breeding is continuous throughout the year but reaches a peak in the spring, the insects being most numerous from late April to the end of June.

DAMAGE

General: the hypocotyl and cotyledons of seedlings are pitted, and holes are formed in the young leaves, but damage rarely occurs after July. **Conifer seedlings:** damage to the hypocotyl and cotyledons results in stunted seedlings with a brush-like mass of swollen, distorted needles; such seedlings develop into useless, multi-stemmed plants.

Family ONYCHIURIDAE

Springtails without ocelli, but with complex sensory organs on the antennae and with mandibulate mouthparts.

Onychiurus spp.
White blind springtails

Various species of *Onychiurus* (e.g. *O. nemoratus* and *O. stachianus*) cause damage to seedlings, pitting the cotyledons, hypocotyl and roots, and chewing the root hairs and rootlets; attacked seedlings collapse and die, often keeling over at about soil level. On older plants, leaves in contact with the soil may also be holed and skeletonized. Damage occurs on various outdoor and greenhouse plants, including ornamentals. The springtails are abundant in wet soil with a high organic content, and often gain entry to pots and seed boxes if these are placed directly onto infested ground. The pests may also be introduced into containers if unsterilized compost is used. Individuals (up to 3 mm long) are white and stout bodied, with a large head, short antennae and legs, and six abdominal segments; the springing organ is reduced or absent. They breed continuously in favourable conditions. Development from egg to adult takes several months, and the insects undergo several moults even after the adult stage is reached.

Order **ORTHOPTERA** (crickets and grasshoppers)

Family **GRYLLOTALPIDAE** (mole crickets)

Crickets with the fore legs greatly enlarged and modified for burrowing.

Gryllotalpa gryllotalpa (Linnaeus) (**1**)
syn. *G. vulgaris* Latreille
Mole cricket

A large, soil-burrowing insect that sometimes causes damage to greenhouse and outdoor plants, including ornamentals, by biting or gnawing the roots and basal parts of the stems. Damage normally occurs at or just below the soil surface but tends to be indiscriminate. Adults are 35–50 mm long and greyish brown to yellowish brown, coated in fine, velvet-like hairs; the prothorax is elongate, and the front tibiae much enlarged and distinctly toothed. The insect is widely distributed in Europe, and has been introduced to the eastern USA. It is also found in North Africa and Western Asia. Mole crickets are of pest status in southern Europe. However, in countries such as Britain they are a protected species and close to extinction.

Family **GRYLLIDAE** (true crickets)

Body relatively broad and somewhat flattened, the fore wings being held more or less horizontally; antennae longer than body; ovipositor and anal cerci long.

Acheta domesticus (Linnaeus) (**2**)
House cricket

Although of only minor importance, this widely distributed species occurs occasionally in heated greenhouses. The insects hide by day in dark crevices, emerging at night to feed. They then cause damage to the stems, flowers and foliage of plants; house crickets also attack the aerial roots of orchids and other ornamentals. The adult males stridulate, producing their characteristic 'song' by rubbing their fore wings together. Individuals are 15–20 mm long, yellowish brown to greyish brown, and clothed with fine hairs.

1 Mole cricket (*Gryllotalpa gryllotalpa*).

2 House cricket (*Acheta domesticus*).

Order **DERMAPTERA** (earwigs)

Family **FORFICULIDAE**

Forficula auricularia Linnaeus (3)
Common earwig

A useful predator of aphids and various other pests, but also a frequent pest of flowers such as carnation (*Dianthus caryophyllus*), *Chrysanthemum, Cineraria, Clematis, Dahlia, Delphinium* and pansy (*Viola tricolor*); buds and leaves are also attacked. Cosmopolitan. Widely distributed in Europe.

DESCRIPTION
Adult female: 12–14 mm long; chestnut-brown; hind wings, when folded away, projecting beyond elytra; pincers slightly curved. **Adult male:** 13–17 mm long; similar to female but pincers distinctly curved. **Egg:** 1.3 × 0.8 mm; pale yellow. **Nymph:** whitish to greyish brown.

LIFE HISTORY
Adults of both sexes overwinter in sheltered situations in the soil, and mate in the early winter. Eggs are laid in December or January. Each female deposits a batch of up to 100 in an earthen cell, and guards over the eggs until after they hatch in February or March. Earwigs are omnivorous insects, the nymphs feeding throughout the spring and reaching maturity by the early summer; there are four nymphal stages. Overwintered adult females sometimes deposit a second batch of eggs in May or June. Nymphs from these eggs develop from late June or early July to September. Earwigs are nocturnal, hiding by day within damaged flowers, in crimpled leaves, under loose bark and so on. Although occurring mainly outdoors, attacks are sometimes reported in greenhouses.

DAMAGE
Damaged flower petals become ragged, spoiling their appearance; attacks on leaves are usually unimportant, but chewed buds may die, resulting in blind shoots; most damage occurs from June to September.

3 Male common earwig (*Forficula auricularia*).

Order **DICTYOPTERA** (cockroaches and mantids)

Family **BLATTIDAE** (cockroaches)

Distinguished from mantises (family Mantidae) by the unmodified front legs and broad pronotum which partly or completely covers the head.

Blatta orientalis Linnaeus
Common cockroach
This generally common, well-known bakery and warehouse insect often occurs in greenhouses, destroying seeds and seedlings, and also causing damage to the aerial parts of older plants. Individuals hide by day but at night they become active and then move rapidly over the floors and beds of infested houses. Eggs are deposited in groups in purse-like oothecae. These egg cases eventually split open to release young nymphs, the incubation period of the eggs lasting for one or more months according to temperature. The nymphs feed for nine months or more before becoming adults. Breeding is continuous under suitable conditions. Adults are 20–30 mm long, rather flattened, shiny blackish brown, with long, many-segmented antennae, long legs and a pair of anal cerci; wings are poorly developed.

Blattella germanica (Linnaeus) (4)
German cockroach
Infestations of this relatively small (c. 12–14 mm long), yellowish-brown cockroach are sometimes established in heated greenhouses and hot-houses. In common with other species, they feed at night and sometimes cause damage to ornamental plants.

Periplaneta americana (Linnaeus)
American cockroach
A relatively large (38–42 mm long), reddish-brown cockroach; often present in heated greenhouses, where it may cause damage to ornamentals and various other plants. Unlike *Blatta orientalis*, the wings are fully developed and reach beyond the tip of the abdomen.

Periplaneta australasiae (Fabricius)
Australian cockroach
Minor infestations of this cockroach have also become established in heated greenhouses in Europe. Individuals are smaller than the previous species (30–36 mm long), and adults and nymphs are more extensively marked with yellow.

Pycnoscelus surinamensis (Linnaeus)
Surinam cockroach
A typically parthenogenetic cockroach, unusual in retaining its ootheca within a brood sac so that the eggs hatch whilst still within the mother's body. Probably of oriental origin but now cosmopolitan; in Europe infestations occur widely in heated greenhouses, where cultivated plants, including ornamentals, are damaged. Adults (21–23 mm long) are dark brown, with paler wings and a pale band along the front of the pronotum; the wings are fully developed.

4 German cockroach (*Blattella germanica*).

Order **HEMIPTERA** (true bugs)

Family **TINGIDAE** (lace bugs)

Flattened bugs, the pronotum and wings with a netted, lace-like pattern; pronotum usually covering the scutellum.

Corythucha ciliata (Say) (**5–6**)

Platanus lace bug

An important North American pest of plane (*Platanus*) trees, especially oriental plane (*P. orientalis*). Since 1964, it has become established in much of central and southern Europe, where it causes extensive damage to both nursery and mature trees. The pest has been introduced into Australia and Japan, and in 2006 was found for the first time in Britain.

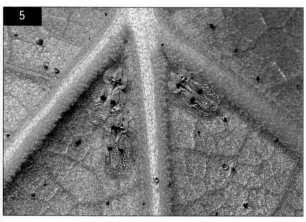

5 Platanus lace bugs (*Corythucha ciliata*).

DESCRIPTION
Adult: 3 mm long; body blackish; wings transparent, with a network of white veins, the fore wings with a dark central patch; antennae brownish white, hairy and with a slightly clubbed apex. **Nymph:** mainly black.

LIFE HISTORY
Adults overwinter under the bark of host trees, usually congregating on the north-west side of the trunks. They emerge in the spring and commence feeding on the new foliage. Eggs are laid on the underside of leaves and hatch shortly afterwards. Nymphs then feed and reach maturity in late June or July. A second generation occurs during the summer, producing adults from late August onwards. These adults then enter hibernation. In particularly favourable, more southerly districts, three generations are reported annually.

DAMAGE
Infested foliage becomes discoloured. Heavily infested leaves drop prematurely, and plant vigour is reduced.

Corythucha arcuata (Say)

Oak lace bug

This North American pest recently appeared in northern Italy in 2000. It has since been found in southern Switzerland and Turkey, and is considered at least a potential threat elsewhere. In Italy, there are three generations annually, and development from egg to adult takes from four to six weeks. As in the case of the previous species, adults form the overwintering stage. Although associated mainly with oak (*Quercus*), the pest has been reported on other plants, including ornamental maple (*Acer*) and rose (*Rosa*).

6 Platanus lace bug (*Corythucha ciliata*) damage to leaf of *Platanus*.

Stephanitis rhododendri (Horváth) (**7–8**)
Rhododendron bug

A locally important pest of *Rhododendron*, especially hybrids of *R. arboreum*, *R. campanulatum*, *R. campylocarpum*, *R. catawbiense* and *R. caucasicum*; probably of North American origin. Infestations occur in various parts of mainland Europe; also found in southern England and Wales.

DESCRIPTION
Adult: 4 mm long; body brownish to blackish; wings with a creamy, net-like venation. **Nymph:** yellowish, ornamented with numerous brown spines.

LIFE HISTORY
Adult bugs first appear in June and may survive on host plants until the early winter. Although fully winged, they are relatively sedentary and do not fly. Once established, therefore, infestations tend on individual plants tend to persist. In the autumn, females deposit eggs along the midrib of young leaves. These eggs overwinter *in situ*, and eventually hatch in the spring. Nymphs then feed on the underside of the leaves, typically occurring in groups of up to 50 individuals. There is a single generation annually.

DAMAGE
This pest causes a yellow mottling of the foliage, the underside of damaged leaves developing a rusty-brown appearance. Heavy attacks lead to extensive discoloration of plants and may cause leaves to wilt. Attacks are most severe on plants growing in sunny, dry situations.

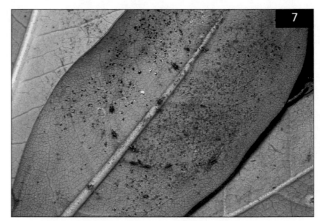

7 Nymphs of rhododendron bug (*Stephanitis rhododendri*) and damage on underside of leaf of *Rhododendron*.

8 Rhododendron bug (*Stephanitis rhododendri*) damage to upper surface of leaf of *Rhododendron*.

Stephanitis takeyai Drake & Maa
Andromeda lace bug

This Japanese species has recently become established in South East England on lily-of-the-valley bush (*Pieris japonica*). It has also been found in parts of mainland Europe, including Germany, Italy, the Netherlands and Poland, and is a problem on ornamental Ericaceae in the USA. Adults and nymphs cause mottling and bronzing of leaves; severely damaged plants may die. The pest, which overwinters in the egg stage, passes through five nymphal stages and completes from two to four generations annually.

Family **MIRIDAE**
(capsid bugs or mirid bugs)

Very active, soft-bodied bugs, with elongate and partly hardened fore wings (termed hemelytra), and long, probing, needle-like mouthparts.

Closterotomus norvegicus (Gmelin)
syn. *Calocoris norvegicus* (Gmelin)
Potato capsid

A generally common capsid on herbaceous weeds; occasionally a minor pest of cultivated *Chrysanthemum* and certain other ornamental Asteraceae. Widely distributed in Europe; also present in Canada.

DESCRIPTION
Adult: 6–8 mm long; hemelytra green, sometimes tinged with reddish brown; pronotum green and straight sided, and typically marked with a pair of black dots. **Nymph:** mainly green to yellowish green, with black hairs.

LIFE HISTORY
Eggs are laid in late July and August in cracks in woody or semi-woody stems of various plants. They hatch in the following May or early June. The nymphs then feed on the buds, growing points, flowers and foliage of herbaceous plants, including clover (*Trifolium*), nettle (*Urtica*) and various members of the family Asteraceae, including chamomile (*Anthemis*), mayweed (*Matricaria*) and thistles (*Carduus* and *Cirsium*). Nymphs reach maturity in late June or July. There is just one generation annually. Attacks on crops tend to occur mainly in weedy situations.

DAMAGE
Adults and nymphs produce necrotic spots which develop into holes; infested young shoots become distorted and may be killed.

Lygocoris pabulinus (Linnaeus) (**9–10**)
Common green capsid

An often abundant pest of trees, shrubs and herbaceous plants, including ornamentals such as *Caryopsis*, *Chrysanthemum, Clematis, Dahlia*, diviner's sage (*Salvia divinorum*), *Forsythia, Fuchsia, Geranium, Hydrangea*, morning glory (*Ipomoea*), *Pelargonium*, rose (*Rosa*) and sunflower (*Helianthus annuus*). Present throughout Europe.

DESCRIPTION
Adult: 5.0–6.5 mm long; bright green, with a dusky-yellow pubescence; pronotum lightly punctured and with moderate callosities; antennae comparatively long. **Egg:** 1.3 mm long; banana-shaped, creamy, smooth and shiny. **Nymph:** light green to bright green; tips of antennae orange-red.

9 Common green capsid (*Lygocoris pabulinus*).

10 Common green capsid (*Lygocoris pabulinus*) damage to leaves of *Prunus*.

LIFE HISTORY
The winter is passed as eggs inserted in the bark of first- or second-year shoots of woody hosts such as crab-apple (*Malus*), currant (*Ribes*), hawthorn (*Crataegus*) and lime (*Tilia*). The eggs hatch from April onwards, and young, very active nymphs then feed on the new foliage. After a few weeks, the nymphs migrate to herbaceous hosts to complete their development. The summer adults occur in June and July. They lay their eggs in the stems of various cultivated herbaceous hosts, including ornamentals, and on weeds such as bindweed (*Convolvulus*), dandelion (*Taraxacum officinale*), dead-nettle (*Lamium*), dock (*Rumex*) and groundsel (*Senecio vulgaris*). Nymphs of the second generation feed on these summer hosts, and reach the adult stage by the autumn. There is then a return migration to woody hosts, where winter eggs are laid. Although most abundant on outdoor hosts, the pest is sometimes introduced into greenhouses during the summer on, for example, infested chrysanthemum plants.

DAMAGE
Foliage at the tips of the new shoots becomes speckled reddish or brownish red, tattered and distorted, and often peppered with small holes. Such symptoms appear on herbaceous hosts from May onwards; in some cases (e.g. clematis and dahlia) flowers are distorted, and in others (e.g. fuchsia, on which damage is often severe) flower buds are aborted.

Lygus rugulipennis Poppius (**11**)
European tarnished plant bug
A common polyphagous pest of annual and herbaceous plants, including African daisy (*Arctosis*), *Chrysanthemum*, *Dahlia*, Michaelmas daisy (*Aster*), nasturtium (*Tropaeolum*) poppy (*Papaver*) and *Zinnia;* particularly troublesome in greenhouses. Widely distributed in Europe.

DESCRIPTION
Adult: 5–7 mm long; robust-bodied, with relatively short antennae; extremely variable in colour, varying from green to yellowish brown, reddish brown or black; apex of hind femora distinctly ringed. **Egg:** 1.0 × 0.25 mm; creamy white, and flask-shaped. **Nymph:** green to brownish, with a pair of black dots dorsally on each thoracic segment.

LIFE HISTORY
Adults overwinter in debris on the ground or in other suitable shelter, and emerge in the following March or April. Eggs are laid during May, most commonly in the buds and stems of wild hosts such as dock (*Rumex*), groundsel (*Senecio vulgaris*) and nettle (*Urtica*). Nymphs then feed on the youngest shoots, and new adults appear from July onwards. Except in more northerly districts, a larger second brood of nymphs is produced and these commonly cause damage to cultivated plants. The next generation of adults eventually appears from September onwards.

DAMAGE
Adults sometimes produce a localized yellowing of the leaves, with brown necrotic spots marking the position of the feeding punctures; attacks on young tissue may result in the leaves or flowers becoming puckered and distorted. Nymphs cause young shoots to become twisted, swollen and often blind.

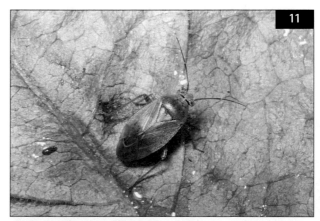

11 European tarnished plant bug (*Lygus rugulipennis*).

Family **CERCOPIDAE** (froghoppers)

Often called 'spittle-bugs'. The nymphs develop on plants within a mass of froth (cuckoo-spit), produced as a fluid from the hind end of the body and through which air bubbles are forced from a special canal by abdominal contractions. Fore wings (elytra) hardened throughout. The hind tibiae of adults bear just a few stout spines (cf. leafhoppers, p. 31).

Aphrophora alni (Fallén) (**12**)
Alder froghopper
A common pest of deciduous trees and shrubs, including alder (*Alnus*), ash (*Fraxinus excelsior*), poplar (*Populus*) and willow (*Salix*). Eurasiatic. Widespread in Europe.

12 Alder froghopper (*Aphrophora alni*).

DESCRIPTION
Adult: 8.0–9.5 mm long; light greyish brown to dark olive-brown, with deep blackish punctures; head and pronotum with a median keel; elytra with a pair of pale patches. **Nymph:** mainly greyish to creamy white, with a distinctive pair of dark spots on the head between the eyes.

LIFE HISTORY
Eggs are laid from July to October, each being deposited close to the ground in the old tissue of host plants. They hatch in the following spring, nymphs then developing within distinctive, round accumulations of spittle, concentrated mainly on the lower parts of plants. Adults appear from late June onwards, ascending host plants to feed on the young tissue.

DAMAGE
The nymphs cause little direct damage. However, young adults produce distinctive calloused rings on the shoots, often weakening the new growth.

Aphrophora salicina (Goeze) (**13–14**)
syn. *A. grisea* Haupt; *A. salicis* (Degeer)
Willow froghopper
A generally common pest of poplar (*Populus*) and willow (*Salix*); often established on ornamental and amenity trees. Widespread throughout Europe; also present in North America.

DESCRIPTION
Adult: 10.0–10.5 mm long; greyish yellow or greenish to olive-brown, finely punctured with black; elytra sometimes with an indistinct whitish basal triangular spot; head and thorax distinctly keeled. **Nymph:** head and thorax reddish brown; abdomen creamy white.

LIFE HISTORY
Eggs, deposited on host plants during the summer or early autumn, hatch in the following spring. The nymphs then feed within large, dripping accumulations of spittle which are often grouped on the young shoots. These nymphal feeding shelters are most obvious in June. Adults appear from late June onwards.

DAMAGE
Feeding by the nymphs affects the vigour and quality of host plants, and the presence of masses of spittle on ornamentals is unsightly.

Philaenus spumarius (Linnaeus) (**15–16**)
syn. *P. leucophthalmus* (Linnaeus)
Common froghopper
nymph = cuckoo-spit bug
Generally abundant on a wide variety of trees, shrubs and low-growing plants; often a minor pest of lavender (*Lavendula*) and various other ornamentals, including barberry (*Berberis*), bellflower (*Campanula*), *Chrysanthemum*, *Coreopsis*, *Geum*, golden-rod (*Solidago virgaurea*), *Lychnis*, *Mahonia*, Michaelmas daisy (*Aster*), *Phlox*, rose (*Rosa*), *Rudbeckia* and many others. Holarctic. Present throughout Europe.

DESCRIPTION
Adult: 5–7 mm long; colour varying from yellowish, greenish or brown to blackish; head bluntly wedge-shaped, with large eyes; elytra convex, and often with dark markings; each hind tibia with an apical ring of spines. **Egg:** 1 mm long; oval. **Nymph:** mainly pale and unicolourous, with dark eyes.

13 Nymphs of willow froghopper (*Aphrophora salicina*).

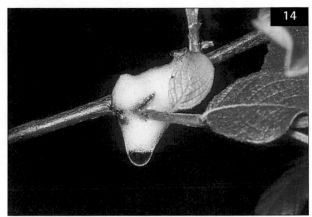

14 Spittle of willow froghopper (*Aphrophora salicina*) on *Salix.*

15 Common froghopper (*Philaenus spumarius*).

16 Nymph of common froghopper (*Philaenus spumarius*).

LIFE HISTORY

The frog-like adults occur mainly from July onwards and are often seen at rest on plants. They are rather sluggish but jump violently into the air if disturbed. Eggs are laid in the stems of host plants during September, typically in batches of up to 30, and hatch in the following May. Sedentary nymphs then feed on the sap of host plants, injecting their needle-like mouthparts into a shoot or leaf vein. Throughout their development, they cover themselves with a protective mass of spittle-like froth. They pass through five nymphal instars, and reach the adult stage in the summer.

DAMAGE

Young shoots on susceptible nymph-infested plants may become distorted and wilted; in some cases flowers are malformed. Adults cause no obvious damage. Ornamentals are disfigured by the presence of the spittle, and this may reduce the marketability of container-grown plants in nurseries.

Family **FLATIDAE** (planthoppers)

Bugs with wings longer than the body and held almost vertically at the sides of the body when in repose; fore wings (elytra) hardened, with a dense network of veins that includes numerous cross-veins.

Metcalfa pruinosa (Say) (**17–18**)
Citrus planthopper

This highly polyphagous North American species first appeared in Europe in the late 1970s. It is now firmly established in southern France, central and northern Italy, and has also been reported elsewhere – e.g. in Switzerland. Infestations occur on a wide range of woody plants, including ornamental nursery stock, roadside hedges and so forth. Nursery plants being raised on sites adjacent to infested wasteland are particularly liable to be attacked.

DESCRIPTION
Adult: 5–8 mm long; body mainly grey; fore wings (elytra) whitish grey to dark grey, marked with black. **Nymph:** whitish to light green, developing distinct wing buds and with conspicuous tufts of white wax at the hind end of the abdomen.

LIFE HISTORY
This pest has one generation per year and overwinters in the egg stage, the eggs being deposited in the woody parts of host plants. The eggs hatch in May; nymphs then feed gregariously on the leaves and shoot tips, amongst masses of mealy wax. The nymphs appear somewhat sedentary, but readily crawl or jump away if disturbed. Adults occur from July to October.

DAMAGE
Heavy infestations cause noticeable discoloration and death of foliage, as well as distortion of the tips of young shoots. In addition, plants are contaminated by wax and by honeydew upon which sooty moulds develop.

17 Citrus planthopper (*Metcalfa pruinosa*).

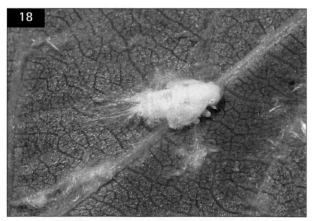

18 Nymph of citrus planthopper (*Metcalfa pruinosa*).

Family **CICADELLIDAE** (leafhoppers)

Adults and nymphs are free-living on the foliage of various plants. The hind legs of the very active adults are adapted for jumping and bear two rows of fine spines along each tibia (cf. family Cercopidae, p. 28). Reliable identification of most species involves examination of genitalia and is a specialist task; some species, however, are distinguishable by characteristic markings on the body or the fore wings (elytra). The following are representatives of those most frequently associated with ornamental plants, and include those often reaching true pest status.

Aguriahana stellulata (Burmeister) (**19**)
Cherry leafhopper
A distinctive, mainly white and generally common leafhopper, occurring on various ornamental trees, including flowering cherry (*Prunus*) and lime (*Tilia*). Infestations are rarely important but leaf discoloration on exotic hosts such as Indian horse chestnut (*Aesculus indica*) is unwelcome.

Alebra albostriella (Fallén) (**20**)
Infestations of this generally distributed and often common leafhopper are found on oak (*Quercus*), and frequently result in foliage damage on young ornamental trees. In mainland Europe, damage also occurs on alder (*Alnus*). Adults are variable in colour but often have the pronotum yellowish, marked with a pair of longitudinal red or orange streaks; the elytra are yellow to hyaline, marked with red or orange. In *Alebra*, unlike all other European genera, the wing venation includes a distinct appendix.

Alebra wahlbergi (Boheman)
This generally common leafhopper is widely distributed in mainland Europe and in the southern half of Britain, causing noticeable damage to the foliage of various trees, especially elm (*Ulmus*), horse chestnut (*Aesculus hippocastanum*), lime (*Tilia*), maple (*Acer*) and sycamore (*Acer pseudoplatanus*). Adults, which occur from July to September, are relatively large (females up to 4.5 mm long); the elytra are mainly pale yellow, sometimes streaked with yellow or orange, with the inner margin usually orange, pink or yellow.

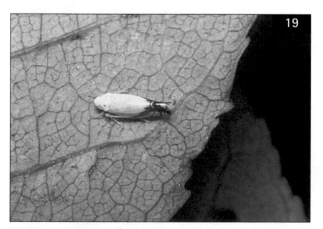

19 Cherry leafhopper (*Aguriahana stellulata*).

20 Adult of *Alebra albostriella*.

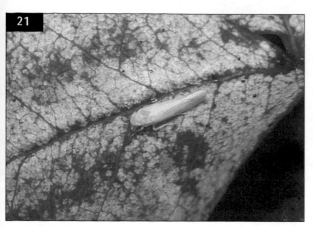

21 Rose leafhopper (*Edwardsiana rosae*).

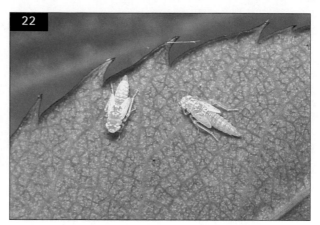

22 Nymphs of rose leafhopper (*Edwardsiana rosae*).

Edwardsiana rosae (Linnaeus) (**21–23**)
Rose leafhopper
A generally abundant and often important pest of rose (*Rosa*); also harmful to various other rosaceous trees and shrubs, including hawthorn (*Crataegus*), rowan (*Sorbus aucuparia*) and whitebeam (*S. aria*). Widely distributed in Europe; also present in North America.

DESCRIPTION
Adult: 3.4–4.0 mm long; mainly pale yellowish, with a yellow abdomen. **Nymph:** translucent to whitish.

LIFE HISTORY
Eggs are deposited in the autumn under the epidermis of young shoots of wild and cultivated rose bushes. The eggs hatch in the following spring. Nymphs then feed on the underside of the leaves, and reach adulthood in late May or early June. Most adults then migrate to summer hosts, such as fruit trees, but some remain on rose and eventually deposit eggs in the leaves. These eggs hatch in late June or early July and a second brood of nymphs develops. Adults appear from mid-August onwards. These, along with newly reared individuals returning to rose from summer hosts, finally deposit the winter eggs.

DAMAGE
Infested leaves become extensively flecked and blanched. Infestations are most harmful in hot, dry weather, and on climbing roses. Severe attacks may cause leaves to turn brown, resulting in premature leaf fall.

Edwardsiana crataegi (Douglas)
syn. *E. oxyacanthae* Ribaut; syn. *Typhlocyba froggatti* (Baker)
This widely distributed species occurs commonly on rosaceous trees and shrubs, and is sometimes damaging to the foliage of crab-apple (*Malus*), flowering cherry (*Prunus*), hawthorn (*Crataegus*), rowan (*Sorbus aucuparia*) and willow-leaved pear (*Pyrus salicifolia*). Adults are similar in appearance to those of *Edwardsiana rosae* but the elytra are darkened along the inner margin of the clavus.

Edwardsiana flavescens (Fabricius) (**24**)
Foliage damage caused by this widely distributed species occurs most frequently on beech (*Fagus sylvatica*) and hornbeam (*Carpinus betulus*), the pale yellow to whitish adults being readily distinguished from the two most common and more strongly marked species associated with these hosts (on beech, *Fagocyba cruenta*, p. 34; on hornbeam, *Typhlocyba bifasciata*, p. 36).

Edwardsiana nigriloba (Edwards) (**25**)
Sycamore leafhopper
This widely distributed leafhopper, which is restricted to sycamore (*Acer pseudoplatanus*), frequently causes extensive damage to the foliage of ornamental trees. Nymphs feed in the spring and early summer, and adults appear in July. The adults (3.6–4.0 mm long) are mainly pale yellowish (with a characteristic black tubercle ventrally on the genital segment); the older nymphs are particularly noticeable, having a characteristic black pattern which contrasts with the pale background of the underside of the leaves.

23 Rose leafhopper (*Edwardsiana rosae*) damage to leaves of *Rosa*.

24 Adults of *Edwardsiana flavescens*.

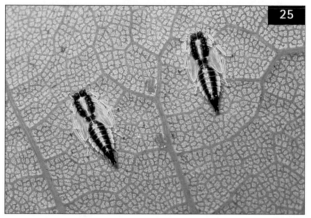

25 Final-instar nymphs of sycamore leafhopper (*Edwardsiana nigriloba*).

26 Chrysanthemum leafhopper (*Eupteryx melissae*).

Eupteryx melissae Curtis (26–27)
Chrysanthemum leafhopper

A generally common species on Lamiaceae, including ornamentals such as diviner's sage (*Salvia divinorum*) and *Phlomis*; other hosts include *Chrysanthemum*, hollyhock (*Alcea rosea*) and tree mallow (*Lavatera arborea*). Attacks are sometimes noted on garden or greenhouse-grown plants during the summer months, infested leaves developing a sickly, flecked appearance. The adults (2.9–3.4 mm long) are whitish or greenish white, with distinctive greenish, blackish or greyish markings.

27 Chrysanthemum leafhopper (*Eupteryx melissae*) damage to leaves of *Salvia*.

28 Adult of *Eurhadina pulchella*.

29 Beech leafhopper (*Fagocyba cruenta*).

30 Rhododendron hopper (*Graphocephala fennahi*).

Eurhadina pulchella (Fallén) (28)

Although associated mainly with oak (*Quercus*), this species also attacks the foliage of other trees, including birch (*Betula*). It is widely distributed and generally common in Europe. Adults (*c.* 4 mm long) are variable in colour, but the elytra are usually yellow with a distinct apical patch. In common with other members of the genus, individuals have a characteristically flattened appearance, the elytra being noticeably broader medially and the head distinctly narrower than the pronotum.

Fagocyba cruenta (Herrich-Schaeffer) (29)
syn. *Typhlocyba douglasi* (Edwards)
Beech leafhopper

Generally distributed and often abundant on beech (*Fagus sylvatica*). Hedges and specimen trees are often disfigured, the foliage becoming extensively flecked with silver. The pest also invades other trees, including oak (*Quercus*), sycamore (*Acer pseudoplatanus*) and whitebeam (*Sorbus aria*). Adults (3.4–4.1 mm long) are usually pale yellowish, with the elytra somewhat darkened towards the inner margin; darker forms also occur.

Graphocephala fennahi Young (30)
syn. *G. coccinea* (Förster)
Rhododendron hopper

Widely distributed on *Rhododendron* in England, particularly in the south, having been introduced from America in the 1930s; also now widely distributed in mainland Europe.

DESCRIPTION
Adult: 8.4–9.4 mm long; head yellow; thorax bluish green, marked with red and yellow; abdomen red above; elytra bluish green, striped with red; hind wings grey. **Nymph:** whitish to yellowish green.

LIFE HISTORY
Eggs are inserted into slits made in scales on the flower buds from August to October. The eggs hatch in the following spring. Nymphs then feed on the underside of leaves from April or May to July or August. Adults and nymphs often occur in considerable numbers on the tips of young shoots. There is just one generation each year.

DAMAGE
Direct feeding has little adverse effect on host plants. However, infested plants are more susceptible to bud blast (a disease caused by the fungus *Pycnostysanus azaleae*), the fungal spores probably gaining easy entry into plant tissue through hopper egg-laying slits.

31 Glasshouse leafhopper (*Hauptidia maroccana*).

32 Adult of *Ribautiana ulmi* alongside cast skin of final-instar nymph.

Hauptidia maroccana (Melichar) (31)
syn. *Erythroneura pallidifrons* (Edwards);
E. tolosana (Ribaut)
Glasshouse leafhopper

A polyphagous, tropical or subtropical species. Well established in the warmer parts of Europe. In northern Europe, infestations occur mainly on greenhouse plants, including *Calceolaria*, *Chrysanthemum*, diviner's sage (*Salvia divinorum*), *Fuchsia*, *Gloxinia*, heliotrope (*Heliotropium*), *Pelargonium*, primrose (*Primula vulgaris*), sweet-scented verbena (*Aloysia citriodora*) and tobacco plant (*Nicotiana*); in favourable areas, also found on outdoor plants such as chickweed (*Stellaria*) and foxglove (*Digitalis purpurea*).

DESCRIPTION
Adult: 3.1–3.7 mm long; mainly pale yellow, with greyish or brownish markings, the latter forming a pair of distinctive, chevron-like marks on the elytra. **Nymph:** whitish.

LIFE HISTORY
Breeding is continuous throughout the year, all stages occurring on the underside of host plants. Eggs are deposited singly in the leaf veins and hatch in about a week at normal greenhouse temperatures. Nymphs feed for about a month, passing through five instars before reaching the adult stage. Adults survive for up to 3 months, each female depositing up to 50 eggs. The duration of the various stages is extended in cool conditions; development is greatly protracted during the winter, when the eggs often take a month or more to hatch and nymphal development lasts for two or more months.

DAMAGE
Growth of heavily infested plants is checked and seedlings killed, but on most hosts damage is limited to specking, silvering or a blanched mottling of the foliage. Leaves are also contaminated by cast nymphal skins.

Ribautiana ulmi (Linnaeus) (32)
A generally common species, associated mainly with elm (*Ulmus*) and causing extensive discoloration of the expanded leaves; attacks also occur on other trees, including hazel (*Corylus*), hornbeam (*Carpinus betulus*), oak (*Quercus*), whitebeam (*Sorbus aria*) and willow (*Salix*). Adults are present from May to November. They breed on the underside of the leaves, and often occur in considerable numbers. Individuals are 3.6–4.4 mm long; the elytra are yellow, with greyish markings on the apical third; the head and pronotum are yellowish, the former with a pair of distinctive back spots and the latter usually with a single spot on the front margin. The active nymphs are yellowish to whitish.

Typhlocyba bifasciata Boheman (**33**)
Hornbeam leafhopper

This common yellow and black leafhopper breeds on elm (*Ulmus*) and hornbeam (*Carpinus betulus*). Infestations occur throughout the summer and lead to noticeable silvering of the leaves. The adults and nymphs often occur in large numbers on the underside of infested leaves, with the cast skins of earlier instars or generations clearly visible. Adults are 3.2–3.7 mm long, with pale yellow elytra, each marked with a pair of black crossbands or with the basal two-thirds mainly or entirely black. The characteristic pattern on the elytra readily distinguishes this species from *Edwardsiana flavescens* (p. 32) and from other leafhoppers associated with elm or hornbeam.

Typhlocyba quercus (Fabricius) (**34***)

Although most often noted on oak (*Quercus*), causing extensive discoloration of the leaves, this generally common species is also associated with various other hosts, including birch (*Betula*), hornbeam (*Carpinus betulus*), smoke bush (*Cotinus coggygria*) and southern beech (*Nothofagus*). Adults are 3.0–3.5 mm long, with the elytra mainly white, marked with greenish yellow to yellowish brown and with prominent brick-red to orange patches. They occur from July to October.

Family **PSYLLIDAE** (psyllids or suckers)

Adults very active, with relatively large, membranous wings and strong hind legs adapted for jumping. Wing venation characteristic; a pterostigma present or absent. Nymphs often flat and scale-like, and usually slow-moving; they produce a white, waxy secretion and masses of sticky honeydew. Most species are free-living but some breed within leaf galls.

Acizzia uncatoides (Ferris & Klyver)
Acacia sucker

An important Australian pest of silk tree (*Albizia*) and wattle (*Acacia*). Now established in many other parts if the world, including southern Europe (e.g. France, Italy, Portugal and Spain) and the USA.

DESCRIPTION

Adult: fore wings hyaline, with grey or brownish markings and pale yellow veins; body varying from pale yellow to green, brownish orange or dark brown, with grey markings. **Egg:** yellow to orange. **Nymph:** mainly pale yellow to yellowish orange or greenish yellow, with a darker head and dark markings on the thorax and abdomen; eyes reddish; antennae short and dark tipped.

LIFE HISTORY

Eggs are laid on buds and the underside of leaves, and hatch shortly afterwards. The flat-bodied nymphs then pass through five instars before moulting into adults. There are several (often up to eight) generations annually, with adults and nymphs overwintering. In addition to wax, the pest produces considerable quantities of honeydew upon which sooty moulds subsequently develop.

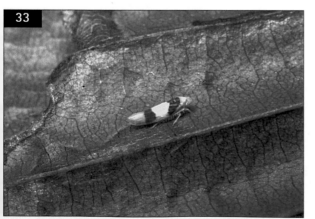

33 Hornbeam leafhopper (*Typhlocyba bifasciata*).

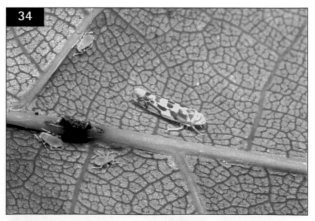

34 Adult of *Typhlocyba quercus*.

DAMAGE

Infested plants are contaminated by wax, sticky honeydew and sooty moulds. Infested leaves and shoots are discoloured and distorted, and the pest also causes premature leaf fall and dieback.

NOTE

Other species of *Acizzia* have been accidentally introduced into Europe. These include *A. acaciaebaileyanae* and *A. jamatonica*. The former is an Australian species, associated with cootamundra wattle (*Acacia baileyana*), but not considered a damaging pest. The latter is an Asian pest of Persian silk tree (*Albizia julibrissin*), causing extensive discoloration of foliage and premature leaf fall. It has been reported from various parts of mainland Europe (including Croatia, France, Italy, Slovenia and Switzerland), and is a particular threat to hosts being raised in nurseries or planted as amenity shade trees.

Cacopsylla buxi (Linnaeus) (35–37)
syn. *Psylla buxi* (Linnaeus)
Box sucker
An often abundant pest of common box (*Buxus sempervirens*). Present throughout Europe; introduced into North America.

DESCRIPTION

Adult: fore wings 2.9–3.7 mm long and shiny, with brownish-yellow veins; body light green to yellowish green. **Nymph:** mainly yellowish green to light green; tips of antennae blackish; eyes pale purplish white.

LIFE HISTORY

There is a single generation each year, with overwintering eggs on the shoots hatching in the spring at about bud-burst. Nymphs then feed on the growing points, sheltered by a loose cabbage-like cluster of cupped terminal leaves which develops around them; these galls are very noticeable, and each measures about 10–20 mm across. One or a few individuals occur within each gall, the nymphs passing through five instars and becoming fully grown within about six weeks. The nymphs produce conspicuous white waxen threads and considerable quantities of honeydew, the latter often spreading over infested foliage. Although adults appear from late April, May or June onwards, mating and egg laying do not occur until the late summer.

DAMAGE

Galls, which remain on bushes long after the causal insects have disappeared, check the growth of new

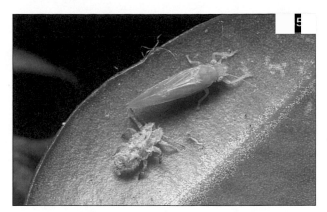

35 Box sucker (*Cacopsylla buxi*) alongside cast skin of final-instar nymph.

36 Nymph of box sucker (*Cacopsylla buxi*).

37 Gall of box sucker (*Cacopsylla buxi*) on *Buxus sempervirens*.

shoots; they also spoil the appearance of hedges. Plants are also disfigured by sticky masses of honeydew and sooty moulds. Although most damage is caused to the shoot tips, nymphal feeding may extend to expanded leaves lower down the twigs, these then becoming blistered, pallid and distorted.

Cacopsylla fulguralis (Kuwayama) (38)
Elaeagnus sucker

In 1999, this eastern Asian species was found in northwestern France on ornamental *Elaeagnus* × *ebbingei* plants. Since then, the pest has spread very rapidly and has appeared in various other parts of Europe, including the Channel Islands, England, Italy, the Netherlands and Switzerland. Infestations appear to be restricted to ornamental species of *Elaeagnus*; they have not been found on cherry elaeagnus (*E. multiflora*) or Russian olive (*E. angustifolia*).

DESCRIPTION
Adult: fore wings 2.5–3.0 mm long, mainly hyaline, clouded with grey; body pale greenish yellow, with brown or greyish-brown markings; eyes reddish. **Nymph:** pale greenish yellow or creamy yellow to brownish, with greyish-brown markings; white waxen plates often protrude from the hind end of the body.

LIFE HISTORY
Adults and nymphs feed on the underside of leaves. They are very active, and move away rapidly if disturbed. Little is known of the biology of this pest, but several generations are completed annually.

DAMAGE
Infested plants are heavily contaminated with honeydew and sooty moulds, which limits photosynthetic activity. The pest also reduces plant vigour, and causes chlorosis, dieback and premature leaf fall. Severe damage to seedlings and established plants is reported.

Cacopsylla melanoneura (Förster) (39)
syn. *Psylla melanoneura* Förster
Hawthorn sucker

Infestations of this widely distributed and generally abundant psyllid are noted occasionally on cultivated hawthorn (*Crataegus*). The greenish (orange- to blackish-marked) nymphs feed on the young shoots during May. Although sometimes present in large numbers, they cause little or no direct damage; however, by producing considerable quantities of honeydew and white waxen threads, the pest often attracts attention on nursery stock, hedges and specimen trees, particularly in dry conditions. Adults are reddish to brown or blackish (with characteristic whitish longitudinal lines on the thorax). They are very active, and occur throughout most of the year. Pine (*Pinus*) trees, upon which they overwinter but do not feed or breed, are also inhabited, as are various other non-host trees and shrubs.

Psylla alni (Linnaeus) (40–42)
Alder sucker

A common pest of alder (*Alnus*). Present throughout northern Europe; also found in North America.

DESCRIPTION
Adult: fore wings 3.8–4.8 mm long, clear but with dark veins; body bright green, later developing brown or red markings. **Nymph:** green, marked with black; antennae, legs and wing pads pale yellowish brown to blackish.

LIFE HISTORY
Eggs are laid in bark crevices during the autumn and hatch at bud-burst in the following spring. Nymphs then feed on the young shoots, each developing within a sticky mass of white waxy 'wool' amongst which beads of honeydew accumulate. The nymphs are very active if disturbed; they then scurry over the leaves or along the shoots and might even drop to the ground. Development is completed in the summer, adults occurring from June or July to October. There is just one generation annually.

DAMAGE
The presence of flocculent masses of 'wool' on plants is unsightly, and infestations may reduce the vigour of new shoots; damage, however, is rarely of significance.

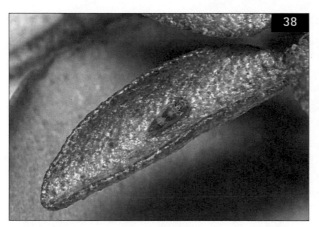

38 Elaeagnus sucker (*Cacopsylla fulguralis*).

39 Nymphs of hawthorn sucker (*Cacopsylla melanoneura*) on *Crataegus.*

40 Alder sucker (*Psylla alni*).

41 Nymph of alder sucker (*Psylla alni*).

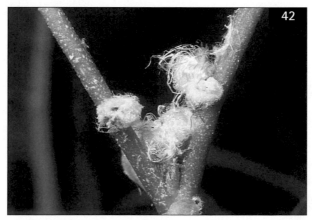

42 Wax-coated nymphs of alder sucker (*Psylla alni*) on *Alnus.*

43 Ash leaf gall sucker (*Psyllopsis fraxini*).

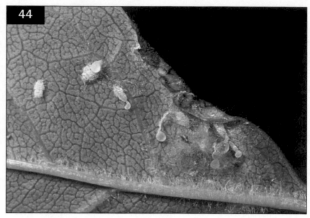

44 Nymphs of ash leaf gall sucker (*Psyllopsis fraxini*).

Psyllopsis fraxini (Linnaeus) (**43–45**)

Ash leaf gall sucker

A common but minor pest of ash (*Fraxinus excelsior*). Eurasiatic. Widely distributed in Europe; introduced into North America.

DESCRIPTION

Adult: fore wings 2.3–3.3 mm long, clear but with brown veins and a blackish apical pattern; body pale bluish green to pale yellow, marked with black. **Nymph:** pale bluish green; eyes red and prominent.

LIFE HISTORY

The winter is passed in the egg stage on dormant shoots. The eggs hatch in the early spring, at about bud-burst. Nymphs then feed on the young expanding leaves, clustered together within rolled leaf edges (galls). They secrete flocculent masses of white, waxy 'wool' and also produce globules of honeydew. The galled tissue turns yellow, changing through red and purplish to brown, and such galls are most conspicuous from summer until leaf fall. Adults appear from June or July onwards, and often shelter within the galls along with any later-developing nymphs. In common with other psyllids, there are five nymphal instars; there is a single generation annually.

DAMAGE

The prominent leaf galls disfigure infested plants. They potentially reduce the value of nursery stock but are otherwise harmless.

45 Gall of ash leaf gall sucker (*Psyllopsis fraxini*) on *Fraxinus*.

Psyllopsis fraxinicola (Förster)

Ash leaf sucker

This widespread and often common species is associated with ash (*Fraxinus excelsior*), but does not produce leaf galls. Adults are greenish or yellowish green, with clear, unpatterned wings; the nymphs are green (cf. *Psyllopsis fraxini*).

Family TRIOZIDAE

Psyllids with a characteristic wing venation and never a pterostigma; associated not only with trees and shrubs but also with herbaceous plants. Nymphs often scale-like and with a fringe of waxen threads.

Trichochermes walkeri (Förster) (46)

Buckthorn sucker

Widespread in mainland Europe, in association with common buckthorn (*Rhamnus cathartica*), and locally common in southern England. Infestations often occur on buckthorns in parks and gardens. Nymphs develop in distinctive, upward leaf-roll galls formed along part of the leaf margin. Such galls are initiated in the spring and enclose nymphs that complete their development in the summer. Adults appear from August onwards. The nymphs are pale greenish yellow, elongate-oval and flattened, with an outer fringe or closely packed hairs, long, narrow wing pads and short antennae. Adults are 2.7–3.1 mm long and brownish, with dark markings and relatively long, thin antennae; the fore wings (3.5–4.0 mm long) are mottled with brown. Damage on infested bushes is often extensive; in severe cases, most if not all leaves are galled.

Trioza alacris Flor (47–48)

Bay sucker

A generally common pest of bay laurel (*Laurus nobilis*), particularly in Mediterranean areas. Commonly associated with container-grown plants, and often established on those adorning pavements in towns and cities; first introduced into Britain in the 1920s, and now widespread in the southern half of England. Also an introduced pest in North and South America.

DESCRIPTION

Adult: fore wings 2.4–3.1 mm long, narrow, and usually clear with yellow veins; body pale whitish yellowish, with brownish or blackish markings; antennae long and slender, with a black tip. **Nymph:** pale yellowish white to greyish white, with greyish or blackish markings; antennae short and mainly pale, with the two apical segments black; body densely covered by tufts of white wax.

46 Galls of buckthorn sucker (*Trichochermes walkeri*) on *Rhamnus.*

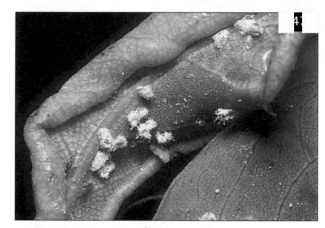

47 Nymphs of bay sucker (*Trioza alacris*).

48 Galls of bay sucker (*Trioza alacris*) on *Laurus.*

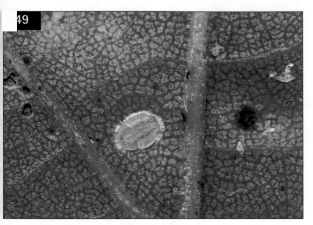

49 Nymph of oak leaf sucker (*Trioza remota*).

50 Oak leaf sucker (*Trioza remota*) damage to leaf of *Quercus.*

51 Galls of pittosporum sucker (*Trioza vitreoradiata*) on *Pittosporum.*

LIFE HISTORY
Adults overwinter on bay laurel, sheltering amongst curled or dense clusters of leaves; they also hide in leaf litter and other debris at the base of host plants. In spring, adults invade the young shoots to feed on the leaves. This causes the leaf edges to curl, and provides suitable sites within which small cluster of eggs are laid. Nymphs feed within these protective galls, excreting considerable quantities of honeydew; the leaf edges of galled leaves become thickened and further curled, changing to yellow, red or brown. Nymphs pass through five instars before attaining the adult stage. In favourable conditions there are two generations annually.

DAMAGE
Foliage damage on heavily infested plants is considerable and specimen plants are commonly disfigured by the prominent galls. Infested plants, young or old, are affected by masses of flocculent wax produced by the nymphs; they also become sticky with honeydew, upon which unsightly sooty moulds develop. Young plants are most susceptible, the leaves often turning brown and dropping prematurely; shoot death also occurs.

Trioza remota Förster (**49–50**)
syn. *T. haematodes* Förster
Oak leaf sucker
A generally common but minor pest of oak (*Quercus*). Widely distributed in central and northern Europe; also found in Japan.

DESCRIPTION
Adult female: fore wings 2.7–3.0 mm long, mainly clear with yellow or brownish veins; body brownish red, with creamy markings. **Nymph:** 1.8–2.0 mm long; broad-bodied and scale-like, with a distinct fringe of setae; light orange, with a pair of orange stripes extending from head to tip of abdomen.

LIFE HISTORY
Adults occur from September to May, overwintering in the shelter of evergreen plants and migrating to oak in the spring. Eggs are laid on the leaves during May. The sedentary nymphs then feed in small pits on the underside of the leaves during the summer, moulting into adults by about September. Adults remain on the host plants into October but then disperse to their overwintering sites.

DAMAGE
Nymphal feeding causes a superficial pitting of the leaves and a noticeable pimpling on the upper surface; however, such damage is not important and does not result in distortion of the leaf blade.

Trioza vitreoradiata Maskell (51)
Pittosporum sucker
A New Zealand pest of *Pittosporum*. Now well established on such plants in the Channel Islands, Cornwall and the Isles of Scilly, where it was first discovered in 1993. The pest has also been found in Ireland.

DESCRIPTION
Adult: 3–4 mm long; body varying from black to yellowish green. **Nymph:** green, flat and scale-like, with a distinct marginal fringe of short waxen filaments.

LIFE HISTORY
The pest overwinters in the adult stage. In spring, eggs are laid on the leaves and hatch in about two weeks. Nymphs then feed for a few weeks, passing through five instars before eventually becoming adults. Up to five generations are completed annually.

DAMAGE
The pest forms blister-like or pit-like galls on the leaves, and young growth is distorted and discoloured. Infested plants are also contaminated by honeydew and sooty moulds.

Family **CARSIDARIDAE**

Psyllids with characteristically flattened, noticeably hairy antennae; mostly of tropical distribution.

Homotoma ficus (Linnaeus) (52–53)
Fig sucker
This tropical, non-indigenous psyllid occurs on common fig (*Ficus carica*) in various parts of Europe, including Austria, England, France, Jersey and Switzerland, but is local and uncommon. The winter is passed in the egg stage, with a single generation annually. Adults (fore wings 3.4–4.3 mm long) occur during the summer and the mainly green, sedentary nymphs develop from early spring to about June. Infestations are not of major importance.

52 Fig sucker (*Homotoma ficus*).

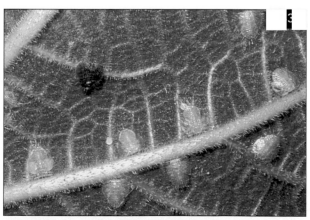

53 Nymphs of fig sucker (*Homotoma ficus*).

54 Eucalyptus sucker (*Ctenarytaina eucalypti*).

55 Nymphs of eucalyptus sucker (*Ctenarytaina eucalypti*).

Family SPONDYLIASPIDAE

A family of psyllids indigenous to Australia; one species has been introduced into Europe on imported *Eucalyptus*.

Ctenarytaina eucalypti (Maskell) (54–55)

Eucalyptus sucker

Established on *Eucalyptus* in various parts of mainland Europe, southern England and the Channel Islands, and often a pest in nurseries.

DESCRIPTION

Adult: fore wings 1.3–1.9 mm long; body mainly dark brown, with the thorax partly orange; fore wings whitish, with yellow veins and a waxy bloom; legs light brown. **Egg:** shiny, pale yellow and pear-shaped. **Nymph:** greyish yellow or orange-brown to blackish; first instar orange.

LIFE HISTORY

The relatively small, robust-bodied adults overwinter on host plants, and deposit their first eggs in clusters on the young growth from February onwards. These eggs hatch in the early spring. Colonies then become established on the young, tender growth, the nymphs producing honeydew and curly strands of whitish wax that are often blown about by the wind. There are two or more overlapping generations annually, but breeding may be continuous on plants growing in protected conditions, all stages of the pest often then occurring together.

DAMAGE

Infestations spoil the appearance of plants, contaminating the new growth with masses of honeydew amongst which the cast nymphal skins accumulate. The shoots also become disfigured by loose masses of waxen 'wool'. Feeding does not distort growth but heavy attacks retard the growth of nursery plants.

Family **ALEYRODIDAE** (whiteflies)

Small, moth-like insects, coated with an opaque, white, waxy powder. The nymphs are flat and scale-like. Development includes a quiescent, non-feeding pupa-like stage, known as a pseudo-pupa.

Aleurothrixus floccosus (Maskell) (**56**)
Woolly whitefly

A polyphagous, subtropical pest of broad-leaved evergreen plants, including ornamentals such as *Citrus*, common fig (*Ficus carica*), fruiting myrtle (*Eugenia*) and *Rhododendron*. Of South American origin, and accidentally introduced into several other countries, including southern Europe and the southern USA. Adults are similar in appearance to other whiteflies but have a yellowish tinge. Eggs are deposited on the underside of leaves in characteristic circles or semicircles and hatch shortly afterwards. The sedentary, light green nymphs then develop rapidly, often forming dense colonies. The nymphs excrete vast amounts of rather viscous honeydew, and infested plants soon become very sticky and blackened following the development of sooty moulds. Third-instar nymphs also secrete long waxen threads that persist throughout the pseudo-pupal stage. As a result, colonies appear to be covered by masses of cottonwool. In favourable conditions breeding continues throughout the year, and there are several overlapping generations annually.

Aleurotuba jelinekii (von Frauenfeld) (**57**)
Viburnum whitefly

This local, southerly distributed species occurs mainly on *Viburnum rotundifolia* and *V. tinus*; it does not invade other species of *Viburnum* but does occur on common myrtle (*Myrtus communis*) and strawberry tree (*Arbutus unedo*). Adults are 1.0–1.5 mm long, with whitish wings and a yellowish-orange body. They occur mainly during June and July, depositing eggs on the younger foliage. Nymphs feed during the summer on the underside of the leaves, and complete their development in the autumn or following spring. The characteristic, scale-like pseudo-pupae (1.0–1.5 mm long) are black, with a whitish fringe and white waxen encrustations on the body. They may be found throughout the winter and spring, and are very conspicuous when infested leaves are turned over. There is just one generation annually. Attacks are relatively harmless but heavily infested plants become contaminated with honeydew, upon which sooty moulds develop.

Aleyrodes lonicerae Walker
syn. *A. fragariae* Walker; *A. rubi* Signoret
Honeysuckle whitefly

Sometimes noted on ornamental honeysuckle (*Lonicera*), snowberry (*Symphoricarpos rivularis*) and many other plants, contaminating the foliage with sticky honeydew upon which disfiguring sooty moulds develop. Several overlapping generations occur each year, the pale yellow, scale-like nymphs and the yellow-bodied, white-winged adults mainly infest the underside of expanded leaves. The adults are whitish, with a grey spot on each fore wing.

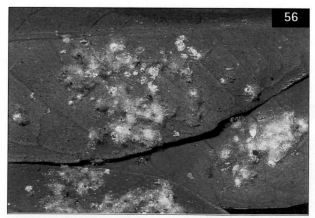

56 Colony of woolly whitefly (*Aleurothrixus floccosus*).

57 Pseudopupa of viburnum whitefly (*Aleurotuba jelinekii*).

Asterobemisia carpini (Koch)
syn. *A. avellanae* (Signoret); *Aleurodes rubicola* (Douglas)
Hornbeam whitefly
Generally common in Europe on hornbeam (*Carpinus betulus*) but also associated with other hosts, including common hazel (*Corylus avellana*) and *Rubus*; various other plants are attacked in mainland Europe, including maple (*Acer*), birch (*Betula*), beech (*Fagus sylvatica*), false acacia (*Robinia pseudoacacia*) and *Tilia* (*lime*). Eggs are laid on the underside of leaves in May and June. The scale-like, pale yellowish to yellowish-green nymphs then feed throughout the summer, completing their development in the autumn. The insects overwinter as pseudo-pupae, from which adults (which are 1.0–1.1 mm long, yellow or orange, with clear white wings) emerge in the spring. Attacks on ornamental plants are usually unimportant, although heavy infestations cause foliage to become sticky with honeydew and subsequently blackened by sooty moulds.

Bemisia hancocki Corbett
A tropical and subtropical whitefly of uncertain status. It infests the underside of the leaves of host plants, including some ornamental species, and has been noted in several parts of Europe, including England where it was fist found on bay laurel (*Laurus nobilis*) in the 1970s. Rightly or wrongly, *Bemisia hancocki* is often regarded merely as a synonym of *Bemisia afer*, a polyphagous pest of cotton, peanut and other crops in Africa and Asia.

Bemisia tabaci (Gennadius)
Tobacco whitefly
As an invasive alien species, this notorious tropical pest has become established in several warmer parts of Europe, including some Mediterranean countries. It occurs from time to time on imported greenhouse plants in northern European countries, such as England where it is a quarantine pest. The insect has a very wide host range but in northern Europe is most likely to occur on ornamentals such as *Gerbera*, *Gloxinia*, *Hibiscus* and poinsettia (*Euphorbia pulcherrima*), and on vegetable crops such as sweet pepper and tomato. Adults and nymphs suck plant sap from the underside of the leaves of host plants, causing slight spotting of the tissue. The insects excrete considerable quantities of honeydew. The lifecycle is similar to that of *Trialeurodes vaporariorum* (see p. 48) but development is slower for any given temperature and may cease completely in the winter, even in heated greenhouses. The pest is best recognized as scattered yellow 'scales' (contrasting with the more clumped distribution and paler 'scales' of *Trialeurodes vaporariorum*); also, eggs are laid singly and randomly on the underside of the leaves. Characteristically, the pseudo-pupae are slightly pointed posteriorly and lack horizontally directed waxen processes from the body wall; in repose, the wings of adults are held in a sloping, roof-like posture. The pest is a notorious virus vector.

Dialeurodes chittendeni Laing
Rhododendron whitefly
This locally distributed, generally uncommon European whitefly is associated with *Rhododendron*, and is sometimes a pest on *R. campylocarpum*, *R. catawbiense*, *R. caucasicum* and *R. ponticum*; other species or cultivars where leaf cuticles are relatively thin or unprotected by hairs or scales are also attacked. Adults are 1.2 mm long, and pale yellow with pure white wings. They occur in June and July and are relatively sluggish, congregating on the underside of the leaves; if disturbed, they fly only a short distance before resettling. Eggs are laid singly on the underside of the leaves of host plants but without any accompanying waxy powder. Nymphs are flat and elliptical, greenish yellow and semitransparent, without a waxy fringe. They feed from mid-July onwards, surviving throughout the winter; they 'pupate' in the following spring, from mid-April onwards. The pseudo-pupae are oval (1.2 × 0.9 mm), pale greenish yellow to whitish yellow, also without a waxy fringe. Because of their shape and colour, nymphs and pseudo-pupae are difficult to see; however, following the emergence of adults, the empty pseudo-pupal cases are more obvious. Light infestations are of little or no significance. However, if attacks are heavy, plant vigour is reduced, infested foliage becoming mottled with yellow, and covered with honeydew and sooty moulds.

58 Colony of citrus whitefly (*Dialeurodes citri*).

59 Colony of phillyrea whitefly (*Siphoninus phillyreae*) on underside of leaf of *Phillyrea.*

Dialeurodes citri (Ashmead) (58)
Citrus whitefly

This polyphagous species is mainly a problem in citrus orchards, but is also found on certain other hosts, including ornamentals such as ash (*Fraxinus excelsior*), *Forsythia*, *Gardenia*, lilac (*Syringa*), Persian lilac (*Melia azedarach*) and privet (*Ligustrum vulgare*). Of East Asian origin, and first noted in Europe (Italy and southern France) in the 1950s.

DESCRIPTION

Adult: 1.2–1.4 mm long; creamy white, coated with white wax. **Egg:** 0.2–0.3 mm long; oval and pale yellow, with a pedicel. **Nymph:** up to 1.5 mm long; pale greenish yellow, flat, oval and scale-like. **Pseudo-pupa:** 1.2–1.5 mm long; flat, oval and scale-like.

LIFE HISTORY

Adults, which first appear in April or May, are short-lived, each surviving for about 10 days. Eggs are laid in batches on the underside of young leaves and hatch 1–4 weeks later, depending on temperature. The scale-like immature stages feed on the underside of leaves for 3–4 weeks; they then enter a quiescent pseudo-pupal stage which lasts anything from 2 to 42 weeks or more. Infestations are often very heavy, with several hundred individuals clustered on a single leaf. The pest overwinters as either third- or fourth-instar nymphs. Under favourable conditions, there are four generations annually.

DAMAGE

Hosts are weakened by heavy, persistent infestations. However, sooty moulds, which accumulate on honeydew excreted by this pest, tend to be more significant; they affect the quality and appearance of ornamentals.

Pealius azaleae (Baker & Moles)
Azalea whitefly

Adults of this widely distributed but local pest of evergreen azalea, especially *Rhododendron mucronatum* and *R. simsii*, appear during the early summer. They congregate on the underside of the leaves, where eggs are laid, and make short flights if disturbed. Nymphs are light green and scale-like. They develop on the underside of the leaves from late summer onwards, overwintering and completing their development in the spring. There is one generation annually. The nymphs excrete considerable quantities of honeydew, and the leaves of infested plants commonly become coated and discoloured by sooty moulds. Infestations are reported from various parts of the world, including Australasia, Europe (e.g. Belgium, England, Germany, France, Italy and the Netherlands) and Japan.

Siphoninus phillyreae (Haliday) (59)
Phillyrea whitefly

A locally common but minor pest of mock privet (*Phillyrea latifolia*); also occurs on ash (*Fraxinus excelsior*). Widely distributed in mainland Europe but absent from northerly areas; in the British Isles confined mainly to southern England. Present in parts of North Africa and Asia.

DESCRIPTION

Adult: 1 mm long; yellow, marked with grey; fore wings mainly white but somewhat smoky basally. **Egg:** 0.3 × 0.1 mm; oblong, with a short stalk; yellow to brownish. **Nymph:** elliptical to oval; brownish, with a fringe of white waxy plates; profusely dusted with white, powdery wax. **Pseudo-pupa:** 1.0 × 0.7 mm; brown to brownish black; elliptical, with a distinctive white waxen rim.

LIFE HISTORY

Adult whiteflies first appear in late May and early June, depositing eggs in clusters on the underside of leaves. The eggs, which are liberally dusted with white waxen powder, hatch in 2–3 weeks. Nymphs then develop on the leaves. Individuals 'pupate' a few weeks later. A second generation of adults appears in August and September. Nymphs of this second generation overwinter, and complete their development in the spring. In favourable, southerly regions of mainland Europe there are three or more generations annually.

DAMAGE

Infestations are usually unimportant, although foliage is contaminated by whitish wax and by honeydew upon which sooty moulds develop.

Trialeurodes vaporariorum (Westwood) (**60**)
Glasshouse whitefly

A tropical and subtropical species, introduced accidentally into northern Europe where it is now a widespread and notorious greenhouse pest. Infestations occur on various ornamentals, including *Asparagus plumosus*, *Begonia*, *Calceolaria*, *Cineraria*, *Coleus*, *Dahlia*, diviner's sage (*Salvia divinorum*), *Freesia*, *Fuchsia*, *Gerbera*, *Hibiscus*, *Pelargonium*, poinsettia (*Euphorbia pulcherrima*), primrose (*Primula vulgaris*), *Solanum*, tobacco plant (*Nicotiana*) and many others; *Chrysanthemum* is a poor host and less commonly attacked. Infestations also occur on seedling trees and shrubs raised under protection and on some, such as London plane (*Platanus hispanica*), growing outdoors.

DESCRIPTION

Adult: 1 mm long; body pale yellow; wings pure white and held relatively flat when in repose. **Egg:** broadly conical, yellowish but soon becoming grey. **Nymph:** light green, oval, flat and scale-like. **Pseudo-pupa:** oval and whitish, with relatively short marginal wax processes and several pairs of long waxen dorsal tubes.

LIFE HISTORY

Infestations occur on the underside of leaves of many cultivated greenhouse plants, as well as on various weeds. Adults usually congregate on the foliage towards the tops of the plants and fly rapidly if disturbed. Each female is capable of depositing over 200 eggs during her lifetime, individuals usually surviving for 3–6 weeks. The eggs are laid in distinctive circular groups on smooth leaves but tend to be more scattered on hairy ones. They darken soon after laying and hatch in about nine days at greenhouse temperatures of 21°C. Newly emerged nymphs crawl over the leaf surface for a short while but soon settle down to feed, inserting their mouthparts into the leaf tissue and remaining completely immobile throughout the rest of their development. They pass through three feeding stages and a non-feeding pseudo-pupal stage; new adults appear about 18 days later. Breeding, which is mainly parthenogenetic, is continuous under favourable conditions, and there are several overlapping generations annually. Between cropping, the pest commonly survives on greenhouse weeds or on outdoor plants in the near vicinity. Adults also hibernate throughout the winter on outdoor weeds, but they are unable to survive if weather conditions are severe.

DAMAGE

Attacked plants are weakened and growth checked, the leaves sometimes becoming spotted with yellow or otherwise discoloured; heavy infestations often lead to premature leaf fall. Plants also become contaminated with sticky honeydew and covered in sooty moulds.

60 Colony of glasshouse whitefly (*Trialeurodes vaporariorum*).

Family **APHIDIDAE** (aphids)

A major group of phytophagous winged (alate) and wingless (apterous) insects in which alternation of generations between winter and summer hosts is commonplace and lifecycles of individual species are often complex. Many species show an alternation of generations, having a primary (winter) host upon which asexual and sexual reproduction occurs and eggs are laid, and a secondary (summer) host where development is entirely parthenogenetic and viviparous. Migration between these alternate hosts is usually achieved following the production of winged forms.

Here, aphids are arranged alphabetically under the following subfamilies: Aphidinae (p. 49 *et seq.*), Callaphidinae, Drepanosiphinae and Phyllaphidinae (p. 72 *et seq.*), Chaitophorinae (p. 78 *et seq.*), Eriosomatinae (p. 80 *et seq.*), Hormaphidinae (p. 87), Lachninae (p. 87 *et seq.*), Mindarinae (p. 93 *et seq.*), Pterocommatinae (p. 93 *et seq.*) and Thelaxinae (p. 94 *et seq.*). These subfamilies were granted full family status in older classification schemes.

Subfamily **APHIDINAE**

Aphids with terminal process of antennae elongate; siphunculi typically cylindrical, tapered or swollen, with or without an apical flange; cauda often triangular, tongue-shaped or finger-shaped.

Acyrthosiphon pisum (Harris) (**61**)

syn. *A. onobrychis* (Boyer de Fonscolombe);
A. pisi (Kaltenbach)
Pea aphid
A generally common pest of certain cultivated legumes (Fabaceae), including sweet pea (*Lathyrus odoratus*). Virtually cosmopolitan. Widespread in Europe. The subspecies *Acyrthosiphon pisum spartii* occurs on wild and cultivated broom (*Cytisus*).

DESCRIPTION

Apterous female: 2.5–4.4 mm long; elongate, light green to yellowish or pinkish, with long antennae, legs and cauda, and very long siphunculi. **Alate:** 2.3–4.3 mm long; similar to aptera. **Nymph:** similar to adult but lightly dusted with wax.

LIFE HISTORY

Adults and eggs overwinter on perennial plants such as clover (*Trifolium*), lucerne (*Medicago sativa*), sainfoin (*Onobrychis vicifolia*) and trefoil (*Lotus*). Colonies develop on these plants in the spring, winged forms migrating to peas and beans from May onwards. Immigrants invade the growing points, and infestations then build up on the shoots and developing pods. The aphids, which readily drop to the ground if disturbed, often form significant infestations in June and July, and colonies usually persist on such hosts into the autumn.

DAMAGE

Most damage is caused to the growing points during June and July, affected leaves becoming distorted and turning yellow. Pea enation mosaic virus, pea leaf roll virus and pea mosaic virus are introduced and spread by this insect.

Acyrthosiphon malvae (Mosley)

syn. *A. pelargonii* (Kaltenbach)
Pelargonium aphid
Infestations of this medium-sized, light green or yellowish aphid sometimes occur on greenhouse-grown *Pelargonium*, often in company with other species such as *Aulacorthum circumflexum* (p. 56). The aphid also infests other ornamentals, including *Cineraria*, but is rarely important. Apterae are 2.5–3.0 mm long, with very long antennae, divergent antennal tubercles, and elongate, tapered siphunculi and cauda; unlike species of *Macrosiphoniella* and *Macrosiphum* (q.v.) the siphunculi are not reticulated at the apex.

61 Colony of *Acyrthosiphon pisum spartii* on *Cytisus*.

Aphis fabae Scopoli (62–63)

Black bean aphid

An often major pest of cultivated beans and ornamental plants. Hosts include guelder-rose (*Viburnum opulus*), mock orange (*Philadelphus coronarius*) and spindle (*Euonymus*), and herbaceous plants such as *Clematis*, *Dahlia*, poppy (*Papaver*), nasturtium (*Tropaeolum*), pot marigold (*Calendula officinalis*) and *Yucca*. Cosmopolitan. Present throughout Europe.

DESCRIPTION

Apterous female: 1.5–2.9 mm long; black to blackish brown or blackish green, often with whitish patches of wax on the abdomen; antennae much shorter than body; siphunculi dark and of moderate length; cauda blunt and finger-shaped. **Alate:** 1.8–2.7 mm long; mainly black, often with noticeable patches of white wax. **Nymph:** black to blackish brown or blackish green, often with distinctive patches of white wax.

LIFE HISTORY

The winter is usually passed in the egg stage on spindle bushes, especially European spindle (*Euonymus europaeus*), but also on other primary (winter) hosts such as guelder-rose and mock orange. Eggs hatch in the early spring, and colonies of aphids then develop on the young leaves and shoots. Winged forms appear in May or June and these disperse to various herbaceous hosts, colonies on the primary host then dying out. Breeding on these secondary (summer) hosts continues throughout the summer, with the frequent production of winged forms and further spread to other summer hosts. Colonies are ant-attended, and are often common in July and August on outdoor ornamentals; colonies may also develop in greenhouses on hosts such as *Chrysanthemum*. In the autumn, there is a return migration to the primary host where, following a sexual phase, winter eggs are eventually laid. In favourable situations, aphids may also survive the winter on herbaceous hosts.

DAMAGE

Primary hosts: spring infestations cause considerable curling of leaves on the new shoots, and spoil the appearance of bushes long after colonies have died out. **Secondary hosts:** leaves may be curled, but infestations on the buds and flower stalks are usually more important as these affect the quality, flowering potential and appearance of plants.

Aphis gossypii Glover (64–65)

Melon & cotton aphid

Generally common on greenhouse ornamentals such as *Begonia*, calla lily (*Zantedeschia aethiopica*), *Chrysanthemum* and *Cineraria*. Virtually cosmopolitan. Widely distributed in Europe.

DESCRIPTION

Apterous female: 1.4–2.0 mm long; very dark green or bluish green, sometimes mottled yellowish green; siphunculi relatively short and dark. **Alate:** 1.1–2.1 mm long; head and thorax dark; abdomen marked with dark spots.

LIFE HISTORY

This subspecies breeds continuously under protection (where it is often associated with members of the Cucurbitaceae, including vegetable crops) but is

62 Colony of black bean aphid (*Aphis fabae*) on *Clematis*.

63 Black bean aphid (*Aphis fabae*) damage to leaves of *Euonymus*.

unlikely to survive the winter outdoors in northern Europe. Other subspecies overwinter as eggs on outdoor hosts, e.g. alder buckthorn (*Frangula alnus*); they inhabit various herbaceous hosts during the summer, including ornamentals such as *Hypericum*.

DAMAGE
Attacked leaves turn yellow and may wilt and die. The appearance of ornamentals is also spoilt by the accumulation of honeydew, development of sooty moulds and presence of cast aphid skins.

64 Colony of melon & cotton aphid (*Aphis gossypii*) on *Hypericum*.

Aphis viburni Scopoli (66)
Viburnum aphid
A common pest of wild and ornamental species of *Viburnum*, including guelder-rose (*V. opulus*). Widely distributed in Europe.

DESCRIPTION
Apterous female: 1.8–3.0 mm long; dark greenish black to brownish black; siphunculi relatively short. **Nymph:** brownish, liberally coated with white wax.

LIFE HISTORY
This species overwinters in the egg stage on the shoots and branches of host plants. The eggs hatch in the early spring, usually from April onwards. Aphids then feed on the underside of leaves on the young shoots. The sexual phase in the lifecycle includes a generation of wingless males. Colonies, which are ant-attended, tend to appear brownish in colour, and often extend onto the inflorescences. They reach their maximum development relatively early in the season and then decline as winter eggs are laid; the aphids have usually disappeared by mid-summer. *Aphis fabae* also infests guelder-rose, but such colonies are longer-lasting and appear black rather than brown.

DAMAGE
Spring infestations curl the leaves of new shoots, the deformed foliage remaining on the bushes long after the aphids have disappeared. The presence of damaged shoots spoils the appearance of ornamental shrubs; *Viburnum carlesii* is most susceptible and may be severely disfigured.

65 Melon & cotton aphid (*Aphis gossypii*) damage to foliage of *Chrysanthemum*.

66 Viburnum aphid (*Aphis viburni*) damage to shoots of *Viburnum carlesii*.

67 Colony of cowpea aphid (*Aphis craccivora*) on *Robinia.*

68 Colony of *Aphis cytisorum* on *Laburnum.*

69 Colony of *Aphis cytisorum sarothamni* on *Cytisus.*

70 *Aphis cytisorum sarothamni* damage to *Cytisus.*

Aphis craccivora Koch (67)
Cowpea aphid

Uncommon in southern England but more numerous in continental Europe, where it is associated with various hosts, especially Fabaceae. The aphids are 1.4–2.0 mm long and shiny black; nymphs are lightly dusted with wax. Dense colonies often develop during the summer on the young shoots of false acacia (*Robinia pseudoacacia*), and infestations are often abundant on such trees in towns and cities. In southern Europe this aphid is an important pest of crops such as lucerne.

Aphis cytisorum Hartig (68)
syn. *A. laburni* Kaltenbach

This greenish-black or blackish aphid infests the leaves and developing pods of *Laburnum* and Spanish broom (*Spartium junceum*). Colonies appear somewhat greyish due to the presence of secreted wax. Infestations are often abundant on established hosts, but are not of economic importance.

Aphis cytisorum sarothamni Franssen (69–70)

This subspecies is very similar to *Aphis cytisorum* but occurs only on wild and cultivated broom (*Cytisus*), affecting the leaves and developing pods. Cast skins accumulate on host plants amongst excreted honeydew, and are often more conspicuous on maturing pods than the live aphids. Although widespread, this aphid is not an important pest.

Aphis farinosa Gmelin (71–72)
Small willow aphid

Dense ant-attended colonies of this widespread and often abundant aphid occur on the young shoots of willow (*Salix*), including ornamental cultivars. Breeding continues from spring to mid-summer, with

71 Colony of small willow aphid (*Aphis farinosa*).

72 Ant-attended colony of small willow aphid (*Aphis farinosa*) on *Salix*.

73 Colony of ivy aphid (*Aphis hederae*).

sexual forms appearing from June or July onwards. The apterous viviparous females (1.7–2.3 mm long) are mottled green and yellow, with mainly pale, moderately long siphunculi. Males are characteristically reddish orange and, owing to their coloration, nymphs of this sex when developing within colonies are sometimes mistaken for predatory midge larvae.

Aphis genistae Scopoli
Genista aphid
Widely distributed but local on cultivated Spanish gorse (*Genista hispanica*), forming dense, greyish colonies on the shoots. Heavy infestations in summer weaken plants and spoil their appearance. Apterae (1.4–2.6 mm long) are blackish-bodied (lightened by greyish wax), with relatively short siphunculi and an elongate cauda.

Aphis hederae Kaltenbach (**73**)
Ivy aphid
This widespread and generally common, dark brown to blackish aphid is most often associated with European ivy (*Hedera helix*) growing on old walls and tree trunks, infestations developing on the young leaves and shoots. Colonies also develop on ornamental ivy plants and on other cultivated Araliaceae. Although commonly overwintering in the egg stage, live aphids will survive the winter if conditions are favourable.

Aphis ilicis Kaltenbach (74)
Holly aphid

Infestations of this widely distributed, greenish-black to reddish-brown or greyish-brown aphid occur on new shoots and young foliage of holly (*Ilex*). Attacks, which result in significant leaf curling, tend to occur only on young plants, on severely pruned bushes or on hedges producing an abundance of new growth. Colonies die out in the absence of young growth (mature foliage apparently being unsuitable for their continued survival). The winter is spent in the egg stage.

Aphis nerii Boyer de Fonscolombe (75)
Oleander aphid

Although associated mainly with oleander (*Nerium oleander*), this entirely parthenogenetic aphid also infests other Apocynaceae, including Chinese dregea (*Dregea sinensis*), as well as plants such as *Citrus* and

74 Colony of holly aphid (*Aphis ilicis*).

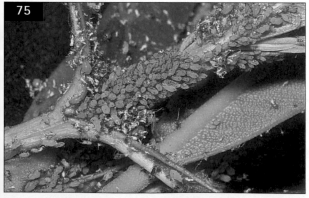

75 Colony of oleander aphid (*Aphis nerii*).

milkweed (*Asclepias*). Of southern European origin but now established in many other warm parts of the world. The aphids occur mainly on the young, succulent growth, and very large colonies develop on the shoots and along the midrib of expanded leaves. Direct damage is of little significance but the aphids produce considerable quantities of honeydew upon which sooty moulds develop, spoiling the appearance of ornamental hosts. The aphids are a particular problem on frequently pruned and heavily fertilized plants that produce an abundance of new growth. Apterae (1.5–2.6 mm long) are bright lemon-yellow (occasionally tinged with green), with dark legs and antennae, a black cauda and relatively long, stout, black siphunculi.

Aphis newtoni Theobald (76–77)
Iris aphid

This dark green to brownish-black aphid is recorded only from various countries in north-western Europe, including England, Germany, the Netherlands and Scandinavia, where it is associated with *Iris*. Large numbers often build up on the lower parts of leaves and, later, on the young inflorescences. Colonies are commonly attended by ants and are sometimes protected by ant-built earthen canopies.

Aphis pomi Degeer (78)
Green apple aphid

Infestations of this generally common aphid often occur on *Cotoneaster*, crab-apple (*Malus*), firethorn (*Pyracantha*), hawthorn (*Crataegus*), medlar (*Mespilus germanica*), rowan (*Sorbus aucuparia*) and other species of *Sorbus*, but they are of greater significance on apple trees. Heavy attacks on ornamentals occur most commonly from June or July onwards. Dense ant-attended colonies develop on the underside of leaves and on the new shoots. The aphids (1.3–2.3 mm long) are bright green or yellowish green, with short antennae and black or dark brown siphunculi. They excrete quantities of honeydew, and foliage and shoots soon become blackened by the accumulation of sooty moulds. Heavy summer infestations cause leaf curl and check the growth of new shoots. Damage to nursery stock may be considerable, but attacks on established trees and shrubs are usually of little or no importance.

Aphis sambuci Linnaeus (79)
Elder aphid

Widespread and often abundant on elder (*Sambucus*), including ornamental forms. In spring, dense, ant-attended colonies develop on the young shoots, infestations being most common on bushes growing in suburban areas. In July there is a migration of winged

76 Colony of iris aphid (*Aphis newtoni*).

77 Earthen shelter around colony of iris aphid (*Aphis newtoni*).

78 Colony of green apple aphid (*Aphis pomi*) on *Cotoneaster*.

79 Colony of elder aphid (*Aphis sambuci*).

forms to the roots of various secondary hosts, including cultivated species of *Dianthus* and saxifrage (*Saxifraga*); colonies on these summer hosts survive until the autumn, when a return migration to elder takes place. The winter is normally passed in the egg stage on elder but it is possible that a few aphids are able to survive on the roots. Apterae on elder are 1.9–3.5 mm long, greyish green to yellowish brown (coated with whitish or greyish wax), with dark, elongate and tapered siphunculi and a bluntly rounded cauda; apterae on summer hosts are smaller, and bluish green in colour.

Aphis schneideri (Börner) (**80**)
Permanent currant aphid

Infestations of this local, but widely distributed, aphid are most numerous on black currant but also occur on the ornamental species golden currant (*Ribes aureum*), causing distortion and tight bunching of the leaves. Adults are 1.2–2.2 mm long and dark green, dusted with bluish-grey wax. Colonies are typically ant-attended, and particularly damaging on young plants.

80 Colony of permanent currant aphid (*Aphis schneideri*) on *Ribes aureum*.

Aphis sedi Kaltenbach (81–82)

Widely distributed on various members of the Crassulaceae, including cultivated species of *Kalanchoe* and stonecrop (*Sedum*). Infestations are often heavy on greenhouse and indoor plants, affected shoots becoming distorted and coated in masses of sticky honeydew. The aphids are small (1.0–1.7 mm long), and mainly dark green to blackish green. They occur in large, ant-attended colonies, particularly on the younger growth; infestations also develop on aerial roots.

Aphis spiraecola Patch (83)

syn. *A. citricola* Patch
Green citrus aphid

A polyphagous pest of East Asian origin; now widely distributed in warmer parts of the world, including southern Europe. The relatively small aphids (1.2–2.2 mm long) are mainly green or yellowish green, with black siphunculi, and are very similar in appearance to *Aphis pomi*. In Asia and North America, *A. spiraecola* has both primary and secondary hosts (the former *Spiraea*; the latter including fruit trees, especially *Citrus*, and many ornamentals). In southern Europe and North Africa, the aphids occur throughout the year on their secondary hosts and the lifecycle does not involve a sexual phase. The aphids are capable of causing considerable damage, including extensive leaf curl; they are also vectors of plant viruses.

Aulacorthum circumflexum (Buckton) (84–85)

Mottled arum aphid

An often abundant, polyphagous pest of indoor and greenhouse ornamentals, including *Asparagus plumosus*, *Begonia*, *Calceolaria*, calla lily (*Zantedeschia aethiopica*), *Cineraria*, *Chrysanthemum*, *Cyclamen*, *Freesia*, *Fuchsia*, *Hippeastrum*, *Lilium*, *Petunia*, primrose (*Primula vulgaris*), rose (*Rosa*), saxifrage (*Saxifraga*), tulip (*Tulipa*) (including dry bulbs in store), violet (*Viola*), and various orchids and ferns; also important as a virus vector. Virtually cosmopolitan. Widely distributed in Europe.

81 Colony of *Aphis sedi* on shoot of a cactus.

82 Colony of *Aphis sedi* on aerial roots of a cactus.

83 Leaf curl caused by green citrus aphid (*Aphis spiraecola*).

84 Mottled arum aphid (*Aulacorthum circumflexum*).

DESCRIPTION
Apterous female: 1.2–2.6 mm long; shiny whitish to greenish white, with blackish markings on the abdomen which often form a horseshoe-shaped pattern; antennae and legs mainly pale; siphunculi and cauda pale and elongate. **Alate:** 1.6–2.4 mm long; similar to aptera. **Nymph:** similar to adult but lacking the black body pattern.

LIFE HISTORY
This species is entirely parthenogenetic. Breeding continues throughout the year if conditions are favourable but the aphids are usually most numerous in the early spring. Winged forms occur occasionally, and these help to spread infestations from plant to plant; along with apterae, they are responsible for transmitting certain plant viruses. Infestations often occur on outdoor plants during the summer, particularly in sheltered situations. The aphids excrete considerable quantities of honeydew.

DAMAGE
Aphids occur on foliage and flowers, and heavy infestations are directly harmful to many kinds of ornamental plants; attacks on cyclamen and certain lilies, e.g. calla lily, are particularly damaging. Infested plants are often soiled by honeydew and sooty moulds; they may also become infected by viruses, such as dahlia mosaic virus, primula mosaic virus and tulip breaking virus, which are commonly transmitted by this species.

Aulacorthum solani (Kaltenbach) (**86**)
syn. *Myzus pseudosolani* (Theobald)
Glasshouse & potato aphid
In unheated situations, a generally common pest of greenhouse-grown crops, including ornamentals such as *Capsicum, Geranium, Pelargonium* and winter cherry (*Solanum capsicastrum*). Also present in summer on various outdoor hosts. Virtually cosmopolitan. Widely distributed in Europe.

DESCRIPTION
Apterous female: 1.8–3.0 mm long; pear-shaped, shiny greenish yellow, with darker patches at base of siphunculi; antennae about as long as body; siphunculi pale with dark tips, long, slender, tapered and distinctly flanged. **Alate:** 1.8–3.0 mm long; head and thorax dark brown to black; abdomen yellowish green, marked with dark brown spots and crossbars. **Nymph:** yellowish green and shiny, with dusky legs and antennae.

LIFE HISTORY
All stages overwinter, including eggs. In spring, infestations spread to various hosts, both indoors and outside. Colonies reach the peak of development in July but largely die out by the autumn. In favourable protected habitats the aphids continue to breed parthenogenetically throughout the year, producing both wingless and winged forms.

DAMAGE
Infestations weaken plants; they also spoil the appearance and quality of hosts.

85 Colony of mottled arum aphid (*Aulacorthum circumflexum*).

86 Glasshouse & potato aphid (*Aulacorthum solani*).

87 Damage on *Prunus* caused by leaf-curling plum aphid (*Brachycaudus helichrysi*).

Brachycaudus helichrysi (Kaltenbach) (87)
Leaf-curling plum aphid

An often serious pest of damson and plum, including various ornamental kinds of *Prunus*, and a summer pest of herbaceous plants such as *Chrysanthemum*, *Cineraria*, forget-me-not (*Myosotis*), Michaelmas daisy (*Aster*) and periwinkle (*Vinca*). Cosmopolitan. Widely distributed in Europe.

DESCRIPTION

Apterous female: 0.9–2.0 mm long; brownish to yellowish green, round and shiny; antennae short; siphunculi pale, short and flanged; cauda tongue-like.

LIFE HISTORY

Eggs are laid in the autumn on various kinds of *Prunus* and hatch in the early spring before bud-burst; in many areas the eggs hatch in November and December. At first, nymphs feed at the base of the buds but, eventually, as the buds open they move onto the new blossoms and foliage. Colonies develop throughout the spring, culminating in the production of winged forms in May. These aphids migrate to summer herbaceous hosts, where colonies develop before a return migration to primary hosts in the autumn. Apterae on secondary (summer) hosts are distinctly smaller and paler than those on primary hosts.

DAMAGE

Primary hosts: infestations cause severe leaf curl, a yellowing of foliage and distortion of new growth, and are of particular significance on nursery stock and young trees. **Secondary hosts:** plants are severely stunted and distorted; leaves also become mottled with yellow, affecting the overall appearance and quality of hosts.

Brachycaudus cardui (Linnaeus) (88)
Thistle aphid

Minor infestations of this generally common aphid sometimes occur in summer on ornamental plants, especially silver ragwort (*Senecio cineraria*). The aphids are similar in appearance to *Brachycaudus helichrysi* but green to shiny brownish black, with distinctive black siphunculi.

Cavariella aegopodii (Scopoli) (89)
Willow/carrot aphid

This generally common aphid overwinters on willow (*Salix*), especially crack willow (*S. fragilis*) and white willow (*S. alba*). Infestations occur in the early spring on the young shoots. Winged forms are then produced, and these migrate in May to various umbelliferous hosts (Apiaceae). Large colonies often develop on the summer hosts from late May to early July, before the production of winged forms and a return migration to the primary, winter hosts. Apterae (1.0–2.6 mm long) are green or yellowish green, with swollen siphunculi and a pair of closely set hairs arising from a prominent tubercle on the eighth abdominal tergite; alatae (1.4–2.7 mm long) are green or yellowish green, with a black patch on the abdomen.

Chaetosiphon tetrarhodum (Walker)
This small, light green to yellowish-green aphid occurs on the young shoots and underside of leaves of rose (*Rosa*) but, unlike many rose-infesting species, the colonies are not ant-attended. Apterae are 1.0–2.6 mm long, with capitate body hairs and a short cauda. Alatae (1.2–2.4 mm long) are greenish, with a blackish head and thorax, and a black mark on the abdomen. In favourable conditions, colonies often persist throughout the year; they are particularly damaging on climbing roses producing an abundance of new growth.

Coloradoa rufomaculata (Wilson)
Small chrysanthemum aphid

This very small, green aphid was introduced from America into northern Europe where it is now a widespread, but usually only minor, pest of greenhouse *Chrysanthemum*. The aphids infest the stems and underside of the leaves, including senescing foliage; they are also vectors of viruses, including chrysanthemum virus B. Apterae are 1.0–1.7 mm long, with spatulate body hairs; the mainly dark siphunculi are very long (longer than the antennae) and swollen close to their apex; the cauda is triangular.

Cryptomyzus korschelti (Börner) (**90–91**)

Infestations of this widely distributed species are associated with alpine currant (*Ribes alpinum*). The aphids cause noticeable discoloration and blistering of the foliage, and damage is often severe. The aphids occur on alpine currant throughout the spring and early summer, having overwintered in the egg stage. Winged forms are produced during the summer, and these migrate to woundwort (*Stachys*), the secondary host, where breeding continues. There is a return migration to alpine currant in the autumn. The aphids are delicate, shiny and rather plump, and whitish to light orange; the siphunculi are relatively long and thin.

Dactynotus tanaceti (Linnaeus)

This uncommon aphid occurs occasionally on greenhouse-grown *Chrysanthemum*, forming colonies on the underside of leaves. Apterae are 2.0–2.5 mm long, dark reddish brown, with long antennae and long, black siphunculi and capitate body hairs.

88 Colony of thistle aphid (*Brachycaudus cardui*) on *Senecio cineraria*.

89 Colony of willow/carrot aphid (*Cavariella aegopodii*) on *Salix triandra*.

90 Galls of *Cryptomyzus korschelti* on *Ribes alpinum*.

91 Colony of *Cryptomyzus korschelti* on *Ribes alpinum*.

92 Colony of tulip bulb aphid (*Dysaphis tulipae*).

93 Colony of hawthorn/carrot aphid (*Dysaphis crataegi*) on *Crataegus.*

94 Gall of hawthorn/carrot aphid (*Dysaphis crataegi*) on leaf of *Crataegus.*

Dysaphis tulipae (Boyer de Fonscolombe) (92)
Tulip bulb aphid

A common pest of stored bulbs and corms, including *Crocus*, *Hippeastrum, Lilium* and snowdrop (*Galanthus nivalis*), but especially Gladiolus, *Iris* and tulip (*Tulipa*); also a virus vector. Virtually cosmopolitan. Widely distributed in Europe.

DESCRIPTION
Apterous female: 1.5–2.5 mm long; pale yellowish, greyish or pinkish brown, dusted with white wax; siphunculi dark, tapered and relatively short; cauda small and triangular. **Alate:** 1.5–2.3 mm long; abdomen with a black dorsal mark and black siphunculi.

LIFE HISTORY
This entirely parthenogenetic aphid occurs on bulbs and corms in store, breeding beneath the dried outer scales throughout the year if conditions are favourable. When infested bulbs or corms are planted out, aphid numbers increase rapidly on the young shoots; later, the flowers, flower spikes and developing seed pods may also be colonized. Colonies are sometimes ant-attended (cf. *Rhopalosiphoninus* spp., p. 70).

DAMAGE
Infestations severely check and distort the growth of young shoots, and heavily infested bulbs and corms may fail to develop. A vector of lily symptomless virus and tulip breaking virus.

Dysaphis crataegi (Kaltenbach) (93–94)
Hawthorn/carrot aphid

This aphid overwinters in the egg stage on hawthorn (*Crataegus*), where spring colonies of blackish aphids lightly dusted with wax inhabit conspicuous, deep-red pseudo-galls on the leaves. Winged aphids later migrate to wild and cultivated carrot (*Daucus carota*), where they initiate dense, ant-attended colonies on the tap root and leaf bases. Further winged forms then return to hawthorn, the primary host, where winter eggs are laid.

Dysaphis sorbi (Kaltenbach) (95)
syn. *D. brevirostris* (Börner)
Rowan aphid

Infestations of this aphid cause distortion of leaves of rowan (*Sorbus aucuparia*), infested shoots becoming contaminated with cast aphid skins, honeydew and sooty moulds. Eggs overwinter on this primary host and hatch in the spring. Colonies of aphids then develop from April onwards, culminating in the production of winged forms in late June or July. There is then a migration to secondary (summer) hosts, such as bell flower (*Campanula*). Apterae (*c.* 2 mm long) on rowan are yellowish brown to yellowish green, with pale siphunculi.

Elatobium abietinum (Walker) (**96–97**)
Green spruce aphid
A major pest of ornamental and forest spruce (*Picea*) trees. Spruces of North American origin are usually more severely affected than Eurasiatic ones. Formerly restricted to North America but now present throughout Europe.

DESCRIPTION
Apterous female: 1.5–1.8 mm long; green, with red eyes; siphunculi long, cauda elongate.

LIFE HISTORY
In parts of mainland Europe which experience harsh winters, a sexual phase occurs in the lifecycle and the aphids overwinter in the egg stage. However, in the British Isles and other areas with less hostile winters, this species is almost entirely parthenogenetic, and wingless adults and nymphs occur on spruce trees throughout the year. These aphids are most numerous from August to early June but details of their development vary considerably from area to area. The aphids feed on the underside of the needles and are very active, often wandering along the shoots. Winged female aphids appear from May onwards, and this brings an end to colony development. These winged aphids produce small numbers of nymphs, which aestivate for several weeks before feeding and completing their development.

DAMAGE
Infestations do not occur on the new growth, but the aphids cause extensive yellowing of the older foliage. Severe attacks lead to distinct bronzing and premature loss of needles. Damage occurs from September onwards and, in mild conditions, continues to increase throughout the winter. In many cases, however, significant damage does not occur until the spring, reaching a peak in May.

95 Rowan aphid (*Dysaphis sorbi*) damage to shoot of *Sorbus aucuparia*.

96 Green spruce aphid (*Elatobium abietinum*) damage to needles of *Picea*.

97 Green spruce aphid (*Elatobium abietinum*) damage to branch of *Picea*.

98 Colony of honeysuckle aphid (*Hyadaphis passerinii*).

99 Honeysuckle aphid (*Hyadaphis passerinii*) damage to flower buds of *Lonicera*.

Hyadaphis passerinii (del Guercio) (98–99)
syn. *H. lonicerae* Börner
Honeysuckle aphid
A generally common pest of honeysuckle (*Lonicera*). Virtually cosmopolitan. Widespread in Europe.

DESCRIPTION
Apterous female: 1.3–2.3 mm long: dark bluish green, with a waxy bloom; antennae and legs black; siphunculi black and swollen; cauda black and elongate. **Alate:** 1.3–2.3 mm long; abdomen green, mottled with darker green, and with a dark patch at the base of the siphunculi.

LIFE HISTORY
Colonies develop on the underside of honeysuckle leaves from the early spring onwards. Winged forms are eventually produced, and these migrate during the summer to umbelliferous hosts (Apiaceae), especially hemlock (*Conium maculatum*). Colonies on honeysuckle then die out. A return migration to honeysuckle, the primary host, takes place in the autumn.

DAMAGE
Heavy attacks affect both shoot growth and flower development. Infested hosts also become sticky with honeydew and fouled by sooty moulds.

Hyadaphis foeniculi (Passerini)
Fly-honeysuckle aphid
Virtually identical to the previous species, with which it is often confused, but associated with fly-honeysuckle (*Lonicera xylosteum*).

Hyalopterus pruni (Geoffroy) (100)
Mealy plum aphid
A cosmopolitan and generally common pest of damson and plum; also inhabits certain other kinds of *Prunus* grown as ornamentals, of which almond (*P. dulcis*) is an example. The winter is passed in the egg stage on host trees. The eggs hatch in the early spring, usually by the white-bud stage, and dense colonies of the light green to bluish-grey, mealy-coated aphids (1.5–2.6 mm long) eventually develop on the leaves and shoots. The aphids produce vast quantities of honeydew and, although they do not cause leaf curl (cf. *Brachycaudus helichrysi*, p. 58), infested plants are disfigured by accumulations of mealy wax and honeydew, and by the subsequent development of sooty moulds. In summer, winged aphids migrate mainly to reed (*Phragmites communis*). A return migration to winter hosts occurs in the autumn.

Idiopterus nephrelepidis Davis
Fern aphid
An American species, now widely distributed in Europe on cultivated ferns, especially ladder fern (*Nephrolepis exaltata*), maidenhair fern (*Adiantum capillus-veneris*), *Polypodium* and *Pteris* growing in heated conditions; infestations also spread to African violet (*Saintpaulia hybrida*), *Cyclamen* and *Streptocarpus*.

DESCRIPTION
Apterous female: 1.3 mm long; dark green to black, with capitate body hairs and whitish antennae; legs long, slender and mainly white; siphunculi white but blackish basally. **Alate:** 1.4–1.6 mm long; black to dark olive-green; wings mottled with dark brown.

100 Colony of mealy plum aphid (*Hyalopterus pruni*) on *Prunus.*

101 Colony of *Illinoia azaleae* on *Rhododendron.*

LIFE HISTORY

Aphids breed parthenogenetically throughout the year, producing winged forms in both spring and summer. Infestations are most common in greenhouses and hot-houses but also persist outdoors in mild districts.

DAMAGE

Infested fern fronds are distorted, and may turn black and die.

Illinoia azaleae (Mason) (**101**)

A common pest of container-grown azaleas (*Rhododendron*), having been introduced into Europe from North America on infested plants. Heavy infestations reduce the vigour of host plants and cause defoliation. Attacks may also occur on open-bedded bushes growing in parks and gardens. Reproduction is typically parthenogenetic and viviparous, and is continuous so long as conditions remain favourable. Adults are medium-sized, spindle-shaped, deep green and shiny, with long, mainly pale, slightly swollen siphunculi.

Illinoia lambersi (MacGillivray) (**102**)

A North American species, now well established on azalea (*Rhododendron*) in parts of Europe, including Denmark, England and the Netherlands. Colonies develop mainly on the new shoots and flower buds. The aphids are similar in appearance to *Illinoia azaleae*, but are larger and either green, pink or yellow, the various colour forms often occurring together.

102 Colony of *Illinoia lambersi* on *Rhododendron.*

Illinoia liriodendri (Monell)

Tulip-tree aphid

This North American species, which is associated with tulip tree (*Liriodendron tulipifera*), was found in France in 1998, and has since been reported elsewhere in Europe, including Britain (southern England and Wales), Germany, Italy and Slovenia. The aphid has also appeared in Japan. Heavy infestations occur on the leaves, and the aphids excrete considerable quantities of honeydew. Apterae (1.8–2.7 mm long) are mainly light green, lightly dusted with white wax, with mainly black but basally pale siphunculi and a yellow cauda.

Liosomaphis berberidis (Kaltenbach) (**103**)
Barberry aphid

This greenish-yellow, orange or pinkish aphid infests barberry (*Berberis*), colonies developing throughout the summer on the young shoots and underside of leaves. Apterae are 2.0–2.5 mm long, with pale legs and antennae, and noticeably swollen siphunculi. Little or no damage is caused.

Longicaudus trirhodus (Walker) (**104–105**)
Compact colonies of this pale yellowish-green (2.0–2.7 mm long) species develop on wild and cultivated columbine (*Aquilegia vulgaris*) during the summer months; meadow-rue (*Thalictrum*) is also a summer host. The aphids are unusual in possessing short siphunculi and a long cauda. Winged aphids, which have an irregular black mark on the abdomen, migrate to rose (*Rosa*), the winter host; small colonies are produced on this woody host in the spring, before a return migration to summer hosts takes place in June.

Macrosiphoniella sanborni (Gillette)
Chrysanthemum aphid

A generally common pest of *Chrysanthemum*. Of eastern Asian origin but now cosmopolitan. Widely distributed in Europe.

DESCRIPTION

Apterous female: 1.0–2.3 mm long; shiny, dark reddish brown to blackish brown; siphunculi black, short and stout; cauda black and elongate, slightly longer than siphunculi. **Alate:** 1.8–2.6 mm long; similar to aptera.

LIFE HISTORY

This species occurs on cultivated chrysanthemum in or near greenhouses, and reproduces parthenogenetically throughout the year. The aphids colonize the underside of leaves, and also infest the buds and flower stems. Winged forms are produced during the summer months, and these help to spread infestations from place to place, but there is no sexual phase in the lifecycle.

DAMAGE

Infested buds and flowers are malformed. The aphids transmit viruses such as chrysanthemum vein mottle virus and chrysanthemum virus B.

Macrosiphoniella absinthii (Linnaeus) (**106**)
syn. *M. artemisiae* (Buckton); *M. fasciata* del Guercio

Infestations of this widely distributed species are sometimes reported on cultivated wormwood

103 Colony of barberry aphid (*Liosomaphis berberidis*) on *Berberis*.

104 Colony of *Longicaudus trirhodus* on leaf of *Aquilegia*.

105 Apterous female and nymphs of *Longicaudus trirhodus* on leaf of *Rosa*.

106 Colony of *Macrosiphoniella absinthii* on *Artemisia*.

107 Apterous female and nymphs of Essig's lupin aphid (*Macrosiphum albifrons*).

108 Colony of Essig's lupin aphid (*Macrosiphum albifrons*) on *Lupinus*.

(*Artemisia*). The aphids are dark reddish brown, partly coated in white wax that forms an attractive and characteristic pattern; the siphunculi and cauda are black. On pale, silvery-looking host plants the dark appearance of the aphids contrasts with the foliage, and colonies are very noticeable; the aphids are ant-attended.

Macrosiphoniella oblonga (Mordvilko)

This widely distributed species occurs throughout the year on protected *Chrysanthemum*, but may also overwinter in the egg stage at the base of plant stems. The aphids feed singly on the underside of older leaves and on the flower stalks. They are easily dislodged from host plants, dropping to the ground if disturbed. In addition to *Chrysanthemum*, infestations occur commonly outdoors on wild hosts such as scentless mayweed (*Tripleurospermum maritimum*). The aphids are 5 mm long, slender-bodied, green to light green, with a darker dorsal longitudinal stripe; the legs are very long and thin, the siphunculi broad and moderately long but without an apical flange; the cauda is blunt and about as long as the siphunculi. Apterous males are reddish brown, with a relatively small body and long legs.

Macrosiphum albifrons Essig (**107–108**)

Essig's lupin aphid

An important North American pest of lupin (*Lupinus*). In Europe, first reported in 1981, in southern England; now widely distributed and spreading within mainland Europe.

DESCRIPTION

Apterous female: 3.2–4.5 mm long; large, light bluish green, with a white, waxy coating; antennae, legs, siphunculi and cauda all long; siphunculi light brown, with darker tips. **Alate:** 3.2–4.5 mm long; similar to aptera but the siphunculi entirely dark. **Nymph:** similar to adult.

LIFE HISTORY

Breeding colonies develop on all parts of host plants, including flower spikes, roots, pods and senescing tissue. The reproductive rate is high and colonies build up quickly. Infestations spread rapidly during the summer, following the production of winged forms. Aphids overwinter in the crowns of mature plants.

DAMAGE

Severest damage is caused to young flowers, with whole spikelets being aborted. Plants may be killed if infestations are heavy.

Macrosiphum euphorbiae (Thomas) (**109**)

syn. *M. solanifolii* Ashmead

Potato aphid

A polyphagous North American species, first introduced into Europe in about 1917 and now cosmopolitan. Infestations occur commonly on ornamental plants, including calla lily (*Zantedeschia aethiopica*), carnation (*Dianthus caryophyllus*) and pink (*D. plumarius*), *Cineraria*, columbine (*Aquilegia vulgaris*), *Dahlia*, *Eucalyptus*, *Freesia*, *Gladiolus*, hollyhock (*Alcea rosea*), *Iris*, pot marigold (*Calendula officinalis*), rose (*Rosa*), snapdragon (*Antirrhinum*), sweet pea (*Lathyrus odoratus*), tulip (*Tulipa*) and valerian (*Valeriana*). Present throughout Europe.

DESCRIPTION

Apterous female: 1.7–3.6 mm long; greyish green to pink, and often shiny; spindle-shaped, with long antennae and legs; siphunculi very long, slender and curved; cauda long and finger-shaped. **Alate:** 1.7–3.4 mm long: similarly coloured to aptera but antennae, head, thorax and siphunculi usually yellowish brown. **Nymph:** long-bodied and pale, with a dark, central, longitudinal stripe and a slight wax coating.

LIFE HISTORY

Although sometimes overwintering in the egg stage on rose and related rosaceous hosts, most populations survive parthenogenetically in protected situations as adults or nymphs. Aphid numbers increase rapidly from early spring onwards, and infestations readily spread to other plants in May and June following the production of winged forms. Attacks often develop on flower stalks, and are most common in unheated greenhouses.

DAMAGE

Heavily infested plants become stunted and distorted. The aphids are capable of transmitting certain plant virus diseases, including freesia mosaic virus and pea enation mosaic virus, but are not important vectors.

109 Colony of potato aphid (*Macrosiphum euphorbiae*) on *Calendula*.

Macrosiphum rosae (Linnaeus) (**110–111**)

Rose aphid

An important and often abundant pest of rose (*Rosa*). Virtually cosmopolitan. Widely distributed in Europe.

DESCRIPTION

Apterous female: 1.7–3.6 mm long; green to pink or reddish brown; spindle-shaped, with long antennae and legs; siphunculi long, black (N.B. siphunculi are pale in other rose-infesting species) and tapered; cauda pale and elongate. **Alate:** 2.2–3.4 mm long; green to pinkish brown, distinctly marked with black along the sides of the abdomen.

110 Rose aphids (*Macrosiphum rosae*).

111 Colony of rose aphid (*Macrosiphum rosae*).

LIFE HISTORY

Overwintered eggs often occur on rose bushes, although adult aphids may survive the winter if conditions are favourable. Substantial colonies build up during the spring and summer, infested shoots often becoming smothered in the aphids. In summer, winged forms spread infestations to other rose bushes; they also migrate to secondary hosts such as holly (*Ilex*), scabious (*Knautia* and *Scabiosa*) and teasel (*Dipsacus fullonum*). Rose bushes remain liable to invasion throughout the summer months, and new colonies are often present throughout the autumn, until their development is curtailed by the onset of cold weather.

DAMAGE

Infestations check the growth of buds and new shoots, and commonly contaminate plants with sticky honeydew upon which sooty moulds develop. Foliage and flowers are also disfigured.

Metopolophium dirhodum (Walker) (**112**)

Rose/grain aphid

Widely distributed on rose (*Rosa*) from autumn to late spring, overwintering in the egg stage. Unlike the main rose-inhabiting species *Macrosiphum rosae* (p. 66), colonies do not persist on rose beyond June, the aphids (and winged migrants) then dispersing to cereals and grasses. *Metopolophium dirhodum* is rarely important on rose. However, in some years it is present in great profusion, and vast clouds of winged migrants are produced; in such circumstances, rose leaves and flower buds become covered in white, cast nymphal skins, giving the superficial appearance of an outbreak of disease. Apterae are elongate (2–3 mm long), and mainly shiny yellowish green, with a darker mid-dorsal longitudinal stripe; the antennae, legs, siphunculi and cauda are mainly pale and relatively long. Alatae (1.6–3.3 mm long) have a uniformly green abdomen.

Microlophium primulae (Theobald) (**113**)

Colonies of this pale yellow to pale yellowish-green aphid sometimes occur on cultivated primrose (*Primula vulgaris*). The aphids congregate on the underside of the leaves amongst cast nymphal skins, which commonly remain attached to the hairy leaf surface. Although aphid numbers are sometimes large, infested foliage is not distorted and attacks are of little or no importance. Apterae are 2.2–2.5 mm long, with long, tapered siphunculi and relatively long legs; alatae are pale yellow, with black abdominal markings.

Myzaphis rosarum (Kaltenbach)

Lesser rose aphid

A widely distributed, small, elongate-oval, somewhat flattened, green to yellowish-green aphid on wild and cultivated rose (*Rosa*), especially climbers. The apterae are 1.0–2.2 mm long, with long, swollen, dark-tipped siphunculi and a long cauda (cf. *Chaetosiphon tetrarhodum*, p. 58). They feed mainly along the midrib on either side of the leaves, and are often hidden from view amongst leaflets that are still furled. Winged females occur during the summer; these are green or yellowish green, with a darker mark on the abdomen. Infestations are usually noticed only when the aphids are unusually abundant. This species is not particularly damaging.

112 Colony of rose/grain aphid (*Metopolophium dirhodum*) on *Rosa*.

113 Colony of *Microlophium primulae* on *Primula*.

Myzus cerasi (Fabricius) and *M. pruniavium* Börner (114)

Cherry blackflies

Generally common pests of cherry, the former (as the sour cherry blackfly) on *Prunus cerasus* and the latter (as the sweet cherry blackfly, which is a more damaging species) on *P. avium*. Eurasiatic; also found in Australasia and North America. Widely distributed in Europe. The two species are morphologically identical, and often considered merely subspecies.

DESCRIPTION

Apterous female: 1.5–2.6 mm long; dark brown to black, and very shiny; front of head emarginate; siphunculi black, moderately long and tapered. **Alate:** 1.4–2.1 mm long; blackish, but abdomen yellowish brown and with a large black dorsal patch. **Nymph:** purplish to blackish.

LIFE HISTORY

Eggs, overwintering on the spurs and young shoots of cherry, hatch in March or April, before the white-bud stage. Colonies of wingless aphids then develop on the underside of the young leaves, and populations build up very rapidly. Winged forms are produced in June, at the climax of colony development, and these fly away to establish colonies on summer hosts such as bedstraw (*Galium*) and speedwell (*Veronica*). Colonies on cherry then decline and eventually die out, although breeding on the primary host often continues throughout July. A return migration to cherry takes place in the autumn, where sexual reproduction takes place and winter eggs are eventually laid.

DAMAGE

On susceptible hosts, colonies of the sweet cherry blackfly cause severe distortion of young shoots and leaves, and affected foliage later turns brown or black; heavy infestations distort growth and may kill shoots. Attacks are most serious on young trees and nursery stock.

Myzus ligustri (Mosley) (115)

syn. *Myzodes ligustri* (Kaltenbach)

Privet aphid

A locally common pest of privet (*Ligustrum vulgare*), and sometimes of importance on garden hedges in suburban areas. Widespread in mainland Europe; in the British Isles most numerous in southern England. Also present in North America.

DESCRIPTION

Apterous female: 1.2–1.5 mm long; shiny yellow to yellowish green; antennae long and green; siphunculi long, yellow to green, with dusky tips, and slightly swollen just beyond the mid-region; cauda yellowish and tapered. **Alate:** 1.5–1.8 mm long; yellowish to yellowish green, with a brown head, brown thoracic lobes and brown abdominal crossbars, the last-mentioned often forming a distinct patch between the dark, moderately long, slightly swollen siphunculi.

LIFE HISTORY

Eggs are deposited on the shoots of privet in early winter and hatch in the spring. Colonies of aphids then develop on the foliage in slightly rolled leaves on the new shoots, reaching the peak of their development from June or July onwards. Males are produced late in the season, and these mate with females (oviparae) which eventually deposit the overwintering eggs.

DAMAGE

Infestations are most common on regularly clipped hedges with an abundance of new growth. Individual leaves become discoloured and curl longitudinally to produce noticeable distortion on the young shoots. Persistent attacks affect the appearance and vigour of hosts, and may lead to leaf death and premature defoliation.

Myzus ornatus Laing (116)

Violet aphid

A generally common, polyphagous pest of herbaceous ornamentals such as African violet (*Saintpaulia hybrida*), azalea (*Rhododendron*), *Begonia*, busy lizzie (*Impatiens*), *Veronica* and violet (*Viola*). Cosmopolitan. Widely distributed in Europe.

DESCRIPTION

Apterous female: 1.0–1.7 mm long; pale brownish yellow or dull green, with paired, blackish markings on the thorax and abdomen; antennal tubercles distinctly convergent; siphunculi pale, cylindrical and moderately long. **Nymph:** light green, with dark red eyes.

LIFE HISTORY

This species is entirely parthenogenetic, and occurs in small numbers on the leaves of various ornamental plants. Under favourable conditions, including greenhouses in northern Europe, breeding continues throughout the year. The aphids produce considerable quantities of honeydew, which coats the foliage and upon which white, cast nymphal skins accumulate.

114 Cherry blackfly (*Myzus cerasi*) damage to leaves of *Prunus avium.*

115 Privet aphid (*Myzus ligustri*) damage to leaves of *Ligustrum.*

116 Violet aphid (*Myzus ornatus*) damage to leaves of *Begonia.*

117 Colony of peach/potato aphid (*Myzus persicae*).

DAMAGE

Direct feeding damage is of little importance, as the aphids do not form dense colonies. However, honeydew is often a problem on pot plants, and the presence of cast nymphal skins on foliage and flowers is unsightly.

Myzus persicae (Sulze) (**117**)
Peach/potato aphid

An often abundant and very polyphagous pest of herbaceous plants, but of greatest significance as a vector of plant virus diseases. Often present on ornamentals such as African violet (*Saintpaulia hybrida*), *Begonia, Calceolaria*, calla lily (*Zantedeschia aethiopica*), carnation (*Dianthus caryophyllus*) and pink (*D. plumarius*), *Chrysanthemum, Fuchsia*, hyacinth (*Hyacinthus orientalis*), morning glory (*Ipomoea*), nasturtium (*Tropaeolum*), *Phlox*, primrose (*Primula vulgaris*), rose (*Rosa*), snapdragon (*Antirrhinum*), sweet pea (*Lathyrus odoratus*), tulip (*Tulipa*), violet (*Viola*), winter cherry (*Solanum capsicastrum*) and various cacti. Cosmopolitan. Widely distributed in Europe.

DESCRIPTION

Apterous female: 1.2–2.5 mm long; light to yellowish green; head with inwardly directed tubercles at base of antennae; siphunculi moderately long, with apical half slightly swollen, pale but dark tipped; cauda triangular. **Alate:** 1.4–2.3 mm long; head and thorax blackish brown; abdomen green to yellowish green and often pinkish, with a dark dorsal patch which includes a pale central mark.

LIFE HISTORY

Although sometimes overwintering in the egg stage on nectarine and peach trees, this aphid more frequently survives as adults and nymphs on herbaceous (secondary) hosts such as outdoor brassicas and protected lettuce. The aphids breed parthenogenetically throughout much of the year. They are very restless and frequently wander over the foodplant, on flower crops often migrating upwards to invade the buds. Winged forms are produced in the summer months, and these readily spread infestations to other hosts. Colonies are never very populous and reach their maximum development in July.

DAMAGE

This aphid is a major vector of plant viruses, including carnation latent virus, chrysanthemum virus B and various mosaic viruses (e.g. carnation, dahlia and orchid mosaic virus). Direct feeding is rarely significant, although the aphids sometimes damage protected crops, causing distortion of leaves, buds and flowers, and stunting of terminal shoots. Although populations are usually small, the mere presence at harvest of aphids on crops affects marketability and is, therefore, unacceptable.

Myzus ascalonicus Doncaster
Shallot aphid

A widely distributed, virtually cosmopolitan species; unknown before 1940 but now an often common pest of herbaceous plants, including ornamentals such as *Gladiolus, Hippeastrum, Lilium* and winter cherry (*Solanum capsicastrum*). The aphids breed asexually throughout the year, and often survive the winter on stored bulbs and corms, and on various greenhouse plants. In mild conditions colonies also survive the winter outdoors. Winged migrants appear in the spring, and these spread infestations to various herbaceous summer hosts. Infestations lead to considerable distortion and malformation of foliage and flower trusses. Further, when previously infested bulbs and corms in store are planted out, new shoots are often weak and noticeably distorted. Apterae (1.1–2.2 mm long) are light brown, greenish brown or yellowish brown, shiny and distinctly convex; the head is emarginate, with slightly convergent prominences, the siphunculi distinctly swollen towards the tip, and the cauda bluntly triangular and barely visible from above.

Pentalonia nigronervosa Coquerel (**118**)
Banana aphid

A tropical and subtropical species, established in various parts of mainland Europe on hot-house palms and certain other monocotyledonous hosts. The aphids are small (1.1–1.8 mm long) and mainly dark brown, with stout siphunculi. Colonies develop on the underside of the leaves and are usually attended by ants.

Rhopalosiphoninus latysiphon (Davidson)
Bulb & potato aphid

Infestations of this cosmopolitan species occur on plant roots and on tulip (*Tulipa*) bulbs in store, building up rapidly during forcing; they also occur on other stored bulbs, corms and tubers, including *Gladiolus*. When developing leaves are attacked they often turn brown and shrivel at the tips. The aphids (1.4–2.5 mm long) are plump and dark olive-green, with shiny black and strongly swollen siphunculi. Unlike *Dysaphis tulipae* (p. 60), which also attacks tulip bulbs, this species is never attended by ants.

Rhopalosiphoninus staphyleae (Koch)
syn. *Hyperomyzus tulipaella* (Theobald)
Mangold aphid

Colonies of this widespread aphid occur during the winter on stored bulbs and corms, including *Crocus*, day-lily (*Hemerocallis*), *Hippeastrum, Lilium* and tulip (*Tulipa*). Aphids are 1.5–2.4 mm long and mainly olive-brown, with dark crossbands along the back; the head and first two antennal segments bear numerous small spines, and the siphunculi are noticeably swollen.

Rhopalosiphum nymphaeae (Linnaeus) (**119**)
Water-lily aphid

A generally common pest of aquatic plants, especially water-lilies (*Nymphaea* and *Nuphar*). Virtually cosmopolitan. Widely distributed in Europe.

DESCRIPTION

Apterous female: 1.6–2.6 mm long; dark olive-green to brown, lightly dusted with whitish wax; siphunculi relatively long and swollen, mainly pale but dark apically. **Alate:** 1.6–2.6 mm long; dark brown to shiny black.

LIFE HISTORY
The winter is passed in the egg stage on blackthorn (*Prunus spinosa*) and various other species of *Prunus*, the primary hosts. Eggs hatch in the following spring and small colonies then develop on the shoots. Winged forms are produced in the early summer and these migrate to various aquatic plants, including arrow-head (*Sagittaria sagittifolia*), false bulrush (*Typha latifolia*) and water-plantain (*Alisma plantago-aquatica*); large colonies often develop on water-lilies, the aphids congregating along the leaf veins and also invading the flowers. In the autumn, there is a return migration to primary hosts, where sexual reproduction occurs and eggs are laid.

118 Colony of banana aphid (*Pentalonia nigronervosa*).

DAMAGE
Primary hosts: if attacks are heavy, young leaves at the tips of the new shoots become crinkled. **Secondary hosts:** heavy infestations cause considerable distortion of the stems and foliage, and also discolour the flowers.

Rhopalosiphum insertum (Walker) (**120**)
syn. *R. crataegellum* (Theobald)
Apple/grass aphid
Minor infestations of this generally common species sometimes occur on crab-apple (*Malus*), *Cotoneaster*, hawthorn (*Crataegus*), Japanese quince (*Chaenomeles japonica*), medlar (*Mespilus germanica*), rowan (*Sorbus aucuparia*) and other species of *Sorbus*, causing leaf distortion. The aphids overwinter as eggs on rosaceous primary hosts, with colonies developing on the shoots in the following spring. Eventually, winged forms are produced and these migrate to the roots of grasses, the secondary hosts. A return migration to primary hosts occurs in the autumn. Apterae (2.1–2.6 mm long) are yellowish green, with darker longitudinal stripes down the body, short, dark-tipped antennae, and short, light green, distinctly flanged siphunculi. Attacks on ornamentals are usually insignificant but may be of some importance on young plants, including nursery stock.

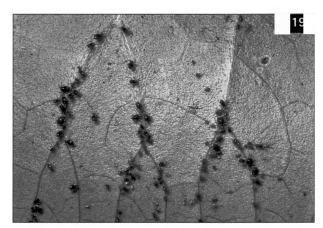
119 Colony of water-lily aphid (*Rhopalosiphum nymphaeae*).

120 Apple/grass aphid (*Rhopalosiphum insertum*) damage to leaves of *Sorbus sargentiana*.

Subfamilies CALLAPHIDINAE, DREPANOSIPHINAE & PHYLLAPHIDINAE

Aphids with terminal process of the antennae of variable length, siphunculi usually stumpy or broadly conical but sometimes pore-like or long and swollen, anal plate often divided into two lobes, and cauda knob-like or rounded.

Betulaphis quadrituberculata (Kaltenbach) (121)

A locally common pest of birch (*Betula*). Widely distributed in Europe.

DESCRIPTION

Apterous female: 1.3–2.0 mm long; greenish or yellowish, sometimes marked with darker green or yellow; abdomen with four rows of hairs along the back and a series along either side, the longer hairs distinctly capitate; siphunculi conical and stumpy; cauda broadly rounded and subtriangular, projecting slightly beyond a deeply cleft subanal plate. **Alate:** 1.5–2.2 mm long; similar to aptera but with capitate hairs restricted to the eighth abdominal segment; larger-bodied individuals have the head, antennae, siphunculi and parts of thorax, legs and abdomen dusky.

LIFE HISTORY

Eggs overwinter on host trees and hatch in the spring. Aphids then feed on the underside of the leaves, individuals of the first generation developing into either winged or wingless forms. Nymphs of subsequent generations, most of which develop into wingless adults, also occur on the underside of the leaves, either singly or in small groups. There is a succession of generations throughout the spring and summer months but dense colonies develop only if conditions are particularly favourable.

DAMAGE

Infestations are usually of minor importance. Any effect on tree growth is slight, even when noticeable colonies develop.

Callaphis juglandis (Goeze) (122)

Large walnut aphid

A generally common but minor pest of walnut (*Juglans*). Colonies develop on the underside of the leaves, the aphids causing slight yellowing and premature leaf fall. Heavy infestations reduce the vigour of young trees but attacks on established plants are of little or no importance. The aphids (2–4 mm long) are bright greenish yellow to yellow, marked with brownish black; the wing veins of alates are dusky bordered.

Callipterinella tuberculata (von Heyden) (123)

syn. *Aphis betularia* (Kaltenbach)

Minor infestations of this widespread but locally distributed species occur on birch (*Betula*), the aphids occurring singly or in small groups on the leaves. Colonies are sometimes noticed in the new foliage of young trees but are not of importance. The aphids (1–2 mm long) are basically greenish, with the body variably patterned by dark sclerites, the darkened areas often being extensive; the siphunculi are small, and the cauda very short and inconspicuous.

121 Colony of *Betulaphis quadrituberculata* on *Betula*.

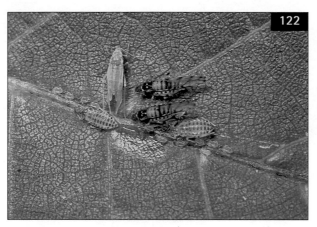

122 Colony of large walnut aphid (*Callaphis juglandis*).

123 Colony of *Callipterinella tuberculata* on *Betula*.

Drepanosiphum platanoidis (Schrank) (**124–125**)
Sycamore aphid

Generally abundant on sycamore (*Acer pseudoplatanus*), but mainly a problem as a copious producer of contaminating honeydew. Widely distributed in Europe; also present in North America.

DESCRIPTION
Alate female: 3.2–4.3 mm long; elongate, light green or greyish green, sometimes tinged with reddish, with darker markings dorsally; antennae with a long terminal process; siphunculi long and swollen; cauda small and rounded. **Nymph:** light green to whitish green, with red eyes; antennae with blackish markings.

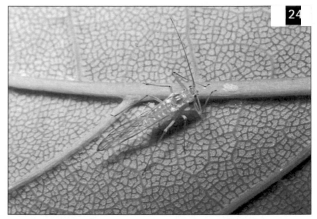

124 Sycamore aphid (*Drepanosiphum platanoides*).

LIFE HISTORY
Overwintered eggs hatch at bud-burst. Small numbers of alate females then develop rapidly on the new shoots and beneath the expanding leaves. All adult forms are winged. In summer, breeding ceases but females continue to survive on the underside of expanded leaves where they may often be found in considerable numbers. Although mainly sedentary they readily become active if disturbed. Breeding recommences in the autumn, when leaves begin to senesce, with populations again increasing rapidly as sexual forms are produced; winter eggs are laid in November or December.

DAMAGE
The main problem is the excreted honeydew, which pollutes host plants and anything beneath the trees.

125 Nymph of sycamore aphid (*Drepanosiphum platanoides*).

126 Lime leaf aphid (*Eucallipterus tiliae*).

127 Nymphs of lime leaf aphid (*Eucallipterus tiliae*).

128 Silver birch aphid (*Euceraphis betulae*).

129 Nymph of silver birch aphid (*Euceraphis betulae*).

Eucallipterus tiliae (Linnaeus) (**126–127**)

Lime leaf aphid

An abundant and widely distributed pest of lime (*Tilia*), but of importance mainly as a producer of contaminating honeydew. Present throughout Europe; also occurs in Central Asia, New Zealand and North America.

DESCRIPTION

Alate female: 1.8–3.0 mm long; black and yellow; wing veins dusky-bordered; siphunculi dark and stumpy; cauda rounded and bearing a small, pigmented, dorsal tubercle. **Nymph:** greenish yellow, marked with several blackish platelets.

LIFE HISTORY

Eggs laid on lime trees in the autumn hatch in the following spring. Colonies of winged aphids then build up on the new shoots and expanding foliage. Such progeny readily spread infestations from tree to tree throughout the spring and early summer. Reproduction slows down in mid-summer but picks up with the initiation of sexual forms in the early autumn.

DAMAGE

Infestations have little direct effect on tree growth. However, the vast quantities of honeydew produced by this species are commonly a nuisance, regularly contaminating cars, garden furniture, pavements and plants beneath infested trees; such contamination is most significant in dry weather. Sooty moulds developing on the honeydew is also a problem.

130 Colony of downy birch aphid (*Euceraphis punctipennis*).

131 Colony of hazel aphid (*Myzocallis coryli*).

Euceraphis betulae (Koch) (**128–129**)

Silver birch aphid

A very common pest of silver birch (*Betula pendula*), including cultivars such as Swedish birch (*B. pendula* 'Dalecarlica'). Infestations frequently develop on trees in gardens and nurseries. Widely distributed in Europe.

DESCRIPTION

Alate female: 3–4 mm long; body elongate, mainly light green or yellowish green and coated in a bluish-white waxy secretion; legs and antennae very long; siphunculi very short; cauda knobbed. **Nymph:** green and waxy, with distinctly dark legs and siphuncular rims.

LIFE HISTORY

This species overwinters in the egg stage on the shoots. The eggs hatch in the spring. Nymphs then develop on the leaves, the first adults appearing in late May or early June. The aphids occur on the youngest, still unfurling leaves and on the underside of fully expanded ones, and infestations persist throughout the spring and summer. Adult aphids are always winged and do not aggregate, typically occurring singly; they are also very active and immediately drop from the foodplant if disturbed. Nymphs either occur singly or in small groups.

DAMAGE

Foliage is not distorted but does become sticky with honeydew; infested plants are also disfigured by sooty moulds.

Euceraphis punctipennis (Zetterstedt) (**130**)

Downy birch aphid

This aphid is virtually identical in appearance and habits to the previous species but breeds on downy birch (*Betula pubsecens*).

Myzocallis coryli (Goeze) (**131**)

Hazel aphid

Generally abundant on both wild and cultivated hazel (*Corylus*), and often a minor pest of nursery stock. Eurasiatic. Present throughout Europe. Also now found in North America and New Zealand.

DESCRIPTION

Alate female: 1.3–2.2 mm long; shiny, whitish to yellowish or light green, with large red eyes; siphunculi stumpy; cauda knob-like and projecting beyond a bilobed subanal plate; wing veins terminate in dusky spots on the wing margins. **Nymph:** pale, with body hairs capitate.

LIFE HISTORY

Overwintered eggs hatch in the spring. Aphids then feed as scattered individuals on the underside of the foliage from May onwards. Breeding continues throughout the summer months, all of the aphids developing into winged forms. Sexual morphs arise in November and oviparous females finally deposit eggs on the shoots.

DAMAGE

Although shoots of infested plants are not distorted, the foliate is disfigured by the accumulation of sticky honeydew and development of sooty moulds.

Myzocallis boerneri Stroyan

This species, presumably of southern European origin, infests chestnut-leafed oak (*Quercus castaneifolia*), Chinese cork oak (*Q. variabilis*), cork oak (*Q. suber*), Lucombe oak (*Q. × hispanica* 'Lucombeana') and Turkish oak (*Q. cerris*). It occurs locally in parks, botanic gardens and arboreta in northern Europe but does not occur on native oaks. The alatae are relatively small (1.3–2.2 mm long) and mainly yellow (cf. *Myzocallis castanicola*).

Myzocallis castanicola Baker

Widely distributed and locally common on oak (*Quercus*) and sweet chestnut (*Castanea sativa*) but not an important pest. Alatae (2.2–2.6 mm long) are deep yellow to brown, with black, bar-like markings and black siphunculi.

Myzocallis schreiberi Hille Ris Lambers & Stroyan

Minor infestations of this local, southerly distributed species occur on holm oak (*Quercus ilex*). The aphids, which are entirely parthenogenetic, occur on the underside of the leaves; they may be found throughout the year.

Phyllaphis fagi (Linnaeus) (**132–133**)
Beech aphid

A generally common pest of beech (*Fagus sylvatica*), and frequently injurious to nursery trees and hedges; infestations also occur on Persian iron wood (*Parrotia persica*). Widespread in Europe; also present in North America.

DESCRIPTION
Apterous female: 2.0–3.2 mm long (dwarf summer form half-size); yellowish green, coated with white flocculent masses of wax; siphunculi pore-like; cauda small, rounded and often inconspicuous.

LIFE HISTORY
Black, oval eggs are deposited on beech twigs during the autumn and hatch in the following spring. Colonies of aphids, which reach the peak of their development by early summer, then develop on the shoots and underside of leaves amongst accumulations of white wax. Winged forms appear after two wingless generations, and these spread infestations from place to place. In mid-summer, dwarf apterae are produced and these undergo a period of aestivation before giving rise to sexual forms. Colonies eventually die out as winter eggs are laid.

DAMAGE
Infested foliage becomes coated with masses of sticky honeydew, upon which sooty moulds develop. Attacked leaves become curled; they may also turn brown around the edges and die prematurely.

Sarucallis kahawaluokalani (Kirkaldy)
syn. *Tinocallis kahawaluokalani* (Kirkaldy)
Crape myrtle aphid

An eastern Asian pest of crape myrtle (*Lagerstroemia indica*), which has been accidentally introduced to various parts of the world (including parts of Europe and North America) where the exotic ornamental tree is now grown.

132 Colony of beech aphid (*Phyllaphis fagi*).

133 Beech aphid (*Phyllaphis fagi*) damage to leaves of *Fagus.*

DESCRIPTION
Adult: 1.2–1.8 mm long; pale yellowish-green aphid, with black markings on the body and wings, and a pair of prominent black tubercles on the back.

LIFE HISTORY
This pest completes several generations annually and, following a sexual phase, overwinters in the egg stage. All adults, other than oviparae, are winged.

DAMAGE
Host trees are contaminated by honeydew and sooty moulds. Although not causing deformation of foliage, infestations result in premature leaf fall.

Takecallis arundicolens (Clarke) (**134–135**)
Bamboo aphid
This eastern Asian species was accidentally introduced into Europe, where it is now widely distributed and sometimes locally abundant on bamboo (e.g. *Bambusa* and *Phyllostachys*). The aphids occur under the expanded leaves but do not appear to be harmful. Adults are medium-sized (1.8–2.8 mm long), whitish or pale yellowish, with pale, dark-ringed antennae, short siphunculi and a small, black, oval cauda. Nymphs are pale with long, capitate body hairs, red eyes and the joints of the antennal segments dark. Reproduction is entirely parthenogenetic and all adults are vivparous alates.

NOTE
Takecallis arundinariae and *T. taiwanus* have also been found on bamboo in Europe.

Tuberculatus annulatus (Hartig)
syn. *Tuberculoides annulatus* (Hartig)
Oak leaf aphid
Widespread and generally abundant on oak (*Quercus*), especially English oak (*Q. robur*), the aphids developing in scattered colonies along the major veins on the underside of the expanded leaves. Infestations often occur on young trees in parks, gardens and nurseries but cause little or no damage, although infested foliage may become contaminated by honeydew and sooty moulds. The aphids (1.4–2.9 mm long) have distinctively banded antennae, stumpy siphunculi and a prominent, knob-like cauda; they vary considerably in colour, being yellow, green, salmon-pink or grey. Viviparous forms of this species, which occur throughout the summer months, are all winged.

134 Bamboo aphid (*Takecallis arundicolens*).

135 Nymph of bamboo aphid (*Takecallis arundicolens*).

Subfamily **CHAITOPHORINAE**

Aphids with body and legs bearing long hairs, terminal process of the antennae long, siphunculi pore-like or stumpy, and cauda knob-like or rounded.

Chaitophorus beuthami (Börner) (136)
Osier leaf aphid

A generally common pest of osier (*Salix viminalis*) and other narrow-leaved willows; often present on ornamental trees. Widely distributed in Europe.

DESCRIPTION

Apterous female: 2.3 mm long; pale greenish, relatively elongate and noticeably hairy; siphunculi stumpy and slightly broadened at the tip; cauda knob-shaped; young individuals more or less colourless. **Alate:** similar to aptera, but with dark abdominal tergites.

LIFE HISTORY

Colonies develop from spring onwards. The aphids feed and breed on the underside of the leaves. They often occur in considerable numbers but usually do not infest the young shoots. Although the aphids excrete honeydew, upon which sooty moulds develop, they are not particularly attractive to ants. By late summer, colonies are often much depleted by predators.

DAMAGE

Damage is unimportant, but heavily infested trees become unsightly following the development of sooty moulds on the aphid-excreted honeydew.

Chaitophorus capreae (Mosley) (137)
Sallow leaf aphid

A widely distributed and very common species on eared willow (*Salix aurita*), grey willow (*S. cinerea*), pussy willow (*S. caprea*) and other broad-leaved willows. The aphids are relatively small (less than 2 mm long), with uninterrupted black abdominal markings; the nymphs are mainly whitish. Colonies develop throughout the summer on the underside of the leaves but have little or no direct effect on their hosts.

Chaitophorus leucomelas Koch (138)
Poplar leaf aphid

Infestations of this generally common species occur on the underside of leaves and on the young shoots of black poplar (*Populus nigra*) and Lombardy poplar (*P. nigra* 'Italica'). However, they are not important pests. Apterae (2.0–2.7 mm long) are dull greenish yellow, with distinctive dark markings; nymphs are paler and

136 Colony of osier leaf aphid (*Chaitophorus beuthami*) on *Salix viminalis*.

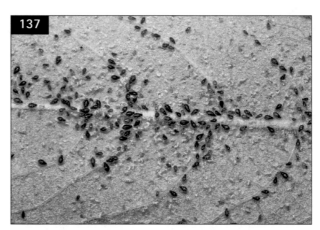

137 Colony of sallow leaf aphid (*Chaitophorus capreae*) on *Salix*.

138 Colony of poplar leaf aphid (*Chaitophorus leucomelas*).

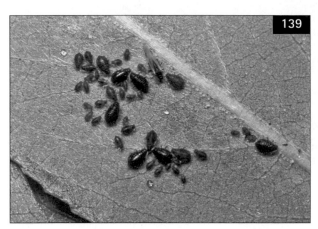

139 Colony of *Chaitophorus niger* on *Salix.*

140 Colony of *Chaitophorus salicti* on *Salix.*

141 Dimorphic summer nymphs of *Periphyllus acericola.*

Chaitophorus niger Mordvilko (**139**)

Although local and generally uncommon, colonies of this species are sometimes found on ornamental willow trees, including crack willow (*Salix fragilis*), purple willow (*S. purpurea*), weeping willow (*S. vitellina* var. *pendula*) and white willow (*S. alba*). The aphids (2.0–2.2 mm long) are distinguished by their overall dark appearance, the tergites forming a solid black carapace.

Chaitophorus saliceti (Schrank) (**140**)

This local species occurs on broad-leaved willows, including common sallow (*Salix atrocinerea*), eared willow (*S. aurita*) and grey willow (*S. cinerea*), and is sometimes found on nursery trees. Individuals are slightly larger and also darker than those of *Chaitophorus capreae* but distinguishable with certainly only by microscopic examination.

Periphyllus acericola (Walker) (**141**)

A generally common species on sycamore (*Acer pseudoplatanus*), infesting both young and mature trees. Widely distributed in Europe.

DESCRIPTION
Apterous female: 3–4 mm long; body light green, oval and hairy; siphunculi stumpy and tapered; cauda broadly rounded. **Nymph:** light green, with long body hairs. **Dimorphic summer nymph:** pale yellow, with long body hairs. **Alate:** 3.0–3.5 mm long; dark-bodied.

LIFE HISTORY
In the spring, colonies of aphids develop on the underside of expanding sycamore leaves. Nymphs of the first generation usually cluster along the major veins towards the leaf base; later, colonies (including winged forms) also develop on the new shoots. The colonies die out in summer, and the species survives by aestivating as first-instar nymphs (dimorphic summer nymphs), which cluster together (cf. *Periphyllus testudinaceus*, p. 80) beneath fully expanded leaves. Activity is resumed in the autumn, and is completed with the production of sexual forms and the deposition of winter eggs.

DAMAGE
Infestations have little effect on tree growth, but spring colonies may be of minor significance on nursery trees.

Periphyllus californiensis (Shinji) (**142**)
Californian maple aphid

This North American species occurs on various species of maple (*Acer*), including downy Japanese maple (*A. japonicum*) and smooth Japanese maple (*A. palmatum*), having been introduced into parts of Europe, including southern England, on infested nursery stock. The aphids (2.3–3.5 mm long) are dark olive-green to brown; the dimorphic nymphs are light green and similar in appearance to those of *Periphyllus testudinaceus* (q.v.), but with just two series of abdominal plates (termed spinal and marginal).

Periphyllus testudinaceus (Fernie) (**143**)
This often abundant, dark brownish-green to blackish (2.0–3.5 mm long) species infests field maple (*Acer campestre*) and various ornamental maples. It is also associated with horse chestnut (*Aesculus hippocastanum*) and sycamore (*Acer pseudoplatanus*). Heavy infestations often develop in the spring on the young growth; later, the aphids occur abundantly beneath the expanded leaves. The aphids produce copious quantities of honeydew, and colonies are ant-attended. The lifecycle follows that of *Periphyllus acericola* but the dimorphic summer nymphs, which are green and fringed with curious leaf-like hairs, aestivate singly along the major leaf veins. Unlike those of *Periphyllus californiensis*, the dimorphic nymphs possess three series of abdominal plates (termed spinal, pleural and marginal).

Subfamily **ERIOSOMATINAE**

Aphids with terminal process of antennae short, eyes reduced to three facets, siphunculi stumpy cones, pore-like or absent, and cauda broadly rounded; body often with groups of well-developed wax glands. Associated with trees and shrubs, often forming galls and sometimes migrating in summer to herbaceous plants or grasses.

Eriosoma lanigerum (Hausmann) (**144**)
Woolly aphid

Common throughout the world on apple, including crab-apple (*Malus*); also a pest of rosaceous ornamentals such as *Cotoneaster*, firethorn (*Pyracantha*), hawthorn (*Crataegus*), Japanese quince (*Chaenomeles japonica*) and Sorbus.

DESCRIPTION

Apterous female: 1.2–2.6 mm long; purplish brown, covered with masses of white, mealy wax; body with numerous wax plates; antennae short; siphunculi pore-like.

LIFE HISTORY

Nymphs, devoid of a wax coating, overwinter in crevices or under loose bark of suitable host plants and become active in March or April. By the end of May breeding colonies, now coated in masses of waxen 'wool', develop on the branches and spurs, mainly around wounds and in splits in the bark. Breeding continues until the autumn, with the production of a small number of winged forms in July. Although a few eggs may be deposited in the autumn, these fail to develop. Thus, completion of the lifecycle is dependent upon the production of nymphs by viviparous, parthenogenetic females.

142 Colony of Californian maple aphid (*Periphyllus californiensis*) on *Acer*.

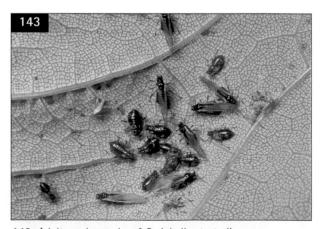

143 Adults and nymphs of *Periphyllus testudinaceus*.

144 Colony of woolly aphid (*Eriosoma lanigerum*).

145 Gall of currant root aphid (*Eriosoma ulmi*) on leaf of *Ulmus*.

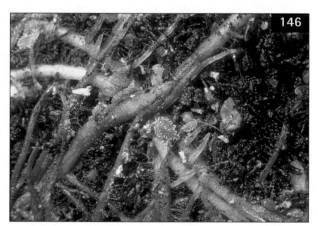

146 Colony of currant root aphid (*Eriosoma ulmi*) on roots of *Ribes*.

DAMAGE

Masses of flocculent wax accumulate on heavily infested plants. This is unsightly and a particular nuisance if infested branches overhang or lie alongside garden paths. The aphids also induce the development of disfiguring galls; these cause new growth to be malformed and the wood of older branches to split.

Eriosoma ulmi (Linnaeus) (145–146)

Currant root aphid

This generally common Eurasiatic species overwinters on elm (*Ulmus*). Overwintered eggs hatch in the spring. Colonies of greyish aphids then develop on the underside of curled leaves, protected by flocculent masses of bluish or whitish wax. Winged forms are reared in June and July; these then migrate to currant and gooseberry bushes, where colonies of greyish or pinkish-grey aphids develop on the roots amongst masses of whitish wax. Winged aphids return to elm in the autumn where, after a sexual phase, eggs are eventually laid. Infested elm foliage is severely curled and shoot growth is checked; however, attacks, although often common in gardens, do not harm established trees. Summer infestations on the roots of ornamental currant, such as flowering currant (*Ribes sanguineum*), affect growth of young plants and are of particular importance on containerized nursery stock.

Kaltenbachiella pallida (Haliday) (**147**)

This widespread species occurs during the summer on various Lamiaceae, including mint (*Mentha*), marjoram (*Origanum vulgare*), thyme (*Thymus*) and woundwort (*Stachys*), where colonies of very small (0.9–1.3 mm long), creamy-white aphids develop on the roots amongst flocculent masses of white wax. Winged forms appear in the late summer and early autumn, and then migrate to elm (*Ulmus*) where, eventually, eggs are laid. The eggs hatch in the following spring. Wingless aphids then invade the unfurling foliage to initiate large, light green leaf galls. Each gall is coated in short whitish hairs, and develops at the base of the midrib as a conspicuous swelling (15–20 mm across) which protrudes both above and below the leaf blade. These galls mature in the early summer. Winged aphids then escape through a stellate opening and eventually establish colonies on secondary (summer) herbaceous hosts. Galls on primary (winter) hosts, although disfiguring leaves of ornamental specimen trees, do not affect plant growth.

Pachypappa tremulae (Linnaeus)
syn. *Asiphum tremulae* (Linnaeus);
Rhizomaria piceae Hartig
Spruce root aphid

A generally common and sometimes abundant pest of spruce (*Picea*) trees, including Norway spruce (*P. abies*) grown for the Christmas market. Often present on the roots of nursery stock, including container-grown material, but also invading the superficial roots of older trees. Present throughout central and northern Europe.

DESCRIPTION
Apterous female [on spruce]: 1.5–2.0 mm long; white to creamy white, with a brownish head, antennae and legs; wax glands are prominent.

LIFE HISTORY
This species overwinters in the egg stage on aspen (*Populus tremula*). Grey poplar (*P. canescens*) also serves as a primary (winter) host. The eggs hatch in spring, and wingless aphids (fundatrices) then develop on the smaller twigs. Their offspring feed on the young shoots, causing a slight bending of the leaf stalks. In June, winged forms appear and these migrate to the roots of spruce trees. Ant-attended colonies then develop on the roots, amongst flocculent masses of white wax. In the early autumn, after one or two wingless generations, winged aphids are produced. These return to winter hosts where, following a sexual phase, winter eggs are eventually laid. Colonies of wingless aphids may persist throughout the year on the roots of spruces.

DAMAGE
Heavy infestations on the roots of spruces may cause death of plants. Damage to winter hosts, however, is of no importance.

Pemphiginus vesicarius (Passerini) (**148–149**)

This species occurs mainly in the Apennines and parts of Asia but is also present in the South Tyrol, from northern Italy into Austria. The aphids inhabit distinctive galls on black poplar (*Populus nigra*). The

147 Gall of *Kaltenbachiella pallida* on *Ulmus*.

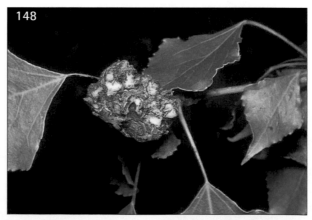

148 Young gall of *Pemphiginus vesicarius* on *Populus*.

galls develop at the base of the young shoots as irregular, brownish growth several centimetres across. Greyish-bodied nymphs develop within the galls amongst flocculent masses of whitish wax. Winged adults eventually escape through small openings (which eventually appear in the wall) and disperse to unknown summer hosts. Vacated galls persist on host trees as distinctive, shrunken, black, woody structures. They are sometimes found on young amenity trees but damage caused by the aphids is of little or no importance.

Pemphigus bursarius (Linnaeus) (**150–151**)
Lettuce root aphid

A generally common gall-forming aphid on Lombardy poplar (*Populus nigra* 'Italica') but most important as a pest of cultivated chicory and lettuce. Present throughout Europe; also found in North America and Australia.

DESCRIPTION
Apterous female: 1.6–2.5 mm long; elongate-oval, yellowish white, with a tuft of white wax posteriorly; small, dark abdominal wax plates clearly visible; siphunculi reduced to mere rings. **Alate:** 1.7–2.2 mm long; abdomen brownish orange.

LIFE HISTORY
This species overwinters in the egg stage on Lombardy poplar. In the spring, wingless females initiate characteristic pouch-like galls by feeding on the leaf stalks. Small colonies of aphids develop within these structures until, in the early summer, winged forms

eventually escape through a beak-like opening. These migrants then invade summer hosts, including *Lactuca*, sow-thistle (*Sonchus*) and various other Asteraceae. Colonies of wingless aphids then build up on the roots until the autumn, when winged forms return to poplar. Following a sexual phase, winter eggs are eventually laid.

DAMAGE
Poplar: autumn populations distort and discolour the foliage but such damage is not of significance; galling in spring is also unimportant.

149 Old gall of *Pemphiginus vesicarius* on *Populus*.

150 Colony of lettuce root aphid (*Pemphigus bursarius*) in gall on *Populus*.

151 Mature gall of lettuce root aphid (*Pemphigus bursarius*) on *Populus*.

Pemphigus phenax Börner & Blunck
syn. *P. dauci* (Goureau)
Carrot root aphid

This species overwinters as eggs on the bark of Lombardy poplar (*Populus nigra* 'Italica'). The eggs hatch in the spring. Young aphids then move to the unfurling leaves where they induce the formation of midrib galls. Each gall becomes an elongate, somewhat wrinkled, reddish swelling (often tinged with yellow laterally), packed with numerous wax-secreting aphids. In summer, winged forms migrate to cultivated and wild carrot (*Daucus carota*), where dense colonies develop on the roots amongst copious amounts of white waxen 'wool'. There is a return migration to poplar in the autumn. Although the leaf galls on poplar trees may attract attention, infestations are not of importance.

152 Galls of poplar/cudweed aphid (*Pemphigus populinigrae*) on *Populus*.

Pemphigus populinigrae (Schrank) (152–153)
syn. *P. filaginis* (Boyer de Fonscolombe)
Poplar/cudweed aphid

Widely distributed in association with black poplar (*Populus nigra*). Eggs overwinter and hatch in the spring. Young aphids then form bulbous, pouch-like galls on the midrib of the leaves. Aphids (fundatrices) initiating the galls are 2.6–2.8 mm long and mainly green to greyish green, characteristically with 4-segmented antennae. Colonies develop amongst a mass of mealy wax, eventually giving rise to mainly brown-bodied alates. These eventually escape from the galls and fly to secondary (summer) hosts such as cudweeds (*Filago* and *Gnaphalium*). There is a return migration to poplar in the autumn.

Pemphigus spyrothecae Passerini (154)
Poplar spiral-gall aphid

Widely distributed in Europe (also recently introduced to North America), and often common on poplar (*Populus*), especially Lombardy poplar (*P. nigra* 'Italica'). Overwintered eggs hatch in the spring, and aphids feed on the leaf stalks which then curl into characteristic, distinctly spiral, pouch-like galls about 10 mm long. The galls, which become packed with aphids, change from green through red to brown as they mature, and are often abundant on ornamental trees. Unlike its close relatives (see above), this species spends its entire lifecycle on poplar.

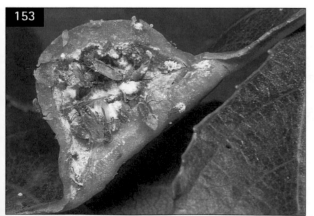

153 Colony of poplar/cudweed aphid (*Pemphigus populinigrae*) in gall on *Populus*.

154 Gall of poplar spiral-gall aphid (*Pemphigus spyrothecae*) on *Populus*.

Stagona pini (Burmeister)
syn. *S. crataegi* Tullgren
Pine root aphid
Small colonies of this generally common and widely distributed pest occur throughout the year on the roots of pine (*Pinus*) trees. Infested parts of the root system become covered in masses of bluish-white waxen 'wool', amongst which may be found the small (1.5–2.0 mm long) greyish-white aphids; individuals possess numerous wax glands but no siphunculi. Attacks are most important on container-grown plants, nursery stock and transplants. Severe infestations cause yellowing of the needles, and wilting and death of plants. Hawthorn (*Crataegus*) is the primary host, but infestations on this plant are not harmful.

155 Young gall of elm leaf gall aphid (*Tetraneura ulmi*) on *Ulmus.*

Tetraneura ulmi (Linnaeus) (**155–157**)
Elm leaf gall aphid
Widespread in north-western Europe, forming large, conspicuous bean-like galls on the expanded leaves of various kinds of elm (*Ulmus*). The galls develop from the upper surface of the leaf blade, each attached by a narrow stalk. The galls, which may exceed 15 mm in height, are initially green but later turn cream and finally brown. Pale yellowish or yellowish-white aphids develop within the galls, amongst whitish masses of flocculent wax. Eventually, in the summer, winged forms are produced, and these finally escape through a conspicuous basal aperture. These aphids migrate to various grasses, where ant-attended colonies develop on the roots. A return migration to the primary host takes place in the autumn. The galls on elm leaves are very conspicuous and often numerous; however, they do not cause significant damage.

156 Mature galls of elm leaf gall aphid (*Tetraneura ulmi*) on *Ulmus.*

157 Old galls of elm leaf gall aphid (*Tetraneura ulmi*) on *Ulmus.*

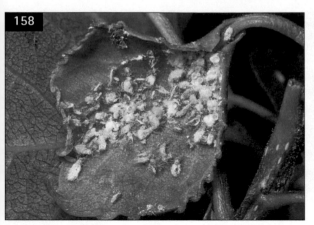

158 Colony of poplar/buttercup aphid (*Thecabius affinis*).

159 Gall of poplar/buttercup aphid (*Thecabius affinis*) on *Populus*.

Thecabius affinis (Kaltenbach) (**158–159**)
syn. *T. ranunculi* (Kaltenbach)
Poplar/buttercup aphid
Generally common on black poplar (*Populus nigra*) and, less frequently, white poplar (*P. alba* 'Pyramidalis'). The summer forms infest the roots of buttercups (*Ranunculus*), primarily creeping buttercup (*R. repens*). Eurasiatic. Widespread in Europe.

DESCRIPTION
Apterous female [on poplar]: 4.0–4.5 mm long; body light green to greyish green, and coated with white waxy powder; head, antennae and legs dark brown; legs short and robust. **Alate [ex: poplar]:** 2.6–3.0 mm long; mainly blackish brown, with a green to purplish-green abdomen; legs long and thin.

LIFE HISTORY
Eggs overwinter on poplar trees and hatch in the spring. Wingless aphids then develop within light green, yellowish-green, pouch-like galls, each formed from a folded leaf. After a further generation of wingless aphids, winged forms are produced; these migrate from July onwards to the roots of buttercups. Subterranean colonies of wingless aphids then develop on these summer hosts, with winged forms appearing in the autumn and returning to poplar in October and November.

DAMAGE
Poplar: galls affect the development of new shoots, disfiguring host plants; however, attacks on established trees are of little or no consequence. **Buttercup:** heavy infestations on the roots produce unsightly masses of white wax and reduce the vigour of host plants; however, attacks on cultivated species are uncommon and rarely important.

Thecabius auriculae (Murray)
Auricula root aphid
This species infests the roots of auricula (*Primula auricula*) and certain other related plants. The aphids feed on the roots, and are particularly damaging on potted plants and on those growing in greenhouses or under other protection. The aphids, which occur on the roots throughout the year, are small (1.3–1.5 mm long), pale yellowish white or pale greenish white, with brownish legs, and are coated with white mealy wax. Winged forms, which are larger (2.5 mm long) and mainly brown to green, also occur. The leaves of infested plants turn yellow, and may become mottled and distorted, with considerable quantities of 'wool' accumulating around the collar and amongst the root system. Severely affected plants may wilt and die.

Subfamily **HORMAPHIDINAE**

A small group of small, scale-like aphids, with siphunculi much reduced or absent.

Cerataphis orchidearum (Westwood)
Orchid aphid

This small, reddish-brown to black aphid breeds parthenogenetically on greenhouse orchids. Apterae are 1.0–1.6 mm long, flattened and scale-like, with very short legs and antennae, short, stubby siphunculi, a knob-like cauda and a marginal fringe of wax. They occur on various kinds of orchid, including tropical climbers (e.g. *Vanilla*); hosts such as *Cattleya*, *Cymbidium*, *Cypripendium*, *Dendrobium* and *Odonotoglossumm* may be damaged but the aphid is usually not of major significance.

Subfamily **LACHNINAE**

Aphids with terminal process of antennae very short; siphunculi usually cone-like and very hairy; cauda broadly rounded.

Cedrobium laportei Remaudière
Small cedar aphid

This pest, first found in North Africa, has become widely distributed in Europe on Atlas cedar (*Cedrus atlantica*), cedar of Lebanon (*C. libani*) and deodar (*C. deodara*). Colonies develop in spring and summer at the base of the needles, and the aphids excrete considerable quantities of honeydew. Infestations result in premature loss of needles and die-back of shoots, and may even lead to the death of host trees. Trees are also disfigured following the development of unsightly sooty moulds. Apterae (1.7–2.0 mm long) are greyish brown and shiny, with pale legs and antennae.

Cinara pilicornis (Hartig) (**160**)
Brown spruce aphid

An often common pest of spruce (*Picea*), including Norway spruce (*P. abies*) grown as Christmas trees; present throughout western Europe.

DESCRIPTION

Apterous female: 2.1–4.7 mm long; greyish brown to brownish, and covered with fine hairs; rostrum short and dagger-like; siphunculi short, arising from pigmented cones; cauda inconspicuous and crescent shaped.

LIFE HISTORY

Overwintered eggs laid on spruce hatch in the early spring, before bud-burst. Colonies of aphids then develop on the underside of the shoots. Winged and wingless forms are produced from May to July, and sexual forms appear from August onwards. Winter eggs are deposited on the twigs during the late summer and autumn.

DAMAGE

Heavy infestations are not directly harmful, although they may cause slight discoloration of the foliage. Sooty moulds develop on honeydew produced by the aphids, and this affects the appearance and, hence, the marketability of nursery stock.

Cinara cedri Mimeur
Cedar aphid

A southerly distributed European species. Well established in Mediterranean areas, where it forms small colonies on the branches of cedar (*Cedrus*) trees. Apterae are 2.5–3.0 mm long, and bright reddish brown, with a pair of blackish longitudinal bands on the body and black siphuncular cones; the body is partly coated in whitish wax. Alates are similar in appearance to apterae, but lack the dark bands; their wings are yellow.

160 Colony of brown spruce aphid (*Cinara pilicornis*) on *Picea*.

Cinara costata (Zetterstedt) (**161**)

Small colonies of this southerly-distributed species are noted occasionally on young Norway spruce (*Picea abies*) trees. Adults (2.7–3.8 mm long) are light brown with darker markings, but often appear greyish due to the presence of secreted wax; the siphunculi are borne on small, dark, widely spaced siphuncular cones (cf. *Cinara pruinosa*). The aphids occur mainly from May to July, but also in the autumn. Colonies are usually attended by ants.

Cinara cuneomaculata (del Guercio)
Larch aphid

This large (2.4–4.6 mm long), dark brown species occurs on larch (*Larix*), including nursery stock, usually establishing small colonies on the shoots. Heavy infestations cause discoloration of foliage. The aphids also produce considerable quantities of honeydew, upon which sooty moulds develop. Wingless aphids occur from May to September. Overwintering eggs are laid in the autumn on young twigs and shoots.

Cinara cupressi (Buckton) and *C. cupressivora* (Watson & Voegtlin)
Cypress aphids

This species complex is associated with Cupressaceae, especially Lawson cypress (*Chamaecyparis lawsoniana*), Leyland cypress (*Cupressocyparis leylandii*) (especially the golden form 'Castlewellan') and Monterey cypress (*Cupressus macrocarpa*); the aphids also occur on *Thuja*. Damage is often caused to nursery stock; hedges and specimen trees in parks and

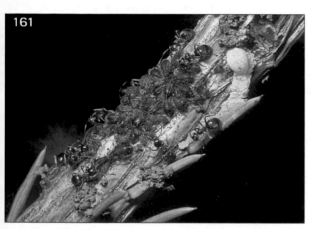

161 Ant-attended colony of *Cinara costata* on *Picea*.

162 Colony of American juniper aphid (*Cinara fresai*) on *Juniperus*.

163 Protective earthen canopy around colony of American juniper aphid (*Cinara fresai*).

164 American juniper aphid (*Cinara fresai*) damage to foliage of *Juniperus*.

gardens are also affected. The aphids, which excrete considerable quantities of honeydew, feed on the foliage, shoots and branches. They cause discoloration and die-back of affected tissue, which reduces plant vigour. The aphids occur from May onwards, although often in only small numbers. Winged forms, which spread infestations, are reared from June to August. Adults (1.8–3.9 mm long) are mainly orange brown to yellowish brown, with blackish markings diverging back from the thoracic region, light grey transverse stripes on the abdomen and a black band between the prominent black siphuncular cones; the rostrum is relatively short (*c*. 1 mm long). The pest is widely distributed in Europe, particularly in southern areas; in Britain attacks are most often reported in the south and west of England.

Cinara fresai (Blanchard) (**162–164**)
American juniper aphid

Dense colonies of this large (2.2–4.2 mm long), dark brownish-grey species occur locally on juniper (*Juniperus*) in southern England from June onwards. This species is readily distinguished from *Cinara juniperi* (see below) by the two broken, divergent black stripes which extend back from the head. The aphids infest the foliage and stems of various ornamental junipers (including *J. chinensis, J. sabina, J. squamata* and *J. virginiana*), protected by earthen shelters constructed by ants. Infestations lead to the development of sooty moulds and noticeable blackening of shoots, affecting the appearance and hence quality of plants. Attacks also cause die-back of shoots and, sometimes, death of trees.

Cinara juniperi (Degeer)
Juniper aphid

A widely distributed species on juniper (*Juniperus*), primarily common juniper (*J. communis*). The aphids are pinkish brown, slightly smaller (2.2–3.4 mm long) than *Cinara fresai*, and have a relatively short rostrum. Also, unlike *Cinara fresai*, infestations are usually restricted to the young, green shoots and needles. Sooty moulds develop on honeydew excreted by the aphids, and this disfigures established ornamentals and nursery stock.

Cinara pinea (Mordwilko) (**165–166**)
Large pine aphid

This common, widely distributed, large-bodied (3.1–5.2 mm long) aphid feeds on Scots pine (*Pinus sylvestris*), infesting young shoots bearing new or one-year-old needles. Adults vary from grey to orange-brown or dark brown, and are present from the spring to

the late autumn or early winter when oviparous forms finally deposit the overwintering eggs. These eggs (*c*. 1.6 mm long) are black and shiny, and are laid in rows along the needles. They hatch in the spring. Colonies of pale aphids then develop on the shoots, individuals clustering together on the wood between the bases of the needles. Breeding continues throughout the summer, including the production of winged forms; later-produced individuals tend to be darker in colour than the spring forms. Colonies are relatively small but produce considerable quantities of honeydew. Although damage to established trees is slight, attacks on nursery trees may cause considerable damage, leading to discoloration of the needles, premature loss of needles and reduced plant vigour.

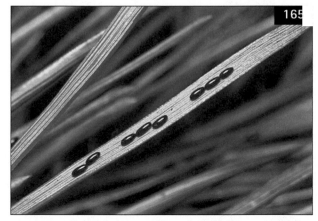

165 Eggs of large pine aphid (*Cinara pinea*).

166 Colony of large pine aphid (*Cinara pinea*) on *Pinus*.

167 Colony of *Cinara pruinosa* on *Picea*.

168 Protective earthen canopy around colony of *Cinara pruinosa*.

Cinara pruinosa (Hartig) (**167–168**)

Colonies of this large (2.4–5.0 mm long), dark green to brown aphid occur on the shoots and branches of spruce (*Picea*) trees, especially Norway spruce (*P. abies*). The rostrum is noticeably long, and the siphuncular cones very large and prominent. In summer, colonies move to the base of the trunk and to the roots, the winter being passed as adults in the soil or as eggs on young twigs. Colonies are ant-attended and often sheltered by earthen canopies constructed by these insects.

Cinara tujafilina (del Guercio)

Thuja aphid

A widely distributed pest of Chilean incense cedar (*Austrocedrus chilensis*), Chinese thuja (*Thuja orientalis*), white cedar (*T. occidentalis*) and certain other Cupressaceae; present in central and southern Europe, and in various other parts of the world. The aphid is parthenogenetic and, in favourable situations,

breeds throughout the year. Colonies develop on the young twigs and underside of the older branches, the aphids often congregating in considerable numbers within bark crevices and in other sheltered places. In summer, colonies also occur on the roots. Apterae (1.7–3.5 mm long) are reddish brown, with a purplish-grey bloom and a pair of black bands extending from the head to the siphunculi; the siphuncular cones are indistinct and linked by a pair of curved, brownish bands. The usually ant-attended aphids produce vast quantities of honeydew, and affected hosts soon become coated in sooty moulds. Infestations spoil the appearance of ornamental trees and, if severe, result in die-back of shoots and branches.

Eulachnus agilis (Kaltenbach)

Narrow spotted pine aphid

A local pest of Scots pine (*Pinus sylvestris*); infestations also occur on Corsican pine (*P. nigra* var. *maritima*). Widely distributed in mainland Europe; in the British Isles found mainly in southern England.

DESCRIPTION

Aptera: 2.2–2.9 mm long; body elongate and spindle-shaped; bright green, with several reddish-brown dorsal platelets, most bearing a short, pointed seta; antennae short, pale greenish brown, the third segment with long setae; siphuncular cones small; legs mainly light green, the tibiae with long dark hairs.

LIFE HISTORY

Colonies build up rapidly during the spring, with both winged and wingless forms present throughout much of the year. Aphid numbers decline during the summer but again reach a peak from late summer onwards. Sexual forms are produced in the autumn and eggs are deposited in the early winter, most placed singly on leaf scars. Individuals arising from the eggs are very long-lived and more prolific than those produced parthenogenetically during the spring and summer.

DAMAGE

Needles in the vicinity of colonies become blackened, following the development of sooty moulds. Heavy infestations on nursery trees cause premature defoliation, with significant losses of older needles.

Eulachnus brevipilosus Börner (**169**)

Narrow green pine aphid

A widespread but generally uncommon species, occurring mainly in pine-growing areas on the needles of Scots pine (*Pinus sylvestris*). Austrian pine (*P. nigra* var. *nigra*) and mountain pine (*P. mugo*) are also

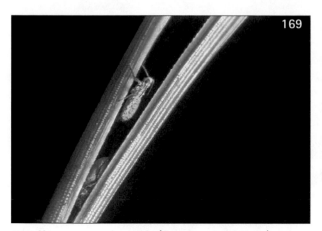

169 Narrow green pine aphid (*Eulachnus brevipilosus*).

170 Eggs of rose root aphid (*Maculolachnus submacula*) on stem of *Rosa*.

attacked. The aphids (1.7–2.2 mm long) are mainly green, with a whitish bloom, and distinguished from *Eulachnus agilis* by the inconspicuous dorsal platelets, the darker appendages and the short setae on the third antennal segment.

Eulachnus rileyi (Williams)
syn. *Protolachnus bluncki* Börner
Narrow brown pine aphid
This widely distributed species feeds mainly on Austrian pine (*Pinus nigra* var. *nigra*), and is recognized by its brownish to orange-red body colour and associated waxy bloom.

Lachnus longirostris (Börner)
A relatively uncommon species on oak (*Quercus*); sometimes noted on ornamentals but rarely if ever troublesome. The aphids are similar in appearance and habits to the following species but are larger (4.5–6.4 mm long) and have inconspicuous siphunculi.

Lachnus roboris (Linnaeus)
This locally common species is associated with oak (*Quercus*), including holm oak (*Q. ilex*), colonies developing on the shoots and smaller branches during the summer months; infestations sometimes attract attention and cause minor damage to garden and amenity trees. The aphids (3.2–3.5 mm long) are shiny black to dark brown, with very long, yellowish-brown legs, and the siphunculi borne on large, blackish, hairy cones; the wings of alates are clouded by brown patches. Colonies are typically ant-attended, and often very large; however, the individual aphids tend to remain relatively well separated and do not cluster tightly together.

171 Protective earthen canopy around colony of rose root aphid (*Maculolachnus submacula*).

Maculolachnus submacula (Walker) (170–171)
syn. *Lachnus rosae* Cholodkovsky
Rose root aphid
An often common pest of rose (*Rosa*), particularly in England, Germany and the Netherlands; also present in North America.

DESCRIPTION
Apterous female: 2.7–3.8 mm long; yellowish brown to dark chestnut-brown or blackish brown; antennae short and blackish brown; legs blackish brown; abdomen with dorsal hairs arising from small plates; rostrum short; siphunculi short and conical; cauda rounded and inconspicuous. **Egg:** 1 mm long; elongate-oval, black and shiny.

LIFE HISTORY

Clusters of overwintering eggs occur on the lower portions of rose stems, and these commonly attract attention when bushes are pruned during the winter or early spring. The eggs hatch in the early spring. Nymphs then migrate to the underground parts of the stems to begin feeding. Ant-attended colonies develop during the spring and summer on the superficial roots and stem bases, and these are often protected by ant-constructed earthen shelters. In summer, winged forms spread infestations to other rose bushes. Aphid numbers decline in the autumn, when eggs are deposited.

DAMAGE

The growth of infested bushes may be checked, and persistent infestations eventually cause the death of plants.

Schizolachnus pineti (Fabricius) (**172**)

Grey pine-needle aphid

Generally common on young pine (*Pinus*) trees, especially Austrian pine (*P. nigra* var. *nigra*), beach pine (*P. contorta*) and Scots pine (*P. sylvestris*). Widespread throughout Europe.

DESCRIPTION

Apterous female: 1.2–3.0 mm long; body broadly oval, dark greyish green, covered with long, fine hairs and coated in white, mealy wax; siphunculi reduced to small cones. **Alate:** 2–3 mm long; greyish green.

LIFE HISTORY

Overwintered eggs hatch in the early spring, and colonies of wingless aphids then develop in rows along the needles. Colonies persist throughout the summer and autumn, and are often attended by ants. Winged forms are produced from May to September. Sexual forms, including winged males, occur in the late autumn and early winter, prior to the deposition of winter eggs. Under favourable conditions colonies of live aphids also survive the winter.

DAMAGE

Infested needles often turn yellow and drop prematurely; this affects the vigour of young plants, and reduces shoot and needle length in the following season.

Trama troglodytes von Heyden

Jerusalem artichoke tuber aphid

Infestations of this large, entirely parthenogenetic aphid sometimes occur on the roots of cultivated Asteraceae, including *Arnica*, *Chrysanthemum maximum*, Michaelmas daisy (*Aster*), sneezewort (*Helenium*) and sunflower (*Helianthus annuus*); they also occur on weeds such as dandelion (*Taraxacum officinale*) and sow-thistle (*Sonchus*). Colonies are ant-attended and occur on the roots throughout the year. Winged forms appear during the summer months and then disperse to new hosts. The aphids reduce the vigour of infested plants and, particularly in dry conditions, cause plants to wilt. Adults (3–4 mm long) are white, and lack siphunculi; characteristically, if disturbed, they vibrate their long hind tibiae.

172 Colony of grey pine needle aphid (*Schizolachnus pineti*) on *Pinus*.

173 Colony of large willow aphid (*Tuberolachnus salignus*) on *Salix viminalis*.

Tuberolachnus salignus (Gmelin) (**173**)
Large willow aphid

An often common pest of willows, including almond willow (*Salix triandra*), crack willow (*S. fragilis*), grey willow (*S. cinerea*), osier (*S. viminalis*) and weeping willow (*S. vitellina* var. *pendula*). Widely distributed, particularly in central and southern Europe; in the British Isles most numerous in southern England.

DESCRIPTION
Apterous female: 4.0–5.4 mm long; blackish brown, with a large tubercle arising from the fourth abdominal tergite; body clothed in fine, grey hairs.

LIFE HISTORY
Dense colonies of this large, conspicuous, long-legged aphid occur on the branches and stems of willow trees from late June onwards but not until much later in some areas. Breeding is entirely parthenogenetic, with populations reaching their maximum in the autumn. In favourable areas, small numbers of apterae survive the winter. Elsewhere, colonies must be re-established in the following year by winged migrants that have been reared further south. Colonies produce vast quantities of honeydew and are constantly visited by sugar-seeking insects such as ants, bees, wasps and various flies.

DAMAGE
The aphids weaken shoot growth, most significant damage occurring on the canes of willows grown for basket-making. Plants, including ornamentals, are also contaminated by masses of sticky honeydew and sooty moulds.

174 Colony of black willow aphid (*Pterocomma salicis*).

Subfamily **MINDARINAE**

A small group of conifer-infesting aphids, with reduced, ring-like siphunculi and a rather long cauda.

Mindarus abietinus Koch
Balsam twig aphid

This light grey species forms small colonies during the summer on the succulent new growth of fir (*Abies*) trees. The aphids secrete considerable quantities of waxen 'wool' and their feeding causes considerable distortion of the foliage. Colonies are sometimes noted on ornamentals but are usually of only minor importance.

Mindarus obliquus (Cholodkovsky)
Spruce twig aphid

Individuals of this species are small, elongate and white. They occur throughout the summer on the tips of new shoots of spruce (*Picea*) trees, and produce masses of bluish-white waxen 'wool'.

Subfamily **PTEROCOMMATINAE**

Relatively large aphids with terminal process of antennae relatively short, siphunculi cylindrical to vase-like (with or without an apical flange), and cauda rounded to broadly tongue-shaped.

Pterocomma salicis (Linnaeus) (**174**)
Black willow aphid

An often common pest of *Salix*, including crack willow (*S. fragilis*), grey willow (*S. cinerea*), osier (*S. viminalis*), pussy willow (*S. caprea*) and white willow (*S. alba*). Holarctic. Present throughout Europe.

DESCRIPTION
Apterous female: 3.5–4.2 mm long; body dull greyish black to black, with greyish or whitish markings, and very hairy; antennae, legs and siphunculi orange-red; antennae short; siphunculi flask-shaped. **Nymph:** pale orange-brown to greenish black, with whitish markings.

LIFE HISTORY
Dense colonies often develop during the summer on the shoots and young stems of host trees, and are commonly attended by ants. The winter is spent in the egg stage in crevices in the bark.

DAMAGE
Colonies on windbreaks and specimen trees or shrubs are disfiguring but of little importance. However,

attacks on osiers grown for basket-making are of some significance; rods may be stunted and the wood stained. New growth may be killed, even when only lightly infested.

Pterocomma populeum (Kaltenbach) (**175**)

A widely distributed aphid on poplar (*Populus*), forming dense colonies on the young branches and twigs. Damage caused, however, is slight. Apterae are greyish green to greyish brown, with a pair of dark spots on each body segment and a light covering of whitish wax.

Pterocomma steinheili (Mordvilko) (**176**)

Colonies of this species, which has a similar lifecycle to that of *Pterocomma salicis*, occur on the young wood of willow (*Salix*); the aphids are distinguished by their reddish-brown to grey body colour.

Subfamily **THELAXINAE**

Aphids with body flattened and eyes reduced to three facets; antennae short, with a short terminal process, siphunculi broad and cone-like, and cauda knob-like or broadly rounded.

Glyphina betulae (Linnaeus) (**177**)

Associated with birch (*Betula*), and often common throughout the summer and autumn months. The dark green, white-marked aphids cluster on the shoot tips and young leaves, and often form dense colonies. Wingless forms are 1.4–1.8 mm long, with short antennae, stout legs, a rounded cauda and small, cone-like siphunculi, and the body coated with short spine-like hairs; they occur most commonly from June to the end of August and again in October and November. Alates, which appear from mid-June to late July, are dark green, with a dark brown to blackish head and thorax; the abdomen bears numerous tubercles and short, truncated siphunculi.

175 Colony of *Pterocomma populeum* on *Populus*.

176 Colony of *Pterocomma steinheili* on *Salix*.

177 Colony of *Glyphina betulae* on *Betula*.

178 Colony of *Thelaxes dryophila* on *Quercus*.

Thelaxes dryophila (Schrank) (**178**)
syn. *T. quercicola* Westwood

Generally common on oak (*Quercus*), and often abundant on young trees. Eurasiatic. Widely distributed in Europe.

DESCRIPTION
Apterous female: 2.2–2.8 mm long; reddish brown to green, with marked segmentation and a pale dorsal line; body flattened and hairy, and lightly dusted with whitish wax; eyes small, antennae short, siphunculi reduced to flat cones, and cauda knob-like.

LIFE HISTORY
Overwintered eggs hatch in the spring. Dense colonies of wingless aphids then develop on the young shoots and underside of oak leaves. Winged aphids arise in the summer, and these readily spread infestations from tree to tree. Sexual forms appear in August, mated females eventually depositing winter eggs. The aphids produce copious quantities of honeydew, and colonies are strongly ant-attended.

DAMAGE
Vigour of heavily infested shoots is reduced, and persistent infestations on young trees are of particular significance.

Family **ADELGIDAE**

Conifer-feeding, aphid-like insects but, unlike true aphids, with short antennal segments, reduced wing venation, no siphunculi and entirely oviparous females; alates have five antennal segments. Lifecycles are very complex, and often involve a variety of different morphs and alteration of host plant.

Adelges abietis (Linnaeus) (**179–180**)
A spruce pineapple-gall adelges

An often abundant pest of spruce (*Picea*), and commonly damaging to ornamentals, including Norway spruces (*P. abies*) grown as Christmas trees. Present throughout the natural range of spruces and widely introduced with such plants to many other areas.

DESCRIPTION
Apterous female (pseudo-fundatrix): yellowish green to light green, with 5-segmented antennae and five pairs of abdominal spiracles. **Alate female (gallicola):** 1.8–2.2 mm long; yellow, with 5-segmented antennae; fore wings 2.5–2.8 mm long.

LIFE HISTORY
This adelgid spends its life on one host, inhabiting characteristic pineapple-like galls on the young shoots of Norway spruce. The galls vary considerably in size, those on the vigorous young shoots usually being the largest, but are most frequently about 15–20 mm long. In this species the number of adult morphs is restricted to just two: wingless pseudo-fundatrices and winged gallicolae. Gallicolae escape from mature galls during the late summer, usually during August and September, and then deposit eggs. These hatch into nymphs which overwinter close to the buds. In the spring the nymphs

179 Gall of spruce pineapple-gall adelges (*Adelges abietis*)

180 Colony of spruce pineapple-gall adelges (*Adelges abietis*).

mature into pseudo-fundatrices. They then deposit pale yellow eggs, usually in batches of about 50, partly covering them with white waxen threads. Nymphs hatching from these eggs are the gall-inducing, gall-inhabiting forms. They feed at the base of the needles, causing a localized swelling which eventually develops into the familiar pineapple-like gall. Each gall contains numerous chambers within which groups of the pale pinkish-orange nymphs develop. This species exhibits only a limited migration, so that heavy populations tend to build up locally on affected trees.

DAMAGE

The galls cause considerable distortion of the shoots but, unlike those formed by *Adelges viridis* (see p. 98), they usually fail to encircle the shoot and do not, therefore, stop further growth. However, galling on trees no more than a few years old may be extensive, rendering young spruces grown as Christmas trees unmarketable.

181 Douglas fir adelges (*Adelges cooleyi*) damage to foliage of *Pseudotsuga.*

182 Douglas fir adelges (*Adelges cooleyi*) on *Pseudotsuga.*

Adelges cooleyi (Gillette) (**181–182**)
Douglas fir adelges

A common pest of conifers, alternating between spruce (*Picea*) and Douglas fir (*Pseudotsuga menziesii*). Originally restricted to western North America but introduced into Europe in the 1930s, where it is now widely distributed.

DESCRIPTION

Alate female (gallicola on spruce): 1.7–2.5 mm long; reddish brown to purplish black. **Alate female [ex: Douglas fir]:** 1.2–1.7 mm long; reddish brown to purplish black.

LIFE HISTORY

On Douglas fir, eggs deposited in the early spring by the overwintering forms hatch at bud-burst. The insects then develop on the needles, amongst conspicuous masses of white, waxen 'wool'. Both winged and wingless forms occur. The former depart in June for spruce trees, where sexual forms are reared. These aphids eventually give rise to overwintering morphs which finally mature in the spring and produce a gall-forming generation. The latter initiate characteristically elongate galls on the shoots, each gall maturing in the summer. Winged forms (gallicolae) then migrate back to Douglas fir, from late summer onwards, where wingless females eventually settle down to overwinter on the underside of the needles.

DAMAGE

Spruce: the shoot galls cause a twisting of the new growth. **Douglas fir:** foliage is yellow-mottled and tree growth is retarded; the foliage also becomes heavily encrusted with masses of white, fluffy wax and discoloured by sooty moulds which develop on the copious quantities of honeydew excreted by the adelgids.

Adelges laricis (Vallot) (183–185)
syn. *A. coccineus* (Ratzeburg)
Larch adelges

An often common pest of larch (*Larix*), including ornamentals, but mainly present on older trees; also gall-forming on spruce (*Picea*). Originally restricted to the central Alps but now widespread throughout Europe; also introduced into North America.

DESCRIPTION
Alate female [ex: larch]: 1.0–1.5 mm long; dark green, with a greyish-green head and thorax; fore wings *c*. 0.5 mm long. **Alate female [ex: spruce]:** 1.9–2.0 mm long; greyish to blackish; fore wings *c*. 0.6–0.7 mm long. **Egg:** greyish black, coated in whitish wax.

LIFE HISTORY
Blackish or purplish-grey nymphs overwinter on the bark of one-year-old larch shoots, often settled close to a bud. They mature in April, appearing blackish grey, and then deposit unprotected clusters of eggs at the base of the leaf spurs. Nymphs developing from these eggs feed on the new foliage, turning into either winged or wingless adults about a month later. The former then disperse to spruce trees, including Sitka spruce (*Picea sitchensis*), while the latter continue to breed on the larch needles, producing vast quantities of waxen 'wool' and sticky honeydew. This honeydew commonly accumulates as large globules amongst the protective canopy of waxen fibres. Individuals migrating to spruces give rise to wingless sexual forms and, in the following spring, gall-forming nymphs develop. Galls formed by this species on spruce are characteristically waxy, creamy and relatively small (*c*. 10–15 mm long). They mature in the summer, and the fully winged gallicolae which emerge, fly back to larches, to settle on the needles and eventually deposit eggs.

DAMAGE
Larch: infested foliage is disfigured with masses of white, waxen 'wool'. The needles become discoloured and distorted, heavily infested trees often appearing blue; attacks may also lead to premature loss of needles and to die-back of shoots. **Spruce:** galls sometimes prevent further shoot growth.

183 Larch adelges (*Adelges laricis*) on *Larix*.

184 Galls of larch adelges (*Adelges laricis*) on *Picea*.

185 Larch adelges (*Adelges laricis*) on *Larix*.

Adelges nordmannianae (Eckstein)
syn. *A. neusslini* (Börner); *A. schneideri* (Börner)
Silver fir migratory adelges

This dark brown to black adelgid is an important pest of young common silver fir (*Abies alba*); heavy infestations also occur on other species, including Caucasian fir (*A. nordmanniana*) and Cilician fir (*A. cilicica*). Nymphs overwinter on the shoots, and eventually mature and deposit clusters of brownish-orange eggs. These hatch at about bud-burst. Nymphs then feed on the needles, and commonly cause significant discoloration, stunting and distortion; severely infested trees may be killed. Infestations may also occur on the stems of host trees but, unlike *Adelges piceae*, with which this species is often confused, very little waxen 'wool' is produced. Several generations may persist on fir-tree hosts but, in summer, winged migrants appear. These disperse to spruce (*Picea*) trees where, in the following year, small, rounded terminal galls (about 10 mm across) are eventually produced. Colonies of this species produce considerable quantities of honeydew.

Adelges piceae (Ratzeburg)
Silver fir adelges

Restricted to silver fir (*Abies*) trees, and most commonly present on species of North American origin, e.g. alpine fir (*A. lasiocarpa*), balsam fir (*A. balsamea*), grand fir (*A. grandis*) and noble fir (*A. procera*). Colonies occur mainly on the stems, the dark brown or blackish insects (unlike *Adelges nordmannianae*, with which – when on common silver fir (*Abies alba*) – this insect is often confused) producing noticeable quantities of waxen 'wool'. Infested tissue often becomes distorted and swollen; in severe cases, growth is significantly affected and host trees may be killed.

Adelges viridis (Ratzeburg) (**186–188**)
syn. *A. laricis* (Hartig)
A spruce pineapple-gall adelges

Generally common on spruce (*Picea*) trees, especially Norway spruce (*P. abies*) and Sitka spruce (*P. sitchensis*), forming pineapple-like galls on the shoots; these are similar to those produced by *Adelges abietis*, but tend to be more elongated and usually completely encircle the shoots, stopping further growth. Winged individuals emerge from the galls in July (earlier than *Adelges abietis*), and then migrate to larch (*Larix*) where eggs, the overwintering stage, are laid. Light green, wingless forms develop on larch from mid-April onwards (earlier than other larch-infesting adelgids), causing a characteristic kinking of the needles. Eventually, light green or yellow, winged forms

are produced and these fly back to spruces, where their eggs hatch into summer, sexual forms. Progeny of the latter overwinter and eventually produce nymphs which initiate the next generation of pineapple galls.

Pineus pini (Macquart) (**189**)
Scots pine adelges

A widespread and common, but minor, pest of Scots pine (*Pinus sylvestris*); other two-needled pines are also hosts. Present throughout Europe, and introduced with the host plant to all other parts of the world.

DESCRIPTION

Apterous female (sisten): 1.0–1.2 mm long; dark brown to dark red, and almost spherical; head and prothorax heavily chitinized; antennae 3-segmented; abdomen with four distinct pairs of spiracles and an ovipositor. **Alate female (sexupara):** 1.0–1.2 mm long; mainly reddish grey; antennae 5-segmented; fore wings 1.4–1.7 mm long, hyaline and with the veins often tinged with red. **Egg:** oval and orange.

LIFE HISTORY

Nymphs, that hatch from eggs deposited in the autumn, overwinter and mature in the following March. They then deposit clusters of eggs, the progeny of which invade the new shoots and eventually develop into either winged or wingless forms. The former migrate to other sites from the end of May to late June, whereas the latter initiate a further wingless generation on the 'parent' plant; all females are oviparous. There may be three or more wingless generations each year. Colonies persist mainly on the bark of the shoots and youngest stems, and the aphids produce masses of white, fluffy 'wool'.

DAMAGE

The presence of waxen 'wool' disfigures host plants and is particularly unsightly on young garden and nursery trees.

186 Gall of *Adelges viridis* on terminal shoot of *Picea.*

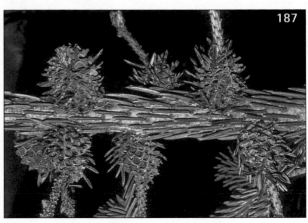

187 Mature galls of *Adelges viridis* on *Picea.*

188 *Adelges viridis* damage to foliage of *Larix.*

189 Colony of Scots pine adelges (*Pineus pini*) on *Pinus.*

Pineus strobi (Hartig) (**190**)
syn. *P. strobus* (Ratzeburg)
Weymouth pine adelges

Essentially similar to *Pineus pini* but slightly smaller and of North American origin, having been introduced into Europe in the mid-1800s. It infests Weymouth pine (*Pinus strobus*) and certain other five-needled pines, and is widely distributed on such hosts in parks and gardens. Overwintering nymphs mature and deposit eggs about a month later than *Pineus pini* and, in consequence, winged forms do not appear until mid-June. A second, wingless generation develops during the summer, and further egg clusters are produced during August. Nymphs appear in August and September, and eventually overwinter on the shoots. Considerable amounts of white, waxen 'wool' are produced on infested stems and younger branches.

190 Colony of Weymouth pine adelges (*Pineus strobi*).

Family **PHYLLOXERIDAE**

A small group of insects, structurally similar to the Adelgidae but alates with just three antennal segments.

Phylloxera glabra (von Heyden) (**191–193**)
Oak leaf phylloxera
A generally common but minor pest of English oak (*Quercus robur*). Widely distributed in Europe.

DESCRIPTION
Apterous female: 0.70–0.85 mm long; yellowish, marked with orange; oval-bodied and scale-like. **Egg:** 0.25 mm long; elongate-oval, yellow and shiny. **Nymph:** similar to adult but smaller.

LIFE HISTORY
The winter is passed in the egg stage in crevices in the bark of oak trees. In spring, nymphs invade the new growth to feed on the underside of the leaves. Once maturity is reached, eggs are deposited in small circles, all female individuals being oviparous. There are several asexual generations throughout the summer. However, in the autumn, smaller-bodied, winged sexual forms are produced. After this sexual phase, mated females eventually deposit the winter eggs.

DAMAGE
Infestations cause yellow and brown spotting of leaves; this sometimes leads to a general browning of the foliage, followed by premature leaf fall. Infestations on established trees are usually of little or no importance but persistent attacks on young trees reduce plant vigour.

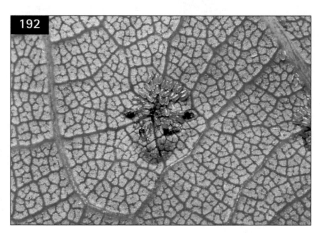

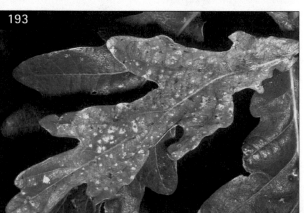

191 Colony of oak leaf phylloxera (*Phylloxera glabra*) on *Quercus*.

192 Eggs of oak leaf phylloxera (*Phylloxera glabra*).

193 Oak leaf phylloxera (*Phylloxera glabra*) damage to leaf of *Quercus*.

Family **DIASPIDIDAE** (armoured scales)

Body of female protected by a hard, scale-like covering (commonly called a 'scale', but also termed a 'test'), formed from wax and cast nymphal skins. The male pupal covering is also termed a 'test'. Owing to the mobility, first-instar nymphs are often known as 'crawlers'.

Abgrallaspis cyanophylli (Signoret)

A tropical, polyphagous species. Often established as a persistent greenhouse pest in temperate regions, including northern Europe where it is most often noted on cacti; infestations also occur on orchids and palms. The female tests are elongate-oval, 1–3 mm long, and mainly yellow to light brown.

Aonidiella aurantii (Maskell)

California red scale

A pest in citrus orchards in various parts of the Mediterranean basin. Also present on various ornamental hosts, including rose (*Rosa*). In cooler regions, sometimes found on greenhouse plants. Widely distributed in tropical and subtropical parts of the world.

DESCRIPTION

Female test: 1.5–2.0 mm across; circular, waxy and transparent. **Adult female:** broadly kidney-shaped, with the abdominal lobes extending back on either side of the pygidium; brownish red and clearly visible through the covering test. **Male test:** 1.2 mm long; oval and dark reddish brown. **Adult male:** 1 mm long; yellowish; winged. **First-instar nymph:** elongate-oval, reddish brown.

LIFE HISTORY

This species is favoured by stable, hot and dry conditions. The adult females are viviparous, and give birth to up to 150 nymphs over a period of about two weeks. The first-instar nymphs invade the stems, but will also settle on the foliage and elsewhere. Development to adulthood involves three nymphal stages and takes from 1 to 2 months. There are several overlapping generations annually, all stages often occurring together.

DAMAGE

Attacked plants are weakened. Infested branches develop distinct lesions and become desiccated, the leaves turning yellow and falling off. Californian red scales often occur amongst colonies of citrus mussel scale (*Lepidosaphes beckii*) (p. 105).

Aspidiotus nerii Bouché (**194**)

Oleander scale

This widely distributed scale insect is a common greenhouse pest in northern Europe, infesting various ornamentals such as *Acacia, Asparagus plumosus*, azalea (*Rhododendron*), *Cyclamen, Dracaena*, oleander (*Nerium oleander*) and palms. In favourable districts, infestations also occur outdoors on hosts such as Japanese laurel (*Aucuba japonica*); in mainland Europe, the pest is often present on decorative container-grown and open-bedded oleanders. The female tests are flat, rounded (1–2 mm across) and whitish, with a central yellow spot; male tests are similar but smaller. The pest often occurs on the stems of host plants but is usually most abundant on the leaves. It occurs throughout southern Europe, North Africa, the United States of America and Australia, but the exact distribution and host range are uncertain owing to frequent confusion with the closely related species *Aspidiotus hederae*.

194 Colony of oleander scale (*Aspidiotus nerii*) on *Nerium.*

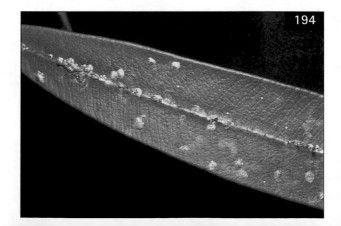

Aulacaspis rosae (Bouché) (**195**)
Rose scale

This widespread species occurs most commonly on non-hybrid rose (*Rosa*), but also attacks blackberry (*Rubus fruticosus*), both sexes developing on the stems. Female tests are 2.0–2.5 mm across, whitish to greyish and more or less oval, with the brownish-yellow nymphal exuviae usually placed eccentrically at the margin; the tiny male tests are elongate (0.8 × 0.3 mm) and white. When clustered on the bark the tests spoil the appearance of host plants, infested wood developing a scurfy appearance. Heavy attacks, which occur most commonly in sheltered sites, also reduce plant vigour. In July or early August, fertilized females lay eggs which they protect with their body. Eventually, orange-coloured first-instar nymphs appear. These nymphs wander over host plants but then become sedentary.

Male nymphs appear in the autumn but female nymphs usually not until the following spring. Individuals of both sexes reach maturity by May or June, when mating occurs. Adult males are winged and have a long caudal spine.

Carulaspis juniperi (Bouché) (**196–198**)
syn. *C. visci* Schrank
Juniper scale

A common pest of cypresses (*Chamaecyparis* and *Cupressus*), juniper (*Juniperus*) and *Thuja*; also associated with Wellingtonia (*Sequoiadendron giganteum*). Widely distributed in Europe, including Belgium, France, Germany and the British Isles, but detailed distribution uncertain owing to confusion with related species.

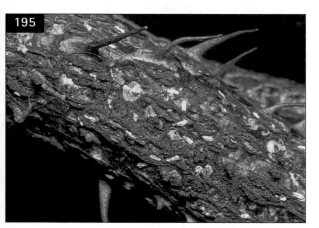

195 Colony of rose scale (*Aulacaspis rosae*) on *Rosa*.

196 Female scale of juniper scale (*Carulaspis juniperi*).

197 Male scale of juniper scale (*Carulaspis juniperi*).

198 Foliage of *Juniperus* heavily infested with juniper scale (*Carulaspis juniperi*).

DESCRIPTION
Female test: 1.0–1.5 mm across; rounded, slightly convex and whitish, with an eccentric yellow spot. **Male test:** 0.5–1.0 mm long; narrow, with a distinct longitudinal rib; mainly white. **Egg:** oval; pale yellowish white. **Nymph:** pale greenish yellow.

LIFE HISTORY
This species infests the foliage, shoots and fruits of various hosts, and is often present in considerable numbers. The insects mature in the late autumn or early winter, when winged males appear and mating takes place. Tests are readily dislodged from the host plant, and the overwintering mated females are frequently taken by insectivorous birds. Eggs are deposited in the following May, the females then dying. The eggs hatch in June. First-instar nymphs then swarm over the host plants before settling down to feed and to complete their development.

199 Nymphs of willow scale (*Chionaspis salicis*) on *Salix.*

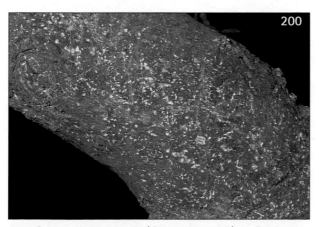

200 Colony of willow scale (*Chionaspis salicis*) on *Fraxinus.*

DAMAGE
Heavy infestations disfigure ornamentals, causing considerable discoloration of the foliage; affected shoots and branches look dull and distinctly unthrifty.

Carulaspis minima (Targioni-Tozzetti)
syn. *C. caerueli* (Signoret)
This more southerly distributed species is virtually indistinguishable from *Carulaspis juniperi*, but slight structural differences are apparent in adults. It is recorded from various parts of Europe; confirmed hosts include *Cupressus*, Lawson cypress (*Chamaecyparis lawsoniana*), pencil cedar (*Juniperus virginiana*) and *Thuja*.

Chionaspis salicis (Linnaeus) (**199–200**)
syn. *C. alni* Signoret; *C. populi* (Baeunsprung)
Willow scale
Locally common on ash (*Fraxinus excelsior*), common alder (*Alnus glutinosa*) and willow (*Salix*); infestations also occur on broom (*Cytisus*), *Ceanothus*, elm (*Ulmus*), flowering currant (*Ribes sanguineum*), lilac (*Syringa*), lime (*Tilia*), maple (*Acer*), poplar (*Populus*), privet (*Ligustrum vulgare*), spindle (*Euonymus*) and winter jasmine (*Jasminum grandiflorum*). Widely distributed in Europe; also present in North America.

DESCRIPTION
Female test: 1.5–2.3 mm long; whitish to waxy yellowish white, and irregularly pear-shaped. **Male test:** 0.5–1.0 mm long; white, elongate and very narrow, with a central and two lateral longitudinal ribs. **Adult male:** 1 mm long; orange to reddish-orange, with a caudal spine and bright yellow legs and antennae; alate and apterous forms occur. **Nymph:** rusty-red, oval and flattened.

LIFE HISTORY
Eggs are laid in August, beneath the body of fertilized females. The females, each covered by their protective test, then die. Eggs remain *in situ* throughout the winter and hatch in May. Newly emerged, red-coloured nymphs then cluster in large, conspicuous groups on the bark of host plants. They soon disperse and settle down to feed, developing to maturity a few weeks later. Adult males occur from late June to mid-July, wingless forms considerably outnumbering those with wings.

DAMAGE
Heavy encrustations impart a whitish appearance to the bark of host plants, and the pest has a deleterious effect on the growth of young trees and shrubs.

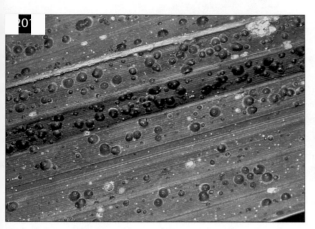

201 Colony of Florida red scale (*Chrysomphalus aonidum*).

202 Citrus mussel scale (*Lepidosaphes beckii*).

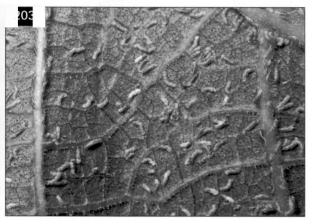

203 Colony of fig mussel scale (*Lepidosaphes conchyformis*).

204 Fig mussel scale (*Lepidosaphes conchyformis*) damage to leaf of *Ficus carica*.

Chrysomphalus aonidum (Linnaeus) (201)
Florida red scale

This tropical and subtropical pest infests *Citrus* and a wide range of other plants, especially palms and other non-deciduous ornamentals. Although the pest occurs outdoors on plants growing in parts of the southern and south eastern Mediterranean basin (e.g. North Africa and the Middle East), in Europe it is normally reported only in greenhouses, hot-houses and so on. The female tests are 2–4 mm in diameter, and dull purplish red with a pale apex; male tests are of similar appearance, but smaller and more elongate. Eggs are deposited by mated females over an extended period and hatch within a few hours into initally active, yellow nymphs. There are several overlapping generations annually, and all stages in the lifecycle often occur together.

Diaspis boisduvalii Signoret
Orchid scale

A common and virtually world-wide pest of greenhouse-grown orchids, especially *Calanthe, Cattleya, Cymbidium* and *Epidendrum*; infestations also occur on palms. Widespread in Europe.

DESCRIPTION

Female test: 2 mm long, flat and oval, yellowish to brownish, and translucent. **Male test:** 0.8–1.0 mm long; elongate, with three distinct longitudinal ribs, and coated with white waxen threads. **Adult female:** yellow to orange-yellow. **Adult male:** orange-yellow and alate, with pale legs and antennae (the latter very long) and a moderately long caudal spine. **Egg:** minute, oval and yellow. **Nymph:** yellow.

LIFE HISTORY

Under suitable conditions, breeding is continuous and infestations build up rapidly, particularly on the underside of the leaves of both young and established monocotyledonous pot plants. Tests of the different sexes are often clustered separately, and dense groupings impart a distinctive waxen appearance to infested parts of plants.

DAMAGE

Infestations disfigure and weaken host plants; the thick masses of white wax associated with male tests are most unsightly.

Lepidosaphes beckii (Newman) (**202**)
Citrus mussel scale

In southern Europe, a pest of *Citrus, Croton, Elaeagnus, Hibiscus* and many other plants; also widely distributed in tropical and subtropical parts of Africa, America and Asia.

DESCRIPTION

Female test: 2–4 mm long; elongate and broadly mussel-shaped; purplish brown. **Male test:** 1 mm long; similar to female test, but more elongate. **Egg:** minute, oval and white. **First-instar nymph:** 1.3 mm long; elongate-oval and flattened; brownish white.

LIFE HISTORY

After mating, each female deposits a batch of 50–100 eggs beneath the shelter of the test and then dies. Following egg hatch, the first-instar nymphs move onto the leaves, shoots and other parts of host plants, where they soon settle down to feed and mature. The lifecycle from egg to adult is completed in 2–4 months, and there are several generations annually.

DAMAGE

Heavy infestations cause premature leaf fall and death of young shoots. The pest often occurs amongst colonies of California red scale (*Aonidiella aurantii*) (p. 101).

Lepidosaphes conchyformis (Gmelin in Linnaeus) (**203–204**)
syn. *L. ficus* (Signoret)
Fig mussel scale

This species occurs outdoors in southern Europe. The tests often cover the underside of the leaves of common fig (*Ficus carica*) and cause a noticeable pale mottling of the foliage, clearly visible from above. The tests are about 2 mm long, relatively narrow, straight or slightly curved and slightly expanded posteriorly. Infested

plants are also reported in northern Europe but here the insect is able to survive only under artificial conditions.

Lepidosaphes machili (Maskell)
Cymbidium scale

In northern Europe, infestations of this scale insect are sometimes noted on the leaves and stems of greenhouse-grown *Cymbidium*. The tests are about 2 mm long, and mussel-shaped but rather narrow. Although disfiguring host plants, they are rarely numerous and are of only minor importance.

Lepidosaphes ulmi (Linnaeus) (**205**)
syn. *Mytilaspis pomorum* (Bouché)
Mussel scale

Generally abundant throughout most of Europe on various trees and shrubs, including fruit trees and ornamentals such as *Ceanothus*, common box (*Buxus sempervirens*), *Cotoneaster*, crab-apple (*Malus*), hawthorn (*Crataegus*), heather (*Erica*), Japanese quince (*Chaenomeles japonica*) and rose (*Rosa*). The female tests are 2.0–3.5 mm long, elongate mussel-shaped and grey to yellowish brown. They often encrust the bark of mature host plants but cause little or no damage. Eggs overwinter, protected by the female tests, and hatch in late May or June. First-instar nymphs then wander over the branches and trunks before settling down to feed. They reach maturity by the end of July, and eggs are laid in late August and September.

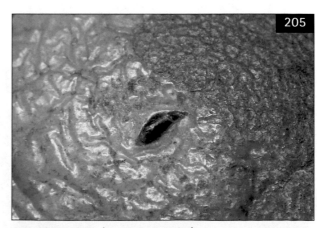

205 Mussel scale (*Lepidosaphes ulmi*).

Pinnaspis aspidistrae (Signoret)
Fern scale

Infestations of this cosmopolitan scale occur commonly on the stems and fronds of greenhouse-grown *Aspidistra*, ferns and palms, causing discoloration of foliage. Female tests are 2.5–3.0 mm long, reddish brown and mussel-shaped. Male tests are *c*. 1 mm long, white, ribbed and elongate; heavily infested foliage often appears white.

Quadraspidiotus perniciosus (Comstock)
syn. *Diaspidiotus perniciosus* (Comstock)
San José scale

An extremely polyphagous pest of trees and shrubs, including alder (*Alnus*), ash (*Fraxinus excelsior*), beech (*Fagus sylvatica*), birch (*Betula*), elm (*Ulmus*), *Eucalyptus*, false acacia (*Robinia pseudoacacia*), hornbeam (*Carpinus betulus*), lime (*Tilia*), maple (*Acer*), poplar (*Populus*), *Sorbus*, walnut (*Juglans*) and willow (*Salix*). Originally found in Asia but now well established in many other warmer parts of the world, including southern Europe. The pest is very destructive, the adults and nymphs imbibing considerable quantities of sap and seriously weakening their hosts. Infestations result in die-back and, often, cause death of plants. The area around feeding sites on leaves and other green tissue frequently turns red. The female tests (*c*. 2 mm diameter) are yellowish, flattened and circular, with a slightly raised eccentric nipple; male tests are smaller and elliptical (1.0 × 0.5 mm). The pest is viviparous, the mature females producing nymphs instead of eggs. There are usually 2–3 generations annually in southern Europe. The pest overwinters in the nymphal stage.

NOTE
Two closely related species, the yellow plum scale (*Quadraspidiotus ostraeformis*) and the yellow pear scale (*Q. pyri*) – both with a more northerly distribution – also occur on ornamental trees and shrubs (especially Rosaceae) but are of only minor significance. Unlike San José scale, their lifecycles include an egg stage and they have just one generation each year.

Unaspis euonymi (Comstock) (**206–207**)
Euonymus scale

A local pest of spindle (*Euonymus*), especially Japanese spindle (*E. japonica*). Widely distributed in southern Europe; also found further north, including coastal areas in southern England.

DESCRIPTION
Female test: 2–3 mm long; brown and mussel-shaped to oyster-shaped. **Male test:** 2.0–2.5 mm long; pale yellow to white and elongate, with three longitudinal ridges.

LIFE HISTORY
There are two generations annually, with eggs hatching and first-instar nymphs swarming on host plants in June and in early September. The nymphs eventually settle down to feed on the foliage and stems of host plants. Male tests are very obvious, and greatly outnumber those of females.

DAMAGE
Attacks cause leaf discoloration and lack of vigour; heavy infestations cause severe decline and, eventually, death of plants.

206 Colony of euonymus scale (*Unaspis euonymi*).

207 Euonymus scale (*Unaspis euonymi*) damage to leaves of *Euonymus*.

Family **COCCIDAE** (soft scales)

Body of female forming a smooth or wax-covered, often tortoise-shaped, scale. Pupal covers of males are known as 'tests'. As in the previous family, first-instar nymphs are often known as 'crawlers'.

Ceroplastes rusci (Linnaeus) (**208**)

Fig wax scale

A polyphagous pest of common fig (*Ficus carica*), common myrtle (*Myrtus communis*), oleander (*Nerium oleander*), *Pittosporum* and various other plants. Of Chinese origin; now widely distributed in the Mediterranean basin, including southern Europe.

DESCRIPTION

Adult female (scale): up to 5 mm long; barnacle-like, formed from eight fused plates; whitish-marble, variably marked with brown. **Male test:** 2.0–2.2 mm long; reddish. **Egg:** 0.32 × 0.23 mm; oval; pale brownish yellow when laid, later becoming reddish brown. **First-instar nymph:** 0.3 mm long; oval-bodied and rusty red in colour.

LIFE HISTORY

After mating, each female deposits many hundreds of eggs over a period of several weeks, typically from May onwards. These eggs remain under her body and hatch 3–4 weeks later. Nymphs pass through three instars before attaining the adult stage, differences between the sexes becoming apparent in the final instar. The scales tend to occur along the major leaf veins or in clusters along the shoots. The pest excretes considerable quantities of honeydew and colonies are often attended by ants. In particularly favourable (e.g. coastal) districts there may be two generations annually; elsewhere, there is just one. The winter is usually passed as second-instar nymphs.

DAMAGE

Infested hosts are weakened, and heavily infested leaves and shoots may be killed. In addition, sooty moulds developing on honeydew excreted by the pest reduces photosynthetic activity of the leaves.

Ceroplastes sinensis del Guercio (**209**)

Citrus wax scale

This minor pest, of Chinese origin, is well established in various Mediterranean areas, including France, Italy and Spain, where it is associated mainly with plants such as Brazilian peppertree (*Schinus molle*), *Citrus*, Japanese holly (*Ilex crenata*), *Rhus* and ornamental *Solanum*.

DESCRIPTION

Adult female (scale): 5–6 mm long; barnacle-like, rectangular in outline and strongly convex; pinkish white when young, but later becoming reddish. **First-instar nymph:** purplish red and oval-bodied.

LIFE HISTORY

Adult females occur on the branches and stems of host plants, eggs (sometimes several thousand per female) being deposited in the late spring. The first-instar nymphs eventually invade the leaves, where they attach themselves to the upper surface and begin feeding. Later, as third-instar nymphs, individuals migrate back to the branches and stems. Here, they become sedentary and then overwinter, eventually attaining the adult stage.

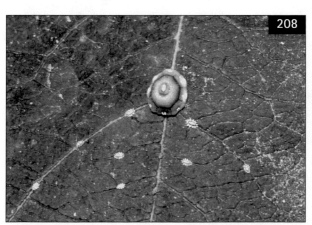

208 Female fig wax scale (*Ceroplastes rusci*) and second-instar nymphs.

209 Citrus wax scale (*Ceroplastes sinensis*).

DAMAGE

Infested shoots and branches may be weakened, but infestations are localized and generally of only minor significance.

Chloropulvinaria floccifera (Westwood) (**210**)
syn. *Pulvinaria floccifera* (Westwood)
Cushion scale

An often common pest of greenhouse-grown *Camellia* and orchids; in favourable areas, infestations also occur outdoors. Other hosts include *Citrus*, holly (*Ilex*), Japanese spindle (*Euonymus japonica*) and *Rhododendron*. Probably of southern European origin but nowadays found in many parts of the world. Widespread in Europe.

DESCRIPTION

Female scale: 2.5 × 2.0 mm; oval and yellowish, surmounting an elongate (10–15 mm long), thin, white ovisac. **Egg:** minute, oval and pinkish.

210 Cushion scale (*Chloropulvinaria floccifera*).

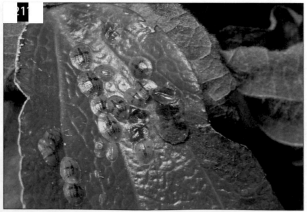

211 Colony of brown soft scale (*Coccus hesperidum*).

LIFE HISTORY

Infestations occur on the foliage and shoots throughout the year with, in suitably warm conditions, a succession of generations annually. Elsewhere, there is just one generation annually, immature scales overwintering and the characteristic ovisacs being produced in April and May soon after mating has occurred. Mature females perish soon after completing their ovisacs and laying eggs, and fall away from the host plant before the eggs hatch.

DAMAGE

Attacked plants are often disfigured by accumulations of flocculent waxen 'wool'.

Coccus hesperidum Linnaeus (**211**)
Brown soft scale

A very common, polyphagous and virtually worldwide pest of greenhouse ornamentals; also attacks outdoor plants growing in favourable situations, including many parts of mainland Europe and southern England. Commonly infested hosts include azalea (*Rhododendron*), bay laurel (*Laurus nobilis*), *Camellia*, *Citrus*, *Clematis*, *Escallonia*, *Geranium*, *Hibiscus*, holly (*Ilex*), ivy (*Hedera*), oleander (*Nerium oleander*), poinsettia (*Euphorbia pulcherrima*), *Stephanotis*, *Viburnum*, weeping fig (*Ficus benjamina*) and various ferns.

DESCRIPTION

Female scale: 3.5–5.0 mm long; very flat and oval; translucent yellow to brown, with an often distinct median longitudinal ridge and rib-like markings.

LIFE HISTORY

This species is viviparous, and usually parthenogenetic, each female producing about a thousand nymphs over a period of 2–3 months. The first-instar nymphs wander over host plants for a few days before settling down to feed on the leaves, with individuals clustered along the midrib and other major veins; the insects often overlap one another, forming dense (often ant-attended) colonies. Breeding is continuous, so long as conditions remain favourable, and the complete lifecycle from birth to maturity occupies about two months at average greenhouse temperatures.

DAMAGE

The scales excrete considerable quantities of honeydew, and foliage beneath infested leaves becomes extensively blackened by sooty moulds. This spoils the appearance of ornamentals and checks growth.

Eulecanium tiliae (Linnaeus) (212)
syn. *E. capreae* (Linnaeus); *E. coryli* (Linnaeus)
Nut scale
A locally common pest of ornamental trees and shrubs, including alder (*Alnus*), *Ceanothus*, *Cotoneaster*, elm (*Ulmus*), firethorn (*Pyracantha*), hawthorn (*Crataegus*), hornbeam (*Carpinus betulus*), horse chestnut (*Aesculus hippocastanum*), lime (*Tilia*), oak (*Quercus*), rose (*Rosa*), spindle (*Euonymus*) and sycamore (*Acer pseudoplatanus*). Widely distributed in Europe; also present in North America.

DESCRIPTION
Female scale: 5–6 mm across; dark chestnut-brown to light brown or greyish brown, and strongly convex. **Male test:** 2.0–2.5 mm long; greyish, elongate-oval. **Adult male:** reddish crimson, with a relatively short caudal spine and a pair of long caudal filaments. **Egg:** pale yellowish white. **Nymph:** pinkish to orange-yellow.

LIFE HISTORY
Adult males emerge in late April and early May, and eggs are deposited by fertilized females about a month later. The eggs hatch towards the end of the summer. First-instar nymphs then wander over host plants before settling down. Young overwintering male and female scales are similar in appearance, and about 1.5–2.0 mm long. However, the more elongate appearance of the former becomes obvious in the spring as development recommences. This species also breeds parthenogenetically.

DAMAGE
Heavy infestations retard growth, and host plants may be killed. Minor attacks are of little or no significance.

Eulecanium excrescens (Ferris) (213)
Wisteria scale
In Asia, this extremely polyphagous species occurs on rosaceous fruit trees and a wide range of ornamental trees and shrubs. It was unknown in Europe until infestations were found in 2001 in Central London, England, on *Wisteria* and certain other ornamentals. The pest also occurs as an introduced species on *Wisteria* in the USA.

DESCRIPTION
Adult female (scale): 13 mm long; globular, strongly concave, with an irregular surface; dark brown to blackish, often dusted with a greyish waxen bloom. **Egg:** 0.5 mm long; pinkish orange. **First-instar nymph:** orange (older nymphs brown, with distinct waxen patches).

212 Colony of nut scale (*Eulecanium tiliae*).

213 Wisteria scale (*Eulecanium excrescens*).

LIFE HISTORY
In England, the pest has just one generation annually. It overwinters in an early nymphal stage, the nymphs eventually reaching maturity in late April or early May. The exceptionally large female scales deposit masses of eggs beneath their bodies, and then die. Following egg hatch, first-instar nymphs emerge *en masse* from beneath the maternal scales and swarm over the foodplant. Initially, these highly mobile nymphs feed on the foliage. Later, they move to the bark of shoots and stems, where they become sedentary and, eventually, overwinter.

DAMAGE
Infested hosts are weakened and soiled by honeydew, upon which sooty moulds develop. Severe infestations may result in the death of host plants.

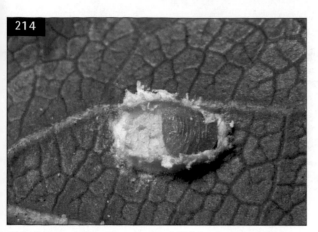

214 Female scale and egg mass of *Lichtensia viburni.*

215 Female scales of *Lichtensia viburni.*

216 Brown scale (*Parthenolecanium corni*).

Lichtensia viburni Signoret (214–215)

Viburnum cushion scale

A locally distributed species, occurring mainly on European ivy (*Hedera helix*) but also associated with *Viburnum tinus*; minor infestations sometimes occur on ornamental plants and hedges in England, Wales and parts of mainland Europe. The female scales are 3–5 mm long, brown and somewhat fleshy. In May, the females mature and produce large numbers of whitish to pale brownish-white eggs. The eggs are contained in prominent (5–6 mm long) ovisacs, composed of sticky masses of white waxen 'wool', which almost completely cover the maternal scales. Eggs hatch in late June. The emerged nymphs then attach themselves to either side of expanded leaves to begin feeding. Development is relatively slow, with the scales maturing in the following spring. Adult males usually appear in late April and May. Male tests are 2.0–2.5 mm long, white and elongate-oval, with distinct longitudinal ridges. This pest produces considerable quantities of honeydew upon which disfiguring sooty moulds develop.

Parthenolecanium corni (Bouché) (216)

Brown scale

A generally common pest of trees and shrubs, including ornamentals such as *Ceanothus*, *Cotoneaster*, crab-apple (*Malus*), *Elaeagnus*, *Escallonia*, firethorn (*Pyracantha*), flowering cherry (*Prunus*), flowering currant (*Ribes* sanguineum), honeysuckle (*Lonicera*), Japanese quince (*Chaenomeles japonica*), *Magnolia*, rose (*Rosa*) and *Wisteria*. Present throughout Europe and in many other parts of the world.

DESCRIPTION

Female scale: 4–6 mm long; more or less oval, very convex and roughened; chestnut-brown and often shiny. **Egg:** minute, oval, whitish and shiny. **Nymph:** oval, flat, pale greenish to orange or light brown.

LIFE HISTORY

In most situations, eggs are laid beneath the female scales in May or June. They hatch from mid-June onwards. First-instar nymphs then move to the young leaves and shoots to begin feeding. Second-stage nymphs, that are also mobile, appear in August. They continue feeding on the young growth, but in the autumn they migrate to the branches and twigs. Here they settle down for the winter, gradually changing colour from greenish to brownish. Activity is resumed in the following spring, individuals then settling permanently and growing rapidly. As development continues, the body becomes increasingly hardened and convex, to form the familiar protective scale. Although most races of this species are parthenogenetic, males occur on some hosts in certain areas. Also, details of the lifecycle vary according to conditions. Under glass, two to three generations may occur annually.

DAMAGE
Infestations disfigure and weaken host plants, and often cause premature leaf fall.

Parthenolecanium persicae (Fabricius) (**217**)
syn. *P. crudum* (Green)
Peach scale
Infestations of this scale insect occur on various ornamental plants, including *Citrus*, common fig (*Ficus carica*), false acacia (*Robinia pseudoacacia*), flowering cherry (*Prunus*), honeysuckle (*Lonicera*) and rose (*Rosa*). In northern Europe, attacks are usually heaviest under glass or on hosts growing against sheltered, sunny walls. Adult female scales (5–6 mm long) are shiny brown, with a distinct anal cleft and a slight longitudinal keel. The nymphs are elongate-oval, yellowish brown to brownish-orange, marked with dark brown, and rather flat; long, straight glass-like strands of silk radiate outwards from the body. Infested stems become coated in scales, which often cluster closely together. The insects excrete considerable quantities of honeydew, and host plants are further contaminated by sooty moulds.

Parthenolecanium pomeranicum
(Kawecki) (**218**)
Yew scale
Similar in appearance and biology to *Parthenolecanium corni* but restricted to yew (*Taxus*). Infestations, which occur mainly on the leaves, are unsightly, particularly when hosts are contaminated by honeydew and sooty moulds. The pest is also directly harmful, and heavy attacks lead to defoliation of hedges and specimen trees. The insect is widely distributed in mainland Europe; in Britain it is most often noted in southern England.

Protopulvinaria pyriformis (Cockerell) (**219**)
Pyriform scale
A pest of *Citrus* and various other plants, including bay laurel (*Laurus nobilis*), castor oil plant (*Ricinus communis*), *Gardenia*, oleander (*Nerium oleander*) and spikenard (*Aralia*). Present in southern Europe, including France, Italy, Portugal and Spain. Found in many other warmer parts of the world, including Africa, America and Asia.

DESCRIPTION
Adult female (scale): 5–7 mm across; mainly light brown and flattened, with a distinctly fan-shaped outline.

217 Peach scale (*Parthenolecanium persicae*).

218 Yew scale (*Parthenolecanium pomeranicum*).

219 Pyriform scale (*Protopulvinaria pyriformis*).

LIFE HISTORY
This species breeds parthenogenetically, with typically two generations annually. Dense colonies occur on the underside of the foliage of infested plants, the scales tending to cluster along the midrib. Copious quantities of honeydew are excreted.

DAMAGE
Infested host plants become extensively blackened by sooty moulds that develop on the excreted honeydew, and this has an adverse effect on photosynthesis and growth.

Pulvinaria hydrangeae Steinweden (**220**)
syn. *Eupulvinaria hydrangeae* (Steinweden)
Hydrangea scale
An invasive pest of ornamental trees and shrubs in various parts of Europe, including Belgium, France, Germany, Hungary, Italy, the Netherlands and Switzerland. Also present in Australasia, Japan and North America. Host plants include *Hydrangea*, lime (*Tilia*), maple (*Acer*), plane (*Platanus*), *Viburnum* and various rosaceous trees and shrubs. The brownish female scales occur on the shoots, branches and trunks, and their white ovisacs develop during the summer. Following egg hatch, first-instar nymphs migrate over the trees before settling down to feed and develop.

Pulvinaria regalis Canard (**221–222**)
Horse chestnut scale
An invasive Asian species, unknown in Europe before the 1960s. Now widely distributed in central Europe and elsewhere (including Denmark, England, France, Germany, the Netherlands and Switzerland). Polyphagous on various trees and shrubs, including bay laurel (*Laurus nobilis*), dogwood (*Cornus*), elm (*Ulmus*), horse chestnut (*Aesculus hippocastanum*), lime (*Tilia*), *Magnolia*, maple (*Acer*), *Skimmia japonica* and sycamore (*Acer pseudoplatanus*).

220 Hydrangea scale (*Pulvinaria hydrangeae*).

221 Horse chestnut scale (*Pulvinaria regalis*).

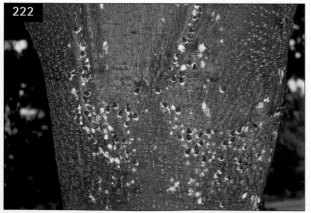

222 Colony of horse chestnut scale (*Pulvinaria regalis*).

223 Woolly currant scale (*Pulvinaria vitis*).

DESCRIPTION

Adult female (scale): 7 mm long; brown; roundly triangular, with a distinct posterior cleft. **Male test:** 3 mm long; similar to female scale but smaller, narrower and lighter brown. **Egg:** minute, oval and whitish. **Nymph:** oval, flat and brownish, with a distinct posterior cleft.

LIFE HISTORY

Unlike most scale insects, this species displays considerable mobility during its developmental stages. Eggs hatch in June to July. First-instar nymphs then migrate to the underside of leaves to begin feeding. In September, when about 2 mm long, the nymphs move onto the twigs where feeding continues throughout the winter; growth, however, is slow. After bud-burst, nymphal development become rapid and size differences soon become obvious between the smaller males (if present) and larger females, with dense colonies developing on suitable hosts. Individuals become mature by May. Before egg laying, the mature females migrate from the twigs to the main branches and trunks. At this time, female scales sometimes fall from host trees. They may then be found on nearby low-growing plants or other objects. Eggs, up to 2,000 per female, are deposited beneath the maternal scales, mainly on the branches and trunks of host plants but also on walls, pavements and elsewhere. Although males occur, females are capable of reproducing parthenogenetically and usually form the bulk of populations.

DAMAGE

Although some hosts, notably limes and maples, are colonized extensively, the scales appear to have little or no effect on tree growth. Their presence on amenity or ornamental trees, however, is unsightly and commonly causes concern. Infested trees often bear whitish marks on the bark long after the dead scales have fallen away.

Pulvinaria vitis (Linnaeus) (**223**)

Woolly vine scale

A highly polyphagous species, now known to include *Pulvinaria betulae* and *P. ribesiae*, both of which have been considered separate entities. Commonly infested ornamentals include alder (*Alnus*), birch (*Betula*), *Cotoneaster*, flowering currant (*Ribes sanguineum*), hawthorn (*Crataegus*), ornamental vine (*Vitis vinifera*) and willow (*Salix*). Holarctic. Widely distributed in Europe.

DESCRIPTION

Adult female (scale): 5–7 mm long; heart-shaped to oval or circular, strongly convex and saddle-like; dark brown and wrinkled. **Male test:** 2 mm long; elongate-oval and brownish, with black markings and distinctly paler ridges. **Adult male:** 1.5 mm long; delicate, pinkish red, with brownish legs and antennae, a short caudal spine and a pair of long caudal filaments; winged. **Egg:** 0.3 mm long; pale purplish red. **First-instar nymph:** 0.5 mm long; light purplish brown to orange-yellow, flat, elongate-oval, with a distinct anal cleft.

LIFE HISTORY

Adults appear in September or October. The short-lived males die after mating. Females, however, are longer-lived and eventually overwinter. In early spring, the surviving females begin to feed and grow, becoming distinctly convex and darker in colour, while their backs gradually harden to form a protective scale. From mid-April or May onwards, when fully mature, each female spins a white, cushion-like ovisac within which, over a period of 2–3 weeks, 1,000 or more eggs are deposited. The female then dies. The bulk of the egg mass forces the scale away from the substratum, with the hind end tilted upwards. The presence of the pest then becomes particularly obvious as strands of waxen 'wool' from beneath the scales are wafted about by the wind, often covering many of the shoots and branches. Eggs hatch from late May or early June onwards. The young, first-instar nymphs, that typically appear in swarms, then wander over the young shoots and leaves. Eventually, the nymphs disperse to the one-year-old wood where they settle down to continue their development. When feeding, this pest excretes considerable quantities of honeydew upon which sooty moulds may develop. There are three nymphal instars, and the adult stage is reached in the autumn. Although there is one generation annually, details of the lifecycle vary according to conditions. Also, populations may include both sexual and asexual races.

DAMAGE

Heavy infestations, which occur on the leaves and branches, and may extend down the main stems onto the roots, severely weaken host plants. Waxen 'wool', honeydew and sooty moulds also contaminate foliage and other parts of host plants, and have a detrimental effect on photosynthesis.

Saissetia oleae (Olivier) (**224**)
Mediterranean black scale

A generally abundant subtropical species, attacking a wide range of trees and shrubs, including ornamentals such as ivy (*Hedera*), oleander (*Nerium oleander*), olive (*Olea europaea*) and *Pittosporum*. Originally Palaearctic, but now virtually cosmopolitan. Widely distributed in southern Europe.

DESCRIPTION
Adult female (scale): up to 5 mm long and 3 mm wide; very convex, dark brown to black, with ridges forming a distinctive H-shaped pattern on the back. **Egg:** minute, oval, whitish to pale brownish yellow. **First-instar nymph:** pinkish, elongate-oval and dorsoventrally flattened.

LIFE HISTORY
Infestations occur on leaves, shoots and branches, each mature female depositing up to 2,000 or more eggs. The eggs hatch 2–3 weeks later. First-instar nymphs occur mainly on the young shoots and underside of leaves. Under ideal conditions the complete lifecycle lasts for 3 or 4 months, and there are two generations annually. However, in the more northerly parts of its outdoor range the pest completes just one generation. Although males are known to occur, reproduction is mainly parthenogenetic.

DAMAGE
Heavily infested hosts are weakened, and shoots and leaves may wither. Also, considerable quantities of honeydew are excreted by the developing nymphs, and this quickly allows sooty moulds to become established. This is both debilitating and unsightly.

Saissetia coffeae (Walker) (**225**)
syn. *S. hemisphaericum* (Targioni-Tozzetti)
Hemispherical scale

An often abundant scale in greenhouses. Infestations occur on a wide variety of hosts, including *Asparagus plumosus*, *Begonia*, carnation (*Dianthus caryophyllus*), *Ficus*, oleander (*Nerium oleander*) and *Stephanotis*, and on various ferns and orchids. Infestations rarely cause significant damage, particularly if plants are otherwise healthy and well-tended, but hosts are often contaminated by honeydew upon which sooty moulds develop. Eggs and nymphs occur at all times of the year, even in unheated greenhouses. Reproduction is entirely parthenogenetic. Each female deposits up to 2,000 eggs in a tight cluster beneath her body and then dies. The female scales are 2–3 mm across, strongly convex and reddish brown to blackish; oval forms are larger than rounded ones, sometimes exceeding 4 mm in length.

224 Mediterranean black scale (*Saissetia oleae*).

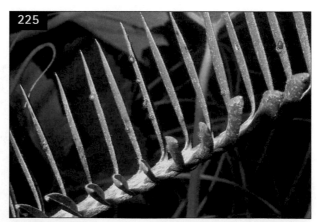

225 Hemispherical scale (*Saissetia coffeae*) on *Encephalartos*.

Family **ERIOCOCCIDAE** (felt scales)

Cryptococcus fagisuga Lindinger (**226–227**)
syn. *C. fagi* (Baerensprung)
Beech scale

A generally abundant pest of beech (*Fagus sylvatica*). Widely distributed in Europe; accidentally introduced into North America in the late 1800s, where it has since become a significant pest.

DESCRIPTION

Adult female (scale): 0.75–1.0 mm long; hemispherical and pale lemon-yellow, coated with white flocculent threads of wax. **Egg:** 0.15 mm long; pale yellow. **First-instar nymph:** 0.25 mm long; lemon-yellow, with three pairs of legs and 5-segmented antennae.

LIFE HISTORY

This species is parthenogenetic and has a single generation. Eggs are laid from June to August, hidden under the protective waxen 'wool' in small string-like clusters of up to eight. They hatch 6–7 weeks later. The very active first-instar nymphs either burrow beneath the remains of adjacent dead scales or swarm over the trunk and branches before settling down in suitable bark crevices. They then overwinter before moulting to the entirely sedentary second nymphal stage. These second-instar nymphs eventually moult to the adult stage, individuals usually reaching maturity in early June. The scales produce considerable quantities of white, waxen 'wool', and often cover considerable areas of the main trunks and branches. On heavily infested sections of bark, the scales commonly overlap the shrivelled, black remains of previous generations.

DAMAGE

Scale-encrusted bark disfigures ornamental trees, but infestations usually have little or no direct effect on the growth or well-being of their hosts. However, in some regions (e.g. Denmark and North America) attacks are associated with pathogenic fungi (*Nectria* spp.) and lead to the decline and eventual death of affected trees.

Eriococcus spurius (Modeer)
syn. *Coccus ulmi* (Linnaeus); *Gossyparia spurius* Modeer
Elm scale

A widely distributed species, associated with elm (*Ulmus*), sometimes forming noticeable encrustations on amenity trees in towns and cities. The adult females (*c.* 2. mm long) are dark red and oval, with a short pair of anal papillae. They occur during the summer on the bark of the trunks and main branches amongst

flocculent masses of whitish wax, each insect depositing about 250 eggs before dying. The eggs hatch from late August onwards, first-instar nymphs migrating onto the leaves to feed. Second-instar nymphs eventually overwinter, resting in sheltered places on the bark of host trees. In the following spring, the nymphs invade the young shoots and foliage. Later, they move onto the trunk and larger branches where they settle down to mature. There is one generation annually.

226 Beech scale (*Cryptococcus fagisuga*).

227 Colony of beech scale (*Cryptococcus fagisuga*) on *Fagus*.

228 Colony of ash scale (*Pseudochermes fraxini*) on *Fraxinus.*

Pseudochermes fraxini (Kaltenbach) (**228**)
syn. *Apterococcus fraxini* (Newstead)
Ash scale

A local insect on ash (*Fraxinus excelsior*) trees, with colonies occurring beneath white, felt-like masses of wax on the stems of small trees and branches of larger ones. Adult males appear from October to early November, and crawl over the bark in search of developing females. The males (0.9 mm long) are wingless, and mainly orange to orange-yellow, with black eyes. Mated orange-red females deposit eggs in the spring and then die. The eggs hatch in mid-June. Reddish first-instar nymphs then swarm over the bark before settling down to feed and develop. Infestations develop rapidly on trees placed under stress by the removal of nearby shelter, and often occur on isolated amenity trees in towns and cities.

Family **PSEUDOCOCCIDAE** (mealybugs)

Typically elongate-oval insects, with poorly developed 5- to 9-segmented antennae but well-developed legs; body covered in a flocculent or mealy waxen secretion. Females elongate-oval and woodlouse-like; males rare and often unknown.

Balanococcus diminutus (Leonardi)
syn. *Trionymus diminutus* Leonardi;
T. calceolariae (Maskell)
New Zealand flax mealybug

An Australasian species, associated mainly with New Zealand flax (*Phormium tenax*) but also attacking *Cordyline australis*. Introduced on such plants to other parts of the world, including northern Europe; often noted in south-western England.

DESCRIPTION
Adult female: 4–5 mm long; grey to purplish grey or dark red, dusted with a white, waxen secretion that forms long filamentous processes. **Nymph:** light grey, dusted with white wax.

LIFE HISTORY
Eggs of this parthenogenetic species are laid in masses within protective ovisacs formed mainly on the leaf blades. The first-instar nymphs are very mobile, particularly if arising from eggs deposited in unfavourable situations; later instars and adults are more or less sedentary, the insects often clustering together at the leaf bases. There are several generations annually, and all life stages often occur together.

DAMAGE
Infested plants are contaminated by vast quantities of honeydew; in addition, masses of flocculent wax spoil the appearance of plants. Attacked hosts are weakened and may be killed.

Nipaecoccus nipae (Maskell)
Palm mealybug

Unsightly infestations of this tropical mealybug sometimes occur in temperate regions on greenhouse-grown palms, individuals tending to settle on the leaf bases. Adults are about 3.5 mm long, salmon-pink in colour, with discrete creamy-white waxen cones distributed over the rather convex body.

229 Citrus mealybug (*Planococcus citri*).

230 Colony of glasshouse mealybug (*Pseudococcus viburni*) on *Cyperus*.

Planococcus citri (Risso) (**229**)

Citrus mealybug

An often abundant pest on greenhouse ornamentals, developing rapidly under conditions of high humidity and temperature, with up to eight generations annually. Heavy infestations commonly develop on *Hippeastrum*, cacti, ferns, orchids, vines and many other plants. Adults are 3–4 mm long and pinkish, coated with whitish wax; the peripheral and caudal waxen processes are characteristically short and stout.

Pseudococcus viburni (Signoret) (**230**)

syn. *P. affinis* (Maskell); *P. latipes* Green; *P. obscurus* Essig

Glasshouse mealybug

A widely distributed tropical or subtropical species, generally common in Europe on a wide variety of greenhouse ornamentals and house plants.

231 Citrophilus mealybug (*Pseudococcus calceolariae*).

DESCRIPTION

Adult: 4 mm long; body pinkish, covered with white mealy wax; caudal processes about half as long as body.

LIFE HISTORY

Eggs are laid in batches within a white, waxy, cottonwool-like sac. After eggs hatch, nymphs eventually disperse over infested plants, often congregating in the axils of buds, at the base of leaves and within leaf sheaths and curled leaves. There are several generations each year and, under favourable conditions, breeding is continuous.

DAMAGE

Adults and nymphs suck the sap of host plants, impairing growth and, sometimes, leading to defoliation. Infested plants are also disfigured by the presence of clumps of white wax and by accumulations of honeydew and sooty moulds.

Pseudococcus calceolariae (Maskell) (**231**)

syn. *P. fragilis* (Brain); *P. gahani* (Green)

Citrophilus mealybug

Unlike most mealybugs, this species is indigenous to northern Europe. In favourable districts, infestations occur outdoors on various plants, including *Ceanothus*, flowering currant (*Ribes sanguineum*), false acacia (*Robinia pseudoacacia*), *Forsythia*, juniper (*Juniperus*) and *Laburnum*; greenhouse ornamentals are also attacked. Adults are 3–4 mm long and broadly oval, with the body partly coated with white, mealy wax; the caudal processes are short and relatively thick.

232 Long-tailed mealybug (*Pseudococcus longispinus*).

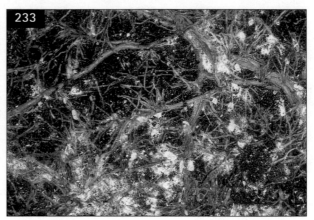

233 Colony of root mealybug (*Rhizoecus falcifer*) on roots of *Lavendula*.

Pseudococcus longispinus (Targioni-Tozzetti) (**232**)

Long-tailed mealybug

Infestations of this common glasshouse mealybug often occur on cacti, lilies, orchids, vines and many other ornamentals. Individuals are relatively small (adults, *c*. 2.5 mm long), with very long caudal processes. Colonies produce considerable quantities of honeydew.

Rhizoecus falcifer (Künckel de Herculais) (**233**)

syn. *R. terrestris* (Newstead)

Root mealybug

Dense subterranean colonies of this and certain other closely related species often develop on the roots of (mainly) greenhouse-grown ornamentals, surrounded by fragile masses of white wax. In common with container-grown plants infested with root aphids, the colonies often occur at the periphery of the root ball. The mealybugs are elongate (1.0–2.3 mm long), with distinct, 5-segmented, geniculate antennae and a pair of very short, waxy anal appendages; although greenish yellow, the insects appear white due to the mealy coating of wax. Reproduction is parthenogenetic, the mature females depositing their white, translucent eggs in batches inside cottonwool-like ovisacs; there is a succession of generations throughout the year. The foliage of attacked plants becomes dull, heavily infested plants making poor growth and eventually wilting.

Family **MARGARODIDAE**

A small group of mainly tropical insects; females typically mobile, with well-developed legs and 6- to 13-segmented antennae.

Icerya purchasi (Maskell) (**234**)

Cottony cushion scale

Originally an Australian species, now a serious pest of citrus plants in various parts of the world; infestations also occur on ornamentals such as *Acacia*, broom (*Cytisus*), *Mimosa* and rose (*Rosa*). Attacks are common in citrus-growing regions of southern Europe, and may also occur occasionally on greenhouse-grown ornamentals and house plants further north. Females are about 5 mm long, brown and oval, each surmounting a large (*c*. 10 mm long), white, distinctively grooved ovisac which contains several hundred eggs.

234 Cottony cushion scale (*Icerya purchasi*).

Order **THYSANOPTERA** (thrips)

Family **THRIPIDAE**

Thrips with a flattened body, narrow and pointed wings, and a saw-like ovipositor. Development includes an egg and two nymphal stages, followed by single non-feeding propupal and pupal stages.

Anaphrothrips orchidaceus Bagnall
Yellow orchid thrips
An introduced thrips, sometimes infesting greenhouse-grown orchids in northern Europe, producing brown patches on the green tuberous tissue. Adults (1.3 mm long) are yellow-bodied, with a pair of brown stripes behind the eyes; antennae are 8-segmented and mainly brown; the fore wings are light brown, but pale basally, and the legs yellow.

Dendrothrips ornatus (Jablonowski) (**235**)
Privet thrips
Infestations of this locally common thrips occur on the upper surface of leaves of lilac (*Syringa*) and privet (*Ligustrum vulgare*), causing noticeable silvering and distortion; in continental Europe, this species is also reported on alder (*Alnus*) and lime (*Tilia*). There are two or more generations each year, the thrips occurring on host plants from April to November. Adults (0.9–1.1 mm long) are brown, with short, 6-segmented antennae and three white crossbands on each fore wing.

Echinothrips americanus Morgan
Poinsettia thrips
Since the 1990s, this pest (which is of eastern USA origin) has become established in various parts of mainland Europe, particularly on greenhouse ornamentals. The thrips has also been intercepted on plants exported to England. Various woody and herbaceous plants are attacked, including busy lizzie (*Impatiens*), *Chrysanthemum, Dieffenbachia, Fatsia, Hibiscus, Homalomena*, peace lily (*Spathiphyllum*), *Philodendron*, poinsettia (*Euphorbia pulcherrima*) and *Syngonium*. The insects, which breed mainly if not entirely parthenogenetically, are about 1.5 mm long and dark brown to blackish with pale wing bases. The nymphs are whitish and translucent. Characteristically, both adults and nymphs often congregate in numbers on the upper surface of infested leaves.

235 Privet thrips (*Dendrothrips ornatus*) damage to leaf of *Ligustrum*.

236 Western flower thrips (*Frankliniella occidentalis*) damage to flower of *Chrysanthemum*.

237 Flecking on petals of *Chrysanthemum*, caused by western flower thrips (*Frankliniella occidentalis*).

Frankliniella occidentalis (Pergande) (**236–237**)
Western flower thrips

An important polyphagous thrips of American origin. Now widely distributed in Europe, originally having been introduced on imported *Chrysanthemum* cuttings in the 1980s. Ornamentals commonly infested in northern Europe include *Achimenes*, African violet (*Saintpaulia hybrida*), busy lizzie (*Impatiens*), cape primrose (*Streptocarpus hybrida*), *Chrysanthemum*, *Cineraria, Cyclamen, Gerbera, Gloxinia, Pelargonium* and vervain (*Verbena hybrida*). An important vector of plant virus diseases. Attempts to eradicate the pest from Europe were unsuccessful.

DESCRIPTION
Adult: 2 mm long; pale yellow to brownish yellow; antennae 8-segmented; distinguishable with certainty from closely related species only by microscopical examination. **Nymph:** translucent to golden yellow; eyes reddish. **Propupa:** white, with short wing cases. **Pupa:** white, with long wing cases.

LIFE HISTORY
Although most frequently associated with chrysanthemum, attacks also become established on many other greenhouse plants. In favourable situations, infestations also occur outdoors. The thrips infest both surfaces of the leaves but tend to occur more commonly on the underside, causing a noticeable scarring and excreting tell-tale specks of black frass, a typical symptom of thrips attack. They also occur commonly beneath bud scales and between the calyx and petals of open blooms. Breeding is continuous whilst conditions remain favourable, the complete lifecycle occupying about 2–3 weeks at temperatures of 20–30°C. Pupation occurs in the soil.

DAMAGE
Thrips cause silvering and spotting of tissue as well as noticeable blanching, especially on petals. Significant damage is caused by a relatively small number of individuals. Infestations also lead to noticeable distortion of host plants. The effects of tomato spotted wilt virus, transmitted by this pest to various hosts (including *Chrysanthemum* and *Gloxinia*), are often severe.

Frankliniella intonsa (Trybom)
Flower thrips

This yellowish to reddish-brown, 1.2–2.8 mm long, flat-bodied thrips infests a wide range of ornamentals, and causes minor damage to the foliage and petals. Infestations are most often reported on heather (*Erica*). Unlike the following species, the front projection of the head is triangular and there are eight pronotal bristles.

Frankliniella iridis (Watson)
Iris thrips

Infestations of this widespread species sometimes occur on *Hippeastrum* and *Iris*. Nymphs and adults cause silvering of foliage and pale speckling of flower petals. Female thrips occur throughout the year, and hibernate during the winter months; males and nymphs are present during the summer and autumn. There are several generations each year. Adult females (1.5–1.7 mm long) are dark brown and flattened, with pale yellow (usually reduced) wings; the front projection of the head is rounded, and there are six strong pronotal bristles.

Heliothrips haemorrhoidalis (Bouché)
syn. *H. adonidium* Haliday
Glasshouse thrips

This tropical or subtropical thrips is a frequent greenhouse pest in temperate regions, infesting many ornamental plants including azalea (*Rhododendron*), *Begonia*, calla lily (*Zantedeschia aethiopica*), *Chrysanthemum*, *Fuchsia* and rose (*Rosa*), as well as ferns, orchids, palms and vines. Under favourable conditions, breeding continues throughout the year, but this insect cannot survive northern European winters in unprotected situations. Adults (1.2–1.8 mm long) are dark brown, with an orange tip to the abdomen and pale yellow antennae, legs and wings; the fore wings are narrow, with few and very short setae on the veins; the antennae are 8-segmented, with the terminal segment needle-like; the tarsi are 1-segmented. The yellowish-brown nymphs produce a reddish fluid, that is deposited in large drops on the surface of host plants and upon which brown fungal growths develop. In addition, feeding by adults and nymphs results in a brownish, silvery or whitish spotting of leaves and flowers.

Kakothrips pisivorus (Westwood)
syn. *K. robustus* (Uzel)
Pea thrips

A widespread and common pest of legumes, including sweet pea (*Lathyrus odoratus*), but rarely of significance on ornamentals. Adults occur from May to July or August, infesting and causing a silvering of the foliage, flowers and pods. There is a single generation each year, nymphs developing during June, July or August. When fully fed, the nymphs enter the soil, where they overwinter; new adults appear in the spring. Adults (1.5–2.0 mm long) are blackish brown and flat-bodied, with dark brown, basally clear, fore wings; the antennae are 8-segmented, with the third segment yellow.

Limothrips cerealium Haliday
syn. *L. avenae* Hinds
Grain thrips

This mainly black-bodied thrips often appears in vast numbers during July or August, at about the time that cereal crops are ripening. The adult females may then invade greenhouses, to appear on crops such as *Chrysanthemum*. Damage caused by other thrips (particularly onion thrips, *Thrips tabaci*, p. 123) is often seen on greenhouse-grown ornamentals at the time of invasion, and grain thrips (which breed only on cereals and grasses) is then sometimes mistakenly assumed to be the harmful species. Unlike onion thrips, grain thrips has 8-segmented antennae, the head is longer than broad and the tenth abdominal tergite bears a pair of stout,

spine-like setae. This and other species of *Limothrips* commonly give rise to the term 'thunder-flies'.

Parthenothrips dracaenae (Heeger) (**238**)
Dracaena thrips

This subtropical species is widely distributed in greenhouses in northern Europe on plants such as *Citrus*, *Croton*, *Cycas*, *Dracaena*, *Ficus*, *Howea*, *Stephanotis* and *Tradescantia*. As often noted on ornamental rubber plants (*Ficus elastica*), affected foliage becomes extensively damaged, both the upper and lower surfaces developing distinctive silvery patches. Adults (*c*. 1.3 mm long) are yellow or brown, with a strongly reticulated head and thorax; the antennae are 7-segmented, with the terminal segment needle-like; the fore wings, which lack a costal fringe of hairs, are very broad and hyaline, with a reticulate pattern, and distinctly banded with brown; the legs are brown, with yellow tibiae and tarsi; the tarsi are 1-segmented.

238 Dracaena thrips (*Parthenothrips dracaenae*) damage to leaf of *Ficus elastica*, viewed from above.

Thrips fuscipennis Haliday (**239**)
syn. *T. menyanthidis* Bagnell
Rose thrips

A common outdoor and greenhouse pest of various flowering plants, including rose (*Rosa*) and many other ornamentals; infestations also occur on the young leaves of various trees. Widespread in Europe.

DESCRIPTION
Adult female: 1.2–1.6 mm long; yellowish brown to dark brown; legs brown; fore wings dark greyish brown, paler basally; antennae 7-segmented; comb of setae on hind margin of the eighth abdominal tergite incomplete centrally. **Nymph:** white to pale yellow.

LIFE HISTORY
Outdoors, adult females hibernate and become active in the spring, depositing eggs from May onwards. Nymphs then occur on host plants from May to August or September and males from June to October, with up to four generations annually. In greenhouses, there are a larger number of generations; however, even under heated conditions, there is an obligatory period of diapause from November onwards, with adult females sheltering in cracks and crevices, in debris or in the soil; the thrips reappear in the late winter or early spring, and eggs are deposited from late February onwards.

DAMAGE
Infested tissue becomes discoloured and distorted; on rose, attacks on the developing flowers lead to malformed, brown-streaked petals.

Thrips simplex (Morison)
syn. *Taeniothrips gladioli* Moulton & Steinweden
Gladiolus thrips

An important pest of greenhouse-grown *Gladiolus* and to a lesser extent *Crocus*; less often, *Freesia, Iris* and lilies are also attacked. Probably introduced into northern Europe from southern Africa.

DESCRIPTION
Adult: 1.5 mm long; dark brown; antennae 8-segmented and mainly dark, with the third and basal part of the fourth and fifth segments pale; fore wings brown but pale basally; legs dark brown, with paler tibial apices and tarsi. **Nymph:** yellow to orange.

LIFE HISTORY
In northern Europe, this species generally survives the winter between the scales of stored corms where, if temperatures remain above 10°C, breeding continues unabated. In the spring, when corms are planted, the thrips will be carried upwards on the enlarging stems and leaves; subsequently, they may also invade the developing flowers. Gladioli growing outdoors are sometimes infested during the summer, particularly in hot, dry conditions.

DAMAGE
Crocus: yellowish-brown areas develop beneath the skin of stored corms; new growth from damaged corms is poor and creamy coloured rather than white. **Gladiolus:** infested corms develop rough, greyish-brown surface patches; infestations on developing plants produce yellowish or silvery streaks on the foliage and stems, which may subsequently turn brown. When feeding extends to the flowers, invaded petals develop silvery flecks; if attacks are severe, the petals may also turn brown and die.

239 Rose thrips (*Thrips fuscipennis*) damage to foliage of *Ostrya.*

Thrips tabaci Lindeman (**240–241**)
syn. *Thrips debilis* Bagnell
Onion thrips

An often common pest of cultivated plants, including ornamentals such as *Asparagus plumosus*, *Begonia*, carnation (*Dianthus caryophyllus*), *Chrysanthemum*, *Cineraria*, *Cyclamen*, *Dahlia*, *Gerbera* and orchids. Most damaging in greenhouses; also a virus vector. Cosmopolitan. Widely distributed in Europe.

DESCRIPTION
Adult female: 1.0–1.3 mm long; greyish yellow to brown; antennae 7-segmented and yellowish brown; fore wings pale brownish yellow; comb of setae on hind margin of eighth abdominal tergite complete. **Nymph:** whitish to pale yellowish-orange.

LIFE HISTORY
Under favourable conditions this entirely parthenogenetic species breeds continuously. Eggs are laid in plant tissue and hatch within 1–2 weeks. Two nymphal stages then occur on host plants, fully fed nymphs dropping to the ground about a week later. They then enter the soil, where the transformation to the adult stage takes place, again in about a week. Considerable populations may develop on greenhouse-grown plants; greenhouses are also subject to frequent invasion by outdoor populations. In unprotected situations, several broods occur from May through to winter, and adult females usually overwinter in the soil.

DAMAGE
Affected leaves become flecked extensively with silver, damage typically being associated with black grains of frass; flower petals may also become discoloured and distorted. Attacks are most severe in hot, dry conditions.

Thrips atratus Haliday
Carnation thrips

This generally abundant species occurs throughout the year, breeding on various outdoor plants from May onwards. Adults (1.3–1.8 mm long) are mainly dark brown, with 8-segmented antennae and narrow, brown (basally paler) fore wings. They commonly enter greenhouses from June to September, and may then cause damage to flowers of carnation (*Dianthus caryophyllus*) and pink (*D. plumarius*).

240 Onion thrips (*Thrips tabaci*) damage to leaf of *Dianthus*.

241 Onion thrips (*Thrips tabaci*) damage to flower of *Cyclamen.*

Thrips nigropilosis Uzel
Chrysanthemum thrips

Generally common and sometimes troublesome on greenhouse-grown *Chrysanthemum*, producing characteristic brownish-red marks at the base of leaf stalks and brownish warts on the foliage. Adults occur throughout the year, with several broods of nymphs developing from May to November. The adults are 1.2 mm long, and mainly yellow with brownish markings, pale yellow fore wings and 7-segmented antennae; some adults may have reduced wings. The main outdoor hosts are plantains, especially ribwort plantain (*Plantago lanceolata*).

Family **PHLAEOTHRIPIDAE**

Thrips without an ovipositor, the fore wings with just one vein; eggs hard-shelled, often sculptured, and deposited on plant tissue or in plant debris; development includes two (often brightly coloured or banded) nymphal stages, one propupal stage and two pupal stages.

Gynaikothrips ficorum (Marchal)
Cuban laurel thrips
This tropical species, which occurs in Central America and the southern parts of the USA, attacks various hosts, including *Citrus, Ficus* and *Viburnum.* In recent years, the pest has also become established in southern Europe and the Mediterranean basin on Cuban laurel (*Ficus retusa*). Infested leaves and young shoots are discoloured and distorted, and severely damaged leaves fall prematurely. Breeding is continuous whilst conditions remain favourable. Adults (*c.* 2.5–3.5 mm long) are mainly dark brown to black.

Liothrips vaneeckei Priesner
Lily thrips
A potentially serious pest of lily bulbs in store, particularly in the Netherlands; also present in North America.

DESCRIPTION
Adult: 2.0–2.5 mm long; dark reddish brown; antennae yellow to brown and 8-segmented; legs orange-yellow, long and slender; abdomen without an ovipositor but terminating in a distinct tube; fore wings strap-like, light brown, with a pale base and a dark median band.

LIFE HISTORY
Attacks are limited to stored bulbs, infestations developing between the scales, close to the base plate. There are several generations each year. Adults or final-instar nymphs overwinter.

DAMAGE
Infestations are often persistent and lead to the development of sunken rust-coloured spots towards the base of the outer scales. Damaged scales become soft and the outer ones papery; the latter may also drop off. If infested bulbs are planted out, they produce fewer new scales, but plant development is hardly affected.

Order **COLEOPTERA** (beetles)

Family **SCARABAEIDAE** (chafers)

Chafers are large to very large, often brightly coloured beetles, with tips of the antennae lamellate. The larvae are large and fleshy, with a distinct head, powerful mouthparts, three pairs of strong legs and a swollen tip to the abdomen; they feed in the soil on decaying matter and plant roots, and are most abundant in light, well-drained sites near heathlands or woodlands.

Amphimallon solstitiale (Linnaeus) (**242**)
syn. *A. solstitialis* (Linnaeus)
Summer chafer

This southerly distributed species is sometimes a pest in gardens and nurseries, the larvae feeding on the roots of various herbaceous plants, including ornamentals; roots of nursery trees are also attacked. The yellowish-brown adults (14–18 mm long) occur in June and July, and are active on warm evenings. The mated females deposit several white, oval eggs in the soil, and the eggs hatch about a month later. The larvae are white and elongate (up to 30 mm long), with two divergent rows of spines surmounting the anal slit; the head and legs are light brown. They feed from August onwards, and usually complete their development within two years. Although capable of causing considerable damage, particularly in their second summer, the larvae are usually present in only small numbers.

Cetonia aurata (Linnaeus) (**243**)
Rose chafer

This locally common, southerly distributed chafer is rarely important as a pest, but the adults do sometimes browse on the flowers of ornamental plants, especially honeysuckle (*Lonicera*), *Viburnum* and various members of the family Rosaceae. Damage is reported most frequently on rose (*Rosa*). Adults (14–20 mm long) are metallic golden-green above, purplish red below, with a few irregular silvery markings on the elytra. They occur from late May onwards, depositing eggs in the soil during the early summer. The larvae (up to 30 mm long) feed on rotting wood and decaying plant material for up to two, or sometimes three, years; they are distinguished from other chafer larvae by the reddish head and legs, and by the transverse rows of reddish hairs on the body.

242 Larva of summer chafer (*Amphimallon solstitiale*).

243 Rose chafer (*Cetonia aurata*).

Epicometis hirta (Poda)
syn. *Tropinota hirtella* Linnaeus

A polyphagous pest of trees, shrubs and herbaceous plants, including ornamentals such as rose (*Rosa*). Widespread in the warmer parts of central, eastern and southern Europe, including the Mediterranean basin. Eurasiatic. An introduced species in North America.

DESCRIPTION
Adult: 8–12 mm long; black and distinctly hairy; elytra each with several white markings; pronotum usually extensively punctured. **Larva:** up to 15 mm long; whitish and fleshy; anal segment with two divergent rows of small spines, each row turning abruptly outwards posteriorly.

LIFE HISTORY
Adults emerge in the spring, and then feed avidly on the flowers of various plants. Eggs are deposited in the soil during the early summer and hatch in 1–2 weeks. Larvae feed in the soil on decaying vegetable matter, and are fully grown in about two months. They then pupate, each in an earthen cell. The adult stage is reached by the autumn, but individuals usually remain within the pupal cell until the following spring.

DAMAGE
Adults destroy the petals and other floral parts, often causing extensive damage. They also browse on the leaves.

Hoplia philanthus (Fuessly)
Welsh chafer

A widely distributed and locally common species, particularly in light soils, the larvae sometimes causing damage to lawns and sports turf. The larvae are similar to those of *Serica brunnea* (p. 127) but have characteristically long, stout claws on the front tarsi. Adults (7–11 mm long) are mainly black with reddish-brown elytra. They occur from May to July.

Melolontha melolontha (Linnaeus) (**244**)
syn. *M. vulgaris* Fabricius
Cockchafer

An often common pest of corms, roots, tubers and underground parts of the stems of various plants, including alpines, herbaceous ornamentals and nursery trees and shrubs. Eurasiatic. Widely distributed in Europe.

DESCRIPTION
Adult: 20–30 mm long; chestnut-brown, the head and thorax darker, and partly coated in whitish hairs which often rub off; elytra each with five longitudinal lines; abdomen terminating in a blunt, downwardly directed spine; antennae with six (female) or seven (male) lamellae. **Egg:** oval, whitish or yellowish. **Larva:** up to 35 mm long; body white, with the last segment somewhat translucent and darkened by the underlying gut contents; head and legs brown and shiny; anal slit transverse and wavy, surmounted by two more or less parallel longitudinal rows of spines. **Pupa:** 25–35 mm long; whitish to brown.

LIFE HISTORY
Adults occur in May or June. They are nocturnal, feeding at night on the buds, flowers or foliage of various trees and shrubs, and are frequently attracted to light. After a few weeks, the females burrow into the soil to a depth of 15–20 cm and then lay their eggs, typically in small batches of 12–30. The eggs swell considerably during an extended period of incubation, usually hatching after about four weeks. The larvae then attack plant roots, feeding for up to three years and passing through three clearly defined instars. Pupation takes place during the third summer in an earthen cell 60 cm or more below the surface. Adults are produced about six weeks later, but they do not emerge from the soil until the following spring.

DAMAGE
Adults make holes in the leaves of various trees and shrubs. However, significant damage is usually restricted to buds of roses, deep cavities being excavated in the sides. The larvae destroy much of the root system of host plants, seriously restricting growth; in severe cases, plants wilt and die. Larval damage is most likely to occur on plants in recently broken-up grassland or pasture.

244 Cockchafer (*Melolontha melolontha*).

Phyllopertha horticola (Linnaeus) (**245–246**)
Garden chafer

The colourful, metallic bluish-green and reddish-brown adults (7–11 mm long) of this widely distributed chafer are particularly common in light-soiled grassland areas. They occur mainly in May and June, flying during the daytime in warm, sunny weather. The larvae are relatively small (up to 15 mm long when fully grown) and characterized by the two parallel rows of spines above the anal slit; also, in repose they hold the head close to the anal segment. They feed on roots of various plants, especially grasses, from June or July onwards, becoming fully grown by the autumn and eventually pupating in the following spring. Although the adults feed on the leaves, flowers and fruits of many garden plants, and the larvae may occasionally damage the roots of herbaceous plants, this species is not an important pest of ornamentals. However, larval damage to lawns and sports turf is often extensive.

Serica brunnea (Linnaeus) (**247**)
Brown chafer

A locally common woodland insect, sometimes causing damage to nursery trees, including Norway spruce (*Picea abies*) grown as Christmas trees, but not an important pest of ornamentals. The larvae (up to 18 mm long) are creamy white, with a yellowish-brown head, numerous reddish body hairs and the anal slit surmounted by a horizontal arc of spines. They feed on plant roots, and usually complete their development in two years. The adults (7–11 mm long) are mainly reddish brown; they occur from June to August.

245 Garden chafer (*Phyllopertha horticola*).

246 Larva of garden chafer (*Phyllopertha horticola*).

247 Larva of brown chafer (*Serica brunnea*).

Family **BUPRESTIDAE**

A family of mainly tropical beetles, with relatively few European representatives. Adults (1–12 mm long) are more or less elongate, and usually distinctly metallic in appearance.

Larvae (sometimes known as 'flat-headed borers') are dorsoventrally flattened, often with a very large prothoracic segment into which much of the head is retracted. The larvae are mainly wood-borers, and feed in galleries excavated beneath the bark or rind of host plants.

Agrilus aurichalceus Redtenbacher

Raspberry jewel beetle
larva = rose stem girdler
A pest of rose (*Rosa*) in mainland Europe from France eastwards. An introduced species in the USA.

DESCRIPTION
Adult: 4.5–7.0 mm long; mainly metallic olive-green to metallic greenish-blue; elytra elongate, with a sinuous outline, and distinctly swollen between the middle and apex.

LIFE HISTORY
Larvae feed singly within the stems of host plants, each burrowing upwards just beneath the epidermis from the point at which the egg was deposited. At first, a relatively tight spiral gallery is formed, which causes a localized swelling (pseudo-gall) about a centimetre long; later, the spiralling becomes more 'open' and the gallery straightens before terminating in an elongated, flask-shaped pupal chamber a few centimetres above the level of the pseudo-gall. Typically, the larval gallery remains filled with frass, as there are no external openings through which this might be expelled. Pupation occurs in the spring. Adults are active from May to July.

DAMAGE
Larva-infested plants are progressively weakened. Also, adults browse on the leaves and cause noticeable loss of tissue.

Agrilus sinuatus (Olivier) (**248**)

Pear jewel beetle
Associated mainly with pear and other rosaceous fruit trees, but also attacking ornamental crab-apple (*Malus*), hawthorn (*Crataegus*) and *Sorbus*. Widely distributed in central Europe and currently extending its range northwards. Also present in North Africa and an introduced pest in the USA.

DESCRIPTION
Adult: 7–11 mm long; coppery red to purplish red; hind margin of pronotum strongly sinuate. **Larva:** up to 22 mm long; body dirty creamy white and dorsoventrally flattened, with the thoracic region noticeably enlarged; tip of abdomen with a pair of pointed, forceps-like projections.

LIFE HISTORY
Adults usually emerge in early June, but their appearance may be delayed by cool, wet weather. At first they browse on the foliage of host trees, before eventually mating. Larvae feed beneath the bark of trees from July onwards, each forming a long, zigzag-shaped gallery in a small branch; galleries up to 1 m in length are reported on some hosts. The larvae usually pupate in the following May. However, unfavourable conditions significantly delay their development, and individuals may then necessarily pass through two winters.

DAMAGE
Infestations are particularly damaging on young hosts, rendering trees liable to attack by secondary organisms such as bark beetles. Old, vacated galleries become particularly noticeable several years after their formation, each developing into a large and distinctive scar on the bark as the host tree continues to grow. Adults feeding on foliage remove irregular sections from the edges. Such damage is of no significance, although on young leaves the incised areas subsequently become necrotic.

248 Pear jewel beetle (*Agrilus sinuatus*).

Family **ELATERIDAE** (click beetles)

A small group of elongate beetles which, when lying on their backs, are capable of propelling themselves into the air, with an audible click. The soil-inhabiting larvae (commonly known as 'wireworms') are long, thin, more or less cylindrical and tough-skinned, with small thoracic legs and powerful jaws.

Agriotes lineatus (Linnaeus)
Common click beetle
This generally common, often important agricultural pest, also causes damage to horticultural crops, including various ornamentals. The larvae are usually most abundant in grassland areas, and often attack crops growing in recently broken-up pasture. They are distinguished from those of *Athous haemorrhoidalis* by the more cylindrical form and pointed hind end; they take four or five years to reach maturity.

Athous haemorrhoidalis (Fabricius) (**249–250**)
Garden click beetle
A generally common pest of herbaceous plants, including ornamentals such as *Anemone*, carnation (*Dianthus caryophyllus*), *Chrysanthemum*, *Dahlia*, *Gladiolus* and primrose (*Primula vulgaris*); also liable to damage seedling trees and nursery stock. Widespread in Europe.

DESCRIPTION
Adult: 10–13 mm long; head and thorax black, the abdomen reddish brown; antennae black; legs brownish black. **Larva:** up to 30 mm long; shiny and yellowish brown, with a darker head; relatively broad and sub-cylindrical, with top of body bifid.

LIFE HISTORY
Adults occur from mid-May to July, and away from upland areas are more commonly encountered than those of many other elaterids. They often fly during the daytime. Eggs are deposited in groups in moist soil, usually beneath the shelter of vegetation. They hatch in about a month. Larvae then attack the underground parts of plants and also feed on other vegetative matter in the soil. Development is slow, usually extending over four or five years. When fully grown, usually in mid- to late summer, larvae pupate in earthen cells. Adults are produced about a month later but normally remain in their cells until the following spring.

DAMAGE
Wireworms bite through the roots and bore into the base of plants, sometimes causing plants to wilt and die. Bulbs, corms, rhizomes, stolons and tubers are also attacked. On chrysanthemum and other relatively fleshy-stemmed plants, wireworms may also tunnel into the stems well above soil level. Damage is most serious on young plants, and in spring and autumn.

249 Garden click beetle (*Athous haemorrhoidalis*).

250 Larva of garden click beetle (*Athous haemorrhoidalis*).

Family **NITIDULIDAE**

A large family of mainly small beetles, the antennae usually terminating in a 3-segmented club.

Meligethes spp.
Pollen beetles

Adult pollen beetles, mainly *Meligethes aeneus*, are increasingly reported invading garden flowers during the summer months. These small (*c*. 3 mm long), bronzy or greenish-black insects breed mainly on brassica seed crops and weeds such as charlock (*Sinapis arvensis*), but often migrate in numbers to various other flowering plants in search of pollen. In gardens, they are often numerous on ornamentals such as *Alyssum*, *Hypericum*, rose (*Rosa*) and sweet pea (*Lathyrus odoratus*). Although causing little or no direct damage, the presence of the beetles on cut flowers brought indoors may be a nuisance.

Family **BYTURIDAE**

A small group of small, hairy beetles with clubbed antennae; the larvae develop in flowers and fruits of *Rubus*.

Byturus tomentosus (Degeer)
syn. *B. urbanus* (Lindemann)
Raspberry beetle

In spring, following their emergence from hibernation, adults of this important raspberry pest often feed on the flowers of rosaceous trees and shrubs, including hawthorn (*Crataegus*) and flowering cherry (*Prunus*); they also attack flowering shrubs such as lilac (*Syringa*). Although sometimes numerous, the beetles do not cause significant damage and soon depart for *Rubus* hosts upon which they eventually breed. Adults are 3.5–4.5 mm long, yellowish brown, with elongate-oval bodies and short, clubbed antennae.

Family **CERAMBYCIDAE** (longhorn beetles) (**251**)

A group of small to large beetles, usually with very long antennae. Larvae are wood-borers, often with a noticeably swollen prothorax and a relatively small head. Many species are important timber pests, and some (including the following examples) cause damage to ornamental trees.

Aromia moschata (Linnaeus) (**252**)
Musk beetle

Associated with willow (*Salix*), and sometimes a pest of shade trees in mainland Europe. Eurasiatic.

DESCRIPTION
Adult: 20–32 mm long; body mainly coppery to metallic green; antennae and legs bluish black or greenish black; pronotum with a tooth-like projection on each side.

LIFE HISTORY
Adults occur mainly from mid-June to August. They are then to be found in association with willow trees or feeding on nearby flowers, especially Apiaceae. Larvae bore within the wood of host trees, and take two or three years to complete their development.

DAMAGE
Host trees are weakened by persistent attacks, and may eventually die.

Phoracantha semipunctata (Fabricius)
Eucalyptus longhorn beetle
larva = phoracantha borer

A potentially important Australian pest of *Eucalyptus*. Accidentally introduced to many parts of the world; now established in various European countries, including France, Italy, Portugal and Spain.

DESCRIPTION
Adult: 25–30 mm long; mainly dark brown, with an irregular yellowish-white band across the middle of the elytra, and a similarly coloured patch at the tip of each elytron.

LIFE HISTORY
Eggs are laid under loose bark or in bark crevices, in groups of up to 30. They hatch 1–2 weeks later. Larvae then bore into the bark, sometimes grazing on the outer surface before doing so. They tunnel within the cambium, forming very long galleries up to a metre or so in length. Sap typically oozes from sites of

251 Larva of a longhorn beetle (family Cerambycidae).

252 Male and female musk beetles (*Aromia moschata*).

infestation, and this is a clear indication of the presence of larvae. Development is completed in three or more months. Fully fed larvae then pupate in chambers at the end of their feeding galleries. Adults emerge at any time from spring to autumn, and there may be up to three generations annually.

DAMAGE
Larvae tunnel within the trunks and branches of infested trees, often causing the foliage to wilt. Liquid commonly seeps from holes or cracks in the bark, and infested branches if not whole trees may be killed. Although attacks most often occur on trees already under stress, healthy trees are also at risk.

Saperda populnea (Linnaeus)
Small poplar longhorn beetle
larva = small poplar borer
An important pest of *Populus*, especially aspen (*P. tremula*) and hybrid black poplar (*P. serotina*). Infestations are most important on young established trees and transplants, and in nurseries. Widespread and generally common in Europe.

DESCRIPTION
Adult: 9–15 mm long; mainly black, patterned with yellow on head, thorax and elytra; antennae black and of moderate length, most segments with grey hairs basally.

LIFE HISTORY
Adults occur from mid- or late May to July, and feed on the leaves and green shoots of host plants. Eggs are laid singly, each at the base of a deep, horseshoe-shaped incision bitten into the bark of a small stem or branch. Several incisions may be made close together on the same stretch of wood. The plant reacts by producing galligenous tissue, the affected area swelling considerably. Eggs hatch in about two weeks. Larvae then bore into the expanding galligenous plant tissue, and gradually extend their feeding galleries upwards into the pith. The local swellings on host trees are clearly visible in the first summer, and become particularly so in the following year. Pupation takes place close to the surface during the spring of the third year, and adults emerge a few weeks later. Larval survival is dependent upon the continued production of galligenous tissue by the host tree, and in wet summers and in damp sites (when a plant's reaction to attack is less marked) many larvae fail to complete their development.

DAMAGE
Infested stems are gnarled, and are eventually snapped off by the wind. Attacks on nursery stock and recent transplants often result in deformation and death of host plants.

Family **CHRYSOMELIDAE** (leaf beetles)

A very large family of mainly small, leaf-feeding beetles; adults often rounded, shiny and brightly coloured. The family includes flea beetles, whose hind legs have strongly swollen femora and are adapted for jumping. Larvae commonly feed in exposed positions on the leaves of host plants, but many attack plant roots or mine within the leaves, leaf stalks and stems.

Agelastica alni (Linnaeus) (**253–256**)
Alder leaf beetle

An important pest of alder (*Alnus*) in mainland Europe, where infestations are often severe on wild and cultivated plants; also associated with beech (*Fagus sylvatica*), hazel (*Corylus*), hornbeam (*Carpinus betulus*) and lime (*Tilia*).

DESCRIPTION

Adult: 6–8 mm long; rather bulbous, the elytra noticeably expanded towards the hind end; bluish to violet, with black antennae, tibiae and tarsi. **Larva:** up to 11 mm long; cylindrical and mainly black.

LIFE HISTORY

Adults hibernate and emerge in the spring, the females eventually laying eggs in large, scattered groups on the fully expanded leaves. Larvae occur gregariously on both sides of the leaves from June to July, feeding for about three weeks. At first they browse on the upper epidermis, but later they bite out holes between the major veins. Fully grown larvae drop to the ground to pupate, either on or just below the surface, and adults appear 1–2 weeks later. These adults browse on the foliage before overwintering. There is a single generation annually.

253 Alder leaf beetle (*Agelastica alni*).

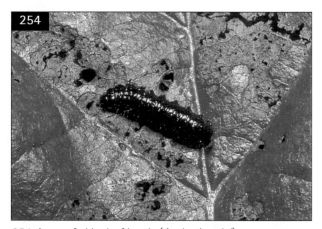

254 Larva of alder leaf beetle (*Agelastica alni*).

255 Larvae of alder leaf beetle (*Agelastica alni*).

256 Alder leaf beetle (*Agelastica alni*) damage to leaves of *Alnus*.

DAMAGE
Attacked foliage becomes peppered with large, irregular holes, weakening and affecting the appearance of host plants. Damage is of particular significance on young trees.

Altica lythri Aubé (**257–259**)
Large blue flea beetle
An occasional pest of *Fuchsia* and certain other ornamentals, including evening primrose (*Oenothera*) and *Potentilla fruticosa*, but associated mainly with wild Onagraceae. Widely distributed in Europe but local.

DESCRIPTION
Adult: 4.5–5.5 mm long; bright blue and rather plump; elytra irregularly punctured and with slightly curved sides. **Egg:** 1.2 × 0.25 mm; whitish orange. **Larva:** up to 12 mm long; head black; body yellowish with numerous darker plates.

LIFE HISTORY
Adults appear from May onwards. They then aggregate on host plants, notably wild hosts such as rose-bay (*Chamaenerion angustifolium*), great willow-herb (*Epilobium hirsutum*) and marsh willow-herb (*E. palustre*). Eggs are laid singly or in small groups on the underside of leaves, and hatch about 2–3 weeks later. The larvae feed on the leaves for several weeks before pupating in the ground. New adults appear in the late summer and then hibernate.

DAMAGE
Infested leaves become notched and covered in numerous holes, affecting the quality and marketability of nursery plants. Heavy infestations cause severe foliage damage, leading to premature leaf fall. Petals of host plants are also damaged.

Altica ericeti (Allard)
Heather flea beetle
A locally common species, associated with heather (*Erica*) and sometimes damaging to such plants in gardens and nurseries. Adults (4–5 mm long) are metallic bluish-green, with distinctly punctured elytra.

257 Large blue flea beetle (*Altica lythri*).

258 Eggs of large blue flea beetle (*Altica lythri*).

259 Larva of large blue flea beetle (*Altica lythri*).

260 Iris flea beetles (*Aphthona nonstriata*) damaging leaf of *Iris.*

261 Willow flea beetle (*Crepidodera aurata*).

262 Poplar flea beetle (*Crepidodera aurea*).

263 Poplar flea beetle (*Crepidodera aurea*) damage to leaf of *Populus.*

Aphthona nonstriata (Goeze) (**260**)
syn. *A. caerulea* (Fourcroy)
Iris flea beetle

Locally common on yellow flag (*Iris pseudacorus*), and sometimes also associated with cultivated irises growing in parks and gardens. The beetles, which occur in the spring and summer, feed on the foliage, forming long whitish markings. Individuals (2.5–3.0 mm long) are mainly blue, with finely punctured elytra; the legs are light brown, with distinctive black hind femora.

Crepidodera aurata (Marsham) (**261**)
syn. *Chalcoides aurata* (Marsham)
Willow flea beetle

A widely distributed and locally common species on willow (*Salix*), affecting both mature trees and nursery stock. Infestations are also increasingly reported on poplar (*Populus*), damage occurring in nurseries, on windbreaks and on isolated trees. Adults, which overwinter in the ground, occur on host plants from May to September, grazing the foliage and producing small but noticeable holes in the expanded leaves; in spring, the beetles will also damage unopened buds. Individuals are 2.5–3.2 mm long and distinguished by the reddish thorax and greenish-black elytra, black or partly black antennae and mainly pale legs.

Crepidodera aurea (Fourcroy) (**262–263**)
syn. *Chalcoides aurea* (Fourcroy)
Poplar flea beetle

This locally common flea beetle is associated mainly with various kinds of *Populus*, including aspen (*P. tremula*), and causes minor damage to the foliage; it also occurs on willow (*Salix*). Adults occur from May to early October. They are similar to those of the previous species but slightly larger (2.7–3.5 mm long), with mainly pale antennae, and the thorax and elytra metallic green with a reddish sheen.

264 Rosemary beetle (*Chrysolina americana*).

265 Larva of rosemary beetle (*Chrysolina americana*).

266 Adult of *Chrysolina polita*.

Chrysolina americana (Linnaeus) (264–265)
Rosemary beetle

In recent years, this southern European beetle has become firmly established in Britain on ornamental lavender (*Lavendula*) and rosemary (*Rosmarinus officinalis*) bushes in private gardens and elsewhere. The beetles aestivate on host plants during the summer, and become active from late August onwards. They then mate and lay eggs, with egg laying continuing throughout the winter months. Eggs hatch in about two weeks. The mainly grey, sac-like larvae (each up to 8 mm long) then browse on the leaves before eventually entering the soil to pupate. New adults appear two or more weeks later, depending on temperature, there being just one generation annually. The rather bulbous adults (*c*. 8 mm long) are metallic green, with longitudinal purple stripes on the thorax and elytra. The pest also breeds on other Lamiaceae, including sage (*Salvia*) and thyme (*Thymus*).

Chrysolina polita (Linnaeus) (266)

Although mainly an inhabitant of damp meadows, adults of this generally common species occur occasionally on lime (*Tilia*) and ornamental willow (*Salix*), causing minor damage to the leaves. Individuals are 6.5–8.5 mm long, broad and convex, with a golden-green, sometimes reddish-flushed head and thorax, and dull-metallic, brownish-red, irregularly punctured elytra (cf. *Chrysomela populi*, p. 136).

Chrysomela populi Linnaeus (**267**)
Red poplar leaf beetle
A locally common pest of *Populus*, including aspen (*P. tremula*); also occurs on willow (*Salix*). Widespread throughout Europe.

DESCRIPTION
Adult: 8–12 mm long; thorax bluish black or greenish; elytra reddish, with a black spot at the extreme tip (cf. *Chrysomela tremula*). **Egg:** 1.0 × 0.56 mm; elongate-oval, pale yellow, yellow to brownish. **Larva:** up to 15 mm long; head black; body creamy white, with a mainly black prothoracic plate and prominent black verrucae.

LIFE HISTORY
Adults occur from May onwards, depositing eggs on the leaves of host plants. They show a particular liking for young trees and are sometimes common in poplar stool beds. Larvae browse on the surface layers of the leaves but later form holes right through the leaf blade. If disturbed, larvae are capable of exuding globules of liquid from a lateral series of tube-like verrucae. Fully fed individuals pupate in the soil, and adults of the next generation appear in about July. A second generation of larvae feed in July and August, and these become adults in September. These beetles eventually hibernate and reappear in the spring.

DAMAGE
Adults and larvae skeletonize the leaves, and infestations often result in extensive damage; young aspen trees are particularly susceptible.

Chrysomela aenea Linnaeus (**268**)
Widely distributed on alder (*Alnus*) and sometimes damaging to the foliage of other trees, including willow (*Salix*). There are two generations annually. Adults are 5–8 mm long and coppery blue or metallic green; the larvae, which often skeletonize the leaves, are up to 10 mm long and blackish to whitish, with prominent black verrucae and a shiny black head and prothoracic plate.

Chrysomela tremula Fabricius
syn. *C. longicollis* (Suffrian)
Adults of this locally distributed beetle are very similar to those of *Chrysomela populi* but are smaller (6–9 mm long) and the body narrower; they also lack the black mark at the tip of the elytra. The pest occurs on aspen (*Populus tremula*) in various parts of Europe, and sometimes causes damage to young cultivated trees.

Crioceris duodecimpunctata (Linnaeus) (**269**)
Twelve-spotted asparagus beetle
A minor pest of asparagus, including ornamental species such as *Asparagus plumosus*. Eurasiatic. Widely distributed in mainland Europe. Also present in North America.

DESCRIPTION
Adult: 5–6 mm long; elytra orange-red, spotted with black; antennae black. **Larva:** up to 8 mm long; mainly yellow, with a black head and two blackish plates on the prothorax.

LIFE HISTORY
Overwintered adults appear in the spring and then attack the young fronds of asparagus plants. Eggs are eventually laid at the tips of the fronds and hatch about 10 days later. Larvae feed on developing berries and eventually pupate in the soil, each in a small chamber. New adults appear 2–3 weeks later. Larvae of a second brood feed in late summer and pupate in the autumn.

DAMAGE
Young adults sometimes cause extensive damage in seedbeds where decorative asparagus fronds are being raised.

Crioceris asparagi (Linnaeus) (**270–271**)
Asparagus beetle
Although mainly of importance as a pest of edible asparagus, this generally common and well-known beetle also attacks ornamental asparagus plants, including *Asparagus plumosus*. Adults and larvae browse on the foliage and, as with the previous species, there are two generations annually. The mainly grey, black-headed larvae (up to 7 mm long) are particularly damaging.

267 Red poplar leaf beetle (*Chrysomela populi*).

268 Adult of *Chrysomela aenea*.

269 Adult twelve-spotted asparagus beetle (*Crioceris duodecimpunctata*).

270 Asparagus beetle (*Crioceris asparagi*).

271 Larvae of asparagus beetle (*Crioceris asparagi*).

Galerucella lineola (Fabricius) (272–275)

Brown willow leaf beetle

An often common pest of willow (*Salix*); most important in stool beds. Widely distributed in Europe.

DESCRIPTION

Adult: 5–6 mm long; mainly yellowish brown, with distinct black marks on the pronotum and shoulders of the elytra; elytra of similar width throughout, with the pubescence shiny and relatively closely set. **Egg:** 1 mm across; rounded, with a roughened surface; pale brownish yellow. **Larva:** up to 8 mm long; head black; body yellowish to black, with numerous black verrucae and plates. **Pupa:** 4.5–5.5 mm long; yellow.

LIFE HISTORY

Adults overwinter in the soil, under dead bark and in various other places, and reappear in the following May.

They favour open, sunny situations and are often numerous in willow beds and on young willow trees. Eggs are laid on the shoot tips or on the upper surface of the leaves, in groups of about 20–25, and hatch about a week later. Larvae appear from about mid-May onwards, feeding gregariously but later becoming solitary. When fully fed, usually after about a month, they enter the soil to pupate; new adults appear about a week later. There are normally two generations annually, with young adults of the final generation appearing in the late summer, but the species is single brooded in northern districts.

DAMAGE

The larvae skeletonize the leaves, and also destroy the buds and shoots, causing considerable damage on heavily infested hosts.

272 Brown willow leaf beetle (*Galerucella lineola*).

273 Eggs of brown willow leaf beetle (*Galerucella lineola*).

274 Larva of brown willow leaf beetle (*Galerucella lineola*).

275 Pupa of brown willow leaf beetle (*Galerucella lineola*).

Galerucella luteola (Müller) (276–278)
syn. *Pyrrhalta luteola* (Müller); *Xanthogaleruca luteola* (Müller)
Elm leaf beetle
An often important and destructive pest of elm (*Ulmus*). Also associated with Caucasian elm (*Zelkova carpinofolia*) and horse chestnut (*Aesculus hippocastanum*). Widely distributed in mainland Europe, Asia Minor and North Africa, but absent from the British Isles; an introduced pest in North America.

DESCRIPTION
Adult: 5–7 mm long; yellowish brown to olive-brown, marked with black on head, thorax and elytra. **Egg:** 1.0 × 0.5 mm; lemon-shaped; pale yellow to brownish white. **Larva:** up to 10 mm long; head black, body yellowish to blackish, marked with numerous black verrucae and plates. **Pupa:** 5–7 mm long; yellowish orange.

LIFE HISTORY
Beetles emerge from hibernation in April and May. They then fly to host plants and begin feeding on the leaves, forming irregular holes between the major veins. Eggs are laid in groups of 10–30 from May onwards, typically in 2–3 rows close to a secondary vein on the underside of a young leaf. Each female is capable of depositing several hundred eggs, and some may successfully lay over a thousand. Eggs hatch in about ten days. The young larvae feed gregariously on the underside of the leaves from May onwards. At the early stages of development larvae are unable to feed on the toughened, older foliage. However, later, when they also become less gregarious, they attack leaves of various ages. Larvae are fully fed in about three weeks. They then pupate amongst leaf litter or in the soil, new adults emerging from early July onwards. Larvae of a second brood complete their development later in the summer. They eventually develop into adults which usually feed briefly before overwintering.

DAMAGE
Adults and larvae cause considerable disfigurement of the foliage, and leaves may be completely destroyed. Heavy infestations affect the growth and vigour of host plants.

276 Elm leaf beetle (*Galerucella luteola*).

277 Larva of elm leaf beetle (*Galerucella luteola*).

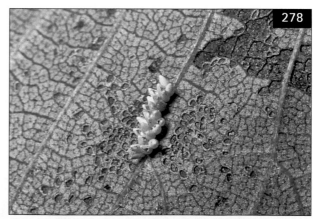

278 Hatched eggs of elm leaf beetle (*Galerucella luteola*).

279 Water-lily beetle (*Galerucella nymphaeae*).

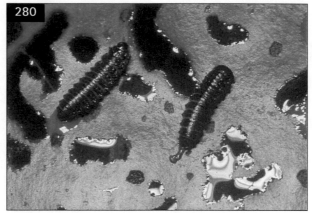

280 Larvae of water-lily beetle (*Galerucella nymphaeae*).

Galerucella nymphaeae (Linnaeus) (279–282)
Water-lily beetle

An often important pest of water-lilies, including yellow water-lily (*Nuphar lutea*) and, especially, white water-lily (*Nymphaea alba*). Present throughout much of Europe, including Britain and Ireland; an introduced pest in North America.

281 Pupae of water-lily beetle (*Galerucella nymphaeae*).

DESCRIPTION
Adult: 6–8 mm long; dark brown to yellowish brown. **Egg:** 0.75 mm across; sub-spherical, pale yellow to yellowish orange, with a finely reticulated surface. **Larva:** up to 9 mm long; dark brown or black, but yellow below. **Pupa:** 5–7 mm long; black and shiny.

LIFE HISTORY
Adult beetles hibernate in the close vicinity of ponds and pools, sheltering amongst vegetation and in dead plant stems. They reappear in the following May or June, to aggregate and feed on the leaves of water-lilies. Eggs are deposited on the upper surface of the leaves in groups of 12–18 and hatch in about a week. At first, the larvae feed in groups, attacking the leaves and grazing away the upper tissue; later, they feed singly and then bite completely through the leaf blades. Flowers may also be damaged. Pupation occurs on the upper surface of the leaves, and young adults of the next generation emerge in July and August. There are normally two generations annually, but there may be three in heated pools and in favourable southerly areas.

282 Water-lily beetle (*Galerucella nymphaeae*) damage to leaves of *Nuphar*.

DAMAGE
Most significant damage is caused by the larvae, attacked leaves becoming disfigured by irregular, elongate holes and, if infestations are heavy, extensively shredded.

Gonioctena viminalis (Linnaeus) (283)
syn. *Phyllodecta viminalis* (Linnaeus)

Minor infestations of this beetle sometimes occur on common sallow (*Salix atrocinerea*), grey willow (*S. cinerea*) and other broad-leaved willows, damage caused by the adults and larvae being similar to that described for related pests such as *Phratora* spp. (p. 144). The adults are 5.0–7.5 mm long and mainly yellowish red, with a black head, black markings on the thorax and 3–5 black spots on each elytron.

Lilioceris lilii (Scopoli) (284–286)
Lily beetle

A destructive and invasive pest of Madonna lily (*Lilium candidum*) and many other Liliaceae, including fritillary (*Fritillaria*) and *Nomocharis saluenensis*. Solomon's seal (*Polygonatum multiflorum*) is also attacked. Of Asian origin and now present in much of Europe. Widely distributed in Britain, and locally abundant. Also present in northern Africa and North America.

DESCRIPTION

Adult: 6–8 mm long; thorax and elytra red; head, antennae, legs and underside of body black; thorax relatively narrow, with a characteristic lateral constriction. **Egg:** 1 mm long; subspherical, red when newly laid but soon turning brown. **Larva:** up to 10 mm long; head black; body orange-red, with small black plates on each abdominal segment; coated with black slime.

LIFE HISTORY

Adults emerge from hibernation in the early spring and may be found in association with host plants from late March or April onwards. Eggs are laid shortly afterwards. They hatch in 7–10 days and the larvae, whose bodies often become coated with slimy black faeces, develop rapidly, feeding ravenously for about two weeks; they then enter the soil to pupate, each in a silken cocoon. Young adult beetles emerge about three weeks later. In favourable conditions, there are two generations annually.

283 Adult of *Gonioctena viminalis.*

284 Lily beetle (*Lilioceris lilii*).

285 Egg of lily beetle (*Lilioceris lilii*).

286 Larva of lily beetle (*Lilioceris lilii*).

287 Adult of *Lochmaea caprea.*

288 Larva of *Lochmaea caprea.*

289 *Lochmaea caprea* larval damage to leaf of *Salix.*

DAMAGE
Leaves, stems, flowers and seed pods are grazed extensively, the tissue of heavily infested plants becoming tattered and often totally destroyed.

Lochmaea caprea (Linnaeus) (287–289)
In mainland Europe a widely distributed and often common pest of willow (*Salix*) (= willow race) and birch (*Betula*) (= birch race). Poplar (*Populus*) and, less frequently, alder (*Alnus*) and hornbeam (*Carpinus betulus*) are also cited as foodplants. Present in the British Isles but not regarded as a pest.

DESCRIPTION
Adult: 4–6 mm long; elongate-oval, the elytra widened posteriorly; head matt black; thorax brownish yellow, marked with black, and strongly pitted; elytra brownish yellow and distinctly sculptured; legs and antennae black. **Egg:** 0.5 mm across; pale yellow to brownish yellow. **Larva:** up to 6 mm long; head black; body light green, with numerous black plates and verrucae.

LIFE HISTORY
Adult beetles emerge from hibernation in early May. They then feed on the buds, leaves and young shoots. Mating takes place in June, eggs being laid in clusters of 10–15, usually just below the surface of the soil immediately beneath host plants. After egg hatch, larvae migrate onto the leaves to feed, usually removing the lower epidermis. Larvae are fully grown in 3–4 weeks. They then pupate in the soil, young beetles appearing from the end of August onwards. These adults feed briefly before overwintering. There is just one generation annually but, because of the extended oviposition period (old adults often surviving throughout the summer), there often appears to be a second brood. This species favours dry, sunny situations.

DAMAGE
Affected leaves are severely disfigured, such damage being most important on young ornamentals and nursery stock.

290 Adult of *Lochmaea crataegi.*

Lochmaea crataegi (Förster) (**290**)

This widely distributed but local species is associated with hawthorn (*Crataegus*), the larvae developing within the fruits and eventually pupating in the soil. The adults are 4–5 mm long, mainly yellowish brown to reddish brown above and blackish below, with orange legs and black antennae. They feed on the leaves from May onwards. They are sometimes found on nursery plants, but damage caused is not of significance.

Oulema melanopa (Linnaeus) (**291–292**)

Cereal leaf beetle

This widely distributed but minor pest of cereals is associated occasionally with ornamental grasses, such as ribbon grass (*Phalaris arundinacea* var. *picta*), the adults and larvae grazing away longitudinal sections of leaf tissue. Such damage is mildly disfiguring but unimportant. The adult beetles (4.0–4.5 mm long) have a black head and antennae, a brownish-orange thorax and metallic, blue-green elytra. They appear in May, the females depositing eggs on the leaves of suitable host plants during June and July. Eggs hatch in about ten days. Larvae, the stage most likely to be noticed on ornamental grasses, are up to 6 mm long, with a brownish-black head and a dirty-yellow body; they are usually coated in slimy black excrement. They feed on the upper surface of the leaves for 3–4 weeks before entering the soil to pupate. Young adults emerge in the late summer and feed briefly before hibernating. There is just one generation annually.

291 Cereal leaf beetle (*Oulema melanopa*).

292 Larva of cereal leaf beetle (*Oulema melanopa*).

Phratora vitellinae (Linnaeus) (293–295)

syn. *Phyllodecta vitellinae* (Linnaeus)

Brassy willow leaf beetle

Generally common on poplar (*Populus*) and willow (*Salix*); often a pest in stool beds and on ornamental trees. Widely distributed in Europe.

DESCRIPTION

Adult: 3.5–5.0 mm long; thorax and elytra metallic bluish, brassy or coppery; punctation on elytra forming striae (cf. *Plagiodera versicolora*, p. 146); legs and antennae black. **Egg:** 1 mm long; elongate-oval, creamy white. **Larva:** up to 6 mm long; whitish to dirty greyish brown, with numerous black plates and verrucae; body rather flat and tapered. **Pupa:** 4–5 mm long; white.

LIFE HISTORY

Adults hibernate under loose bark, or in similar situations, emerging in the following May or June. The beetles are then commonly found on host plants where, after mating, the females eventually deposit batches of eggs on the leaves. The eggs hatch in about two weeks. Larvae then feed in groups on the expanded leaves. When fully grown, usually in 2–3 weeks, the larvae enter the soil and pupate. A second brood of larvae occurs during the late summer or autumn.

DAMAGE

Adults and larvae browse on the surface of leaves, the remaining tissue turning brown. Such damage is often very noticeable and is very unsightly on ornamentals; attacks have little effect on established trees but may check the growth of seedlings and of young plants in stool beds.

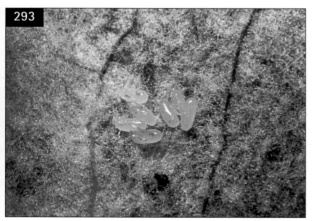

293 Eggs of brassy willow leaf beetle (*Phratora vitellinae*).

294 Larvae of brassy willow leaf beetle (*Phratora vitellinae*).

295 Pupae of brassy willow leaf beetle (*Phratora vitellinae*).

Phratora laticollis (Suffrian) (**296**)
syn. *Phyllodecta laticollis* Suffrian
Small poplar leaf beetle
Locally common on *Populus*, especially aspen (*P. tremula*) and Lombardy poplar (*P. nigra* 'Italica'); also found on grey willow (*Salix cinerea*) and other broad-leaved willows. Adults occur from early spring onwards, at first attacking unopened buds but later feeding on the expanded leaves. Eggs are laid in early June. Larvae (up to 6 mm long) are mainly grey, with numerous black plates and verrucae and a black head. They feed during the summer, at first removing tissue on the underside of the leaves, the upper surface turning brown. Young larvae typically feed gregariously but towards the end of their development individuals often become solitary. Heavy infestations cause extensive skeletonization of the foliage. Fully fed larvae drop to the ground to pupate. Adults, which emerge in the autumn, are 3.5–5.0 mm long and mainly metallic blue with black legs; the head has a characteristic longitudinal depression and the elytra are somewhat elongate. In most areas this species is single brooded. However, in especially favoured districts there may be two generations annually.

Phratora vulgatissima (Linnaeus)
syn. *Phyllodecta vulgatissima* (Linnaeus)
Blue willow leaf beetle
A locally common pest of *Populus* and *Salix*, especially aspen (*P. tremula*), hybrid black poplar (*P. serotina*), Lombardy poplar (*P. nigra* 'Italica') and osier (*S. viminalis*). Often of importance in nursery beds, where young plants may be destroyed, but apparently more restricted in its range than formerly. Adults occur from mid-April or May onwards, eggs being deposited in double rows between two major veins on the underside of the leaves. Eggs hatch in about 1–2 weeks, the larvae then feeding on the leaves for about four weeks before pupating in the soil. Young adults appear in July and August, producing a second generation of larvae which reach adulthood in the autumn. The adults (4–5 mm long) are distinguished from those of *Phyllodecta laticollis* by the slightly narrower body and the blue or green colour; lateral punctation on the elytra is irregular.

Phyllotreta cruciferae (Goeze)
Turnip flea beetle
A generally common pest of brassicaceous plants, including ornamentals such as stock (*Matthiola*) and wallflower (*Cheiranthus cheiri*); also occurs on nasturtium (*Tropaeolum*). Holarctic. Widepsread in Europe.

DESCRIPTION
Adult: 1.8–2.4 mm long; metallic greenish-black and somewhat rounded in outline, with black legs; antennae 11-segmented, mainly black, with the second and third segments yellowish red. **Egg:** 0.3 mm long; yellowish white and oval. **Larva:** up to 6 mm long; head black; body white and narrow, with blackish plates and pinacula on the thoracic and abdominal segments; legs very short.

LIFE HISTORY
Adults are active from early spring onwards, feeding on the leaves and cotyledons of various brassicacous plants. Eggs are deposited in the soil close to host plants, usually in batches of 20–30, and hatch about two weeks later. The larvae attack the roots, feeding externally for about two weeks. Individuals then pupate and new adults emerge two weeks later, usually in July or August. There is just one generation each year.

DAMAGE
Adults bite into the cotyledons and leaves, forming small pit-like blemishes which enlarge as the tissue grows. Attacks are usually most serious in May and June, particularly on seedlings or recent transplants whose growth is retarded by lack of moisture. Larvae destroy the outer tissue of the roots, and heavy attacks affect plant vigour.

296 Larva of small poplar leaf beetle (*Phratora laticollis*).

297 Adult of *Plagiodera versicolora.*

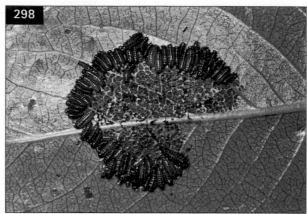

298 Young larvae of *Plagiodera versicolora.*

Plagiodera versicolora (Laicharting) (**297–299**)
This species infests poplar (*Populus*) and willow (*Salix*)
in various parts of Europe. Adults occur mainly from
April or May to September, depositing eggs in groups of
10–30 on the underside of the leaves. The eggs hatch in
about a week. Larvae then feed in groups on the
underside of the leaves. They remove the lower tissue,
the upper surface remaining intact but turning brown.
Young larvae are mainly black and shiny; older
individuals, that feed on either the upper or the lower
surface of leaves, are lighter in colour but have the body
darkened by shiny black verrucae and plates. There are
two main generations annually, larvae occurring from
May onwards, but the developmental stages tend to
overlap. The ladybird-like adults are 3.0–4.5 mm long,
somewhat flattened, and metallic blue or blue-green,
with irregularly punctured elytra and black legs. They
also browse on the foliage of host plants but
damage, although unsightly on ornamentals, is usually
insignificant. Adults overwinter under loose bark,
amongst dry leaves and in other shelter.

Podagrica fuscicornis (Linnaeus) (**300**)
Mallow flea beetle
This species is associated mainly with wild mallows,
including common mallow (*Malva sylvestris*), marsh
mallow (*Althaea officinalis*) and musk mallow
(*M. moschata*). Infestations sometimes also occur
during the summer on ornamental mallows in parks and
gardens, the adults causing extensive damage to the
leaves. Adults are 3–6 mm long, with the head, thorax
and legs pale yellowish red, and the elytra metallic blue
to metallic blue-green.

299 Older larvae of *Plagiodera versicolora* feeding on leaf of
Salix helvetica.

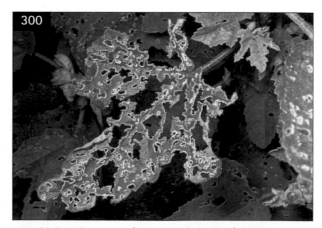

300 Mallow flea beetle (*Podagrica fuscicornis*) damage to
foliage of *Malva.*

Pyrrhalta viburni (Paykull) (**301–303**)

Viburnum beetle

Generally common on wild guelder-rose (*Viburnum opulus*); also a pest of ornamental *Viburnum*. Widespread in central and northern Europe.

DESCRIPTION

Adult: 4.5–6.5 mm long; yellowish brown to light brown; elytra elongate, parallel-sided and regularly rounded posteriorly; body coated with a short, silky pubescence; antennae relatively long, particularly in the male. **Egg:** 0.4 mm across; rounded, dark yellow to brownish. **Larva:** up to 9 mm long; shiny greenish yellow to whitish, with numerous black pinacula and plates; body distinctly swollen, particularly posteriorly.

301 Viburnum beetle (*Pyrrhalta viburni*).

LIFE HISTORY

Overwintered eggs hatch in May. The larvae then feed voraciously on the expanded leaves, becoming fully grown in 4–5 weeks. They then enter the soil to pupate in earthen cells some 30–50 cm below the surface. Adults appear about ten days later, usually in July. After mating, the females, which may survive until September or October, deposit eggs in the tips of one-year-old shoots. Each female is capable of depositing several hundred eggs, and infested bushes are commonly attacked by large numbers of larvae. There is just one generation annually.

302 Larva of viburnum beetle (*Pyrrhalta viburni*).

DAMAGE

Larvae cause considerable damage; they bite out irregular holes in the leaf blades, between the major veins, much of the remaining tissue eventually turning brown. Such depredations are often extensive, affecting both the appearance and vigour of infested plants.

303 Viburnum beetle (*Pyrrhalta viburni*) damage to foliage of *Viburnum*.

Family **RHYNCHITIDAE**

A small group of weevils whose antennae lack a long scape, each segment being of similar length; mandibles dentate.

Neocoenorrhinus aequatus (Linnaeus) (**304**)
syn. *Rhynchites aequatus* (Linnaeus)
Apple fruit rhynchites
A small (2.5–4.5 mm long), reddish-brown to purplish-bronze weevil, associated primarily with hawthorn (*Crataegus*) but also attacking certain other rosaceous ornamental trees and shrubs, including crab-apple (*Malus*), flowering cherry (*Prunus*) and *Sorbus*. The adults feed in the spring on the buds, flowers and foliage, causing slight damage. The females later attack the berries, depositing eggs in small pits bitten into the flesh. Larvae feed in the developing fruits for about three weeks, eventually dropping to the ground to pupate in the soil in earthen cells. There is just one generation annually. Infestations on ornamentals are not important.

Family **ATTELABIDAE** (leaf-rolling weevils)

A group of weevils whose antennae lack a long scape, each segment being of similar length; prothorax much narrower than the abdomen.

Apoderus coryli (Linnaeus) (**305–308**)
Hazel leaf roller weevil
A generally common but minor pest of common hazel (*Corylus avellana*); at least in mainland Europe, attacks also occur on alder (*Alnus*), beech (*Fagus sylvatica*), birch (*Betula*), European hop-hornbeam (*Ostrya carpinifolia*) and hornbeam (*Carpinus betulus*). Present throughout Europe.

DESCRIPTION
Adult: 6–8 mm long; mainly red, with a black, elongate, pear-shaped head, black legs and antennae. **Egg:** 1.0–1.5 mm long; oval, orange. **Larva:** up to 10 mm long; bright orange, with a brown head. **Pupa:** 6–8 mm long; orange.

LIFE HISTORY
Adults emerge in the spring to feed on the leaves of hazel. Eggs are laid in May and June, singly or in small groups, usually in the midrib towards the tip of an expanded leaf. The leaf blade is then severed near the base, the cut extending from one edge to or just beyond the midrib (cf. *Attelabus nitens*, pp. 149–50); the cut tissue then curls laterally to remain suspended from the unsevered part of the leaf blade as a stumpy, cigar-like leaf roll. The larvae develop within these leaf rolls and then pupate, new adults appearing at the end of July or in early August. Larvae of a second generation complete their development in the autumn; they eventually overwinter on the ground, within their fallen habitations, and pupate in the spring.

DAMAGE
Although infestations are sometime established on ornamental plants and larval habitations might be very numerous, damage caused is unimportant.

Attelabus nitens (Scopoli) (**309–310**)
syn. *A. curculionoides* Linnaeus
Oak leaf roller weevil
A locally common but minor pest of oak (*Quercus*); also found on alder (*Alnus*), common hazel (*Corylus avellana*) and sweet chestnut (*Castanea sativa*). Widespread in mainland Europe, northwards to southern Scandinavia; in the British Isles most numerous in the southern half of England.

304 Apple fruit rhynchites (*Neocoenorrhinus aequatus*).

305 Hazel leaf roller weevil (*Apoderus coryli*).

306 Larva of hazel leaf roller weevil (*Apoderus coryli*).

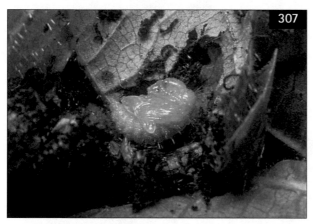

307 Pupa of hazel leaf roller weevil (*Apoderus coryli*).

308 Larval habitation of hazel leaf roller weevil (*Apoderus coryli*) on *Ostrya.*

309 Larva of oak leaf roller weevil (*Attelabus nitens*).

310 Larval habitation of oak leaf roller weevil (*Attelabus nitens*) on *Quercus.*

DESCRIPTION

Adult: 5.0–6.5 mm long; bright red to dark red, with the head, antennae, legs, scutellum and underside of body black. **Egg:** 1.0–1.5 mm long; oval, yellow. **Larva:** head brownish yellow; body yellow to orange-yellow, and strongly C-shaped.

LIFE HISTORY

Adults appear in May and are most commonly associated with young oak trees. In June, mated females deposit eggs singly or in small groups in the midrib of expanded leaves. At the same time, they initiate larval feeding shelters by severing the leaf blade on both sides (cf. *Apoderus coryli*, p. 148) of the midrib at right-angles to the main axis, about a third of the distance from the base (the oviposition point being about 10 mm beyond the line of cut). The freed parts of the leaf blade on either side of the midrib then fold until their surfaces meet, and finally roll up from the tip to produce a characteristic pouch within which larval development takes place. Larvae feed within their shelters for about six weeks, the rolled leaf fragment eventually falling to the ground. Larvae then overwinter and pupate in the spring.

DAMAGE

The unusual larval shelters attract attention but damage to infested trees is not important.

Byctiscus betulae (Linnaeus)
syn. *B. betuleti* (Fabricius)
Poplar leaf roll weevil

A widespread but locally distributed weevil, associated mainly with common hazel (*Corylus avellana*) and poplar (*Populus*), but also attacking alder (*Alnus*), birch (*Betula*), elm (*Ulmus*), ornamental pear (*Pyrus calleryana* 'Chanticleer') and willow (*Salix*). Eggs are laid in leaf veins during late May or June, typically several per leaf. The female also partly severs the leaf stalks, the leaf blade then rolling into a pendulous, cigar-like tube within which larval development takes place. Fully grown larvae pupate in the ground, some adults emerging in the autumn but others not until the following spring. Infestations are sometimes noticed on cultivated plants but are usually of only minor importance. The adult weevils are 6–9 mm long, dark blue, green or red, with distinctly metallic legs; males possess a pair of characteristic, forwardly directed spines on the thorax.

Byctiscus populi (Linnaeus)
Aspen leaf roller weevil

This smaller (5–6 mm long), green or reddish, black-legged weevil (the underside is characteristically bluish black) is associated mainly with young aspen (*Populus tremula*) trees, causing similar damage to the previous species. Infestations occur occasionally on ornamental plants but are of little or no importance.

Deporaus betulae (Linnaeus) (311–312)
Birch leaf roller weevil

An often common but minor pest of young birch (*Betula*) trees; also associated with certain other trees, including alder (*Alnus*), beech (*Fagus sylvatica*), common hazel (*Corylus avellana*) and hornbeam (*Carpinus betulus*). Present throughout Europe.

DESCRIPTION

Adult: 3–5 mm long; black, with rectangular, deeply pitted elytra; hind femora of male greatly thickened. **Larva:** up to 7 mm long; whitish.

311 Female birch leaf roller weevil (*Deporaus betulae*).

312 Larval habitation of birch leaf roller weevil (*Deporaus betulae*) on *Fagus*.

LIFE HISTORY

Adults appear in the spring and then feed on the foliage of various host plants. Egg laying begins from early May onwards, when mated females initiate characteristic larval habitations. Firstly, one or two eggs are deposited in the upper surface of an expanded leaf; the selected leaf is then severed neatly across on both sides of the midrib, just beyond the base, the tissue rolling into an elongate, cone-like tube. The suspended material soon dries out. Cones formed on beech turn rusty brown and are particularly obvious. Larvae are fully grown in about three months. They then enter the soil where they pupate. Adults emerge in the following spring. There is a single generation annually.

DAMAGE

Infestations disfigure host plants and are often extensive, with all leaves on certain shoots affected, but damage is not of economic importance.

Deporaus tristis (Fabricius) (**313**)

Maple leaf roller weevil

This relatively small (3.5–4.0 mm long), bluish-black species is associated mainly with sycamore (*Acer pseudoplatanus*); occasionally, beech (*Fagus sylvatica*) and oak (*Quercus*) are also hosts. The adults feed on the leaves, removing significant amounts of tissue; they also form characteristic, tightly wound leaf rolls within which larval development takes place. Affected leaves are sometimes found on amenity trees in mountainous parts of Europe, from southern Germany southwards, but infestations are not of significance. The weevil is widely distributed in southern Europe.

Family **APIONIDAE**

A group of small, often narrow-bodied, dull-looking weevils; antennae usually not geniculate, although basal segment (scape) may be noticeably elongated.

Rhopalapion longirostre (Olivier)

syn. *Apion longirostre* Olivier

Hollyhock weevil

A destructive pest of hollyhock (*Alcea rosea*). Originally found mainly in south-eastern Europe and Asia Minor, but currently greatly extending its range in Europe. First found breeding in England in 2006. Now widespread and of some significance in the USA, where it was first recorded in 1914.

DESCRIPTION

Adult: 4–5 mm long; mainly black, with a grey pubescence; legs yellowish or reddish yellow; rostrum very long.

LIFE HISTORY

Eggs are laid in early summer in the unopened buds of hollyhock. Larvae then feed on the developing seeds for 4–6 weeks. Fully fed larvae pupate *in situ*, and new adults emerge from the buds shortly afterwards. These adults overwinter, there being just one generation annually.

DAMAGE

Adults feed on buds and leaves, boring holes in the tissue and causing extensive damage.

313 Larval habitation of maple leaf roller weevil (*Deporaus tristis*) on *Acer*.

Family **CURCULIONIDAE** (true weevils and bark beetles)

The main family of weevils, the antennae being geniculate and with a very long basal segment (scape). The larvae are apodous, with a distinct head; in repose, they often adopt a C-shaped posture. Adult weevils typically feed externally on plant tissue. Larvae may also feed externally, but those of most weevils mine within the leaves, stems or roots of host plants.

Bark beetles were formerly separated into the family Scolytidae but are now included in the Curculionidae. They are typically small, cylindrical, wood-boring insects (including some known as ambrosia beetles and some as shot-hole borers) with relatively short but distinctly clubbed antennae. The larvae have an enlarged prothoracic region, giving them a hunch-backed appearance. Development takes place entirely within the host, some species being dependent upon the presence in their galleries of hyphae of ambrosia fungi upon which the larvae feed. Many species of bark beetle attack ornamental trees, but infestations occur mainly on already weakened trees and those growing under stress. Examples cited below are restricted to the genera *Hylesinus, Phloeosinus* and *Scolytus*.

Anthonomus brunnipennis (Curtis) (**314**)
syn. *A. comari* Crotch
A generally common and widely distributed species, associated with wild and cultivated cinquefoil (*Potentilla*). The weevils feed on the foliage and flower petals, producing numerous small holes. Individuals are about 2 mm long and mainly black, with the thorax and elytra covered extensively in fine punctures. Infestations have no significant effect on plant growth. However, damage is unsightly, particularly on petals of opened blossoms.

Barynotus obscurus (Fabricius) (**315**)
Adults of this widely distributed and locally common weevil feed on the flowers of various plants, including ornamentals such as primrose (*Primula vulgaris*), rose (*Rosa*) and violet (*Viola*). Damage occurs from spring to mid-summer, most serious attacks having been reported on rose bushes in England. The weevils are 8–10 mm long and black or brownish, with a coating of round, close-set, yellow (sometimes greenish or coppery) scales; they are superficially similar in appearance to otiorhynchid weevils (pp. 159–63) but the thoracic region is relatively broad, with only a slight constriction between thorax and abdomen; the femora are untoothed.

Barypeithes araneiformis (Schrank) (**316**)
Smooth broad-nosed weevil
Adults of this widely distributed but local weevil are sometimes damaging during the spring and summer in conifer nursery beds, when they destroy the initial shoots and also ring-bark young seedlings. They also attack young deciduous trees, such as birch (*Betula*), horse chestnut (*Aesculus hippocastanum*) and oak (*Quercus*), damaging the buds and severing or forming holes in the expanded leaves. The larvae feed on the roots of herbaceous plants, including various weeds. Adults are 3–4 mm long, shiny brownish yellow to black, with a pointed abdomen.

Barypeithes pellucidus (Boheman) (**317**)
Hairy broad-nosed weevil
This locally distributed weevil is sometimes damaging to young or seedling trees. Adults are distinguished from those of *Barypeithes araneiformis* by the longer, denser and more distinct pubescence.

Cionus scrophulariae (Linnaeus) (**318–320**)
Figwort weevil
A generally common pest of cape figwort (*Phygelius capensis*), mullein (*Verbascum*) and orange ball buddleia (*Buddleja globosa*); also abundant on wild figwort (*Scrophularia nodosa*). Widely distributed in Europe.

DESCRIPTION
Adult: 4.0–4.5 mm long; thorax small, elytra square; body black but largely coated with a purplish-grey pubescence; elytra with a regular pattern of black markings, including two large circular patches; antennae dark red. **Larva:** up to 6 mm long; yellowish, with a black head and prothoracic plate; body coated with greenish-yellow slime.

LIFE HISTORY
Adults occur on host plants from May or June onwards, feeding on the leaves, leaf stalks and flowers. Larvae feed during the summer on the leaves, removing tissue from one surface, the other remaining intact. Fully grown larvae construct more or less spherical, brownish to yellowish-brown, parchment-like cocoons on the stems and leaves, and then pupate. Adults emerge 2–3 weeks later and eventually produce a further brood of larvae. Adults of the final generation enter hibernation to reappear in the spring.

DAMAGE
Attacked plants are disfigured, and leaf tissue infested by larvae may eventually split; flower buds are also destroyed.

314 *Anthonomus brunnipennis* damage to flower of *Potentilla*.

315 Female of *Barynotus obscurus*.

316 Smooth broad-nosed weevil (*Barypeithes araneiformis*) damage to leaf of *Betula*.

317 Hairy broad-nosed weevil (*Barypeithes pellucidus*).

318 Figwort weevil (*Cionus scrophulariae*).

319 Larva of figwort weevil (*Cionus scrophulariae*).

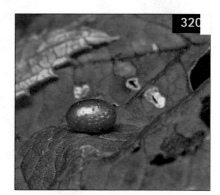

320 Pupal cocoon of figwort weevil (*Cionus scrophulariae*).

321 Mullein weevil (*Cionus hortulanus*).

322 Osier weevil (*Cryptorhynchus lapathi*).

323 Osier weevil (*Cryptorhynchus lapathi*) damage to shoot of *Populus*.

Cionus hortulanus (Fourcroy) (321)
Mullein weevil

This locally common weevil also infests cultivated mullein (*Verbascum*) but has a more restricted host range than the previous species, being found most often on dark mullein (*V. nigrum*). Adults of both species of *Cionus* are similar in appearance but *C. hortulanus* is slightly smaller (3.5–4.0 mm long) and paler in colour, the elytra appearing mainly yellowish grey or greenish grey, interrupted by two black patches and a distinct pattern of small black markings.

Cryptorhynchus lapathi (Linnaeus) (322–323)
syn. *Cryptorhynchidius lapathi* (Linnaeus)
Osier weevil

A locally important pest of poplar (*Populus*), notably hybrid black poplar (*P. serotina*), and willow (*Salix*); also associated with alder (*Alnus*) and, less frequently, birch (*Betula*). Often damaging in nursery beds, including basket-willow beds. Widely distributed in Europe, including the Czech Republic, England, France, Germany, Hungary, Italy, the Netherlands, Poland, Spain, Wales and Yugoslavia; introduced into North America.

DESCRIPTION

Adult: 8–9 mm long; robust; mainly black, interspersed with white or yellowish-white scales, particularly towards the base and at the apex of the elytra. **Egg:** 1 mm long; creamy white. **Larva:** up to 10 mm long; creamy white, with a brown head.

LIFE HISTORY

In subtropical areas and southern Europe, this insect has a one-year lifecycle. In most parts of Europe, however, development usually extends over two years. Adults emerge in the spring, from May onwards. They usually remain hidden beneath host plants during the day and become active in the evening, crawling up the stems to feed on the young shoots. They are very sluggish, dropping to the ground if disturbed; although fully winged, they rarely, if ever, fly. Adults continue to feed throughout the summer and autumn, mated females eventually laying eggs singly in the rods and stools of host plants. Surface wounds, leaf scars and lenticels are often chosen as oviposition sites, the females boring out a cavity before depositing a single egg a few millimetres beneath the surface. Eggs hatch in about three weeks. Irrespective of when eggs are laid, the larvae feed only briefly during the first season; they then enter diapause, still in their first instar, to resume activity in the following spring. The larvae feed within the host tissue for several months, ejecting whitish frass through a hole bored through the bark. They often penetrate to the pith of the branches and trunks of host plants; they also burrow extensively within the rods and stools. Fully fed larvae eventually pupate at the end of their galleries from mid-July onwards. Adults emerge from the pupae in late July or August but do not vacate the pupal cells until the following spring.

DAMAGE

Young terminal shoots are partly severed by the adult weevils, the tips keeling over and eventually turning black. Shoots may also become riddled with holes; such damage is of particular importance in willows intended for basket making; affected rods have to be discarded. The larvae reduce the production of new shoots from infested stools, and are sometimes of importance in nurseries. Heavy attacks cause death of plants.

324 Adult of *Dorytomus taeniatus.*

Dorytomus taeniatus (Fabricius) (**324**)

Generally common on willow (*Salix*) from May onwards. The adults browse on the leaves, removing patches of tissue to expose a network of fine veins; they often feed in groups, but damage caused is insignificant (although noticeable and sometimes extensive). Eggs are laid in the axils of the catkin buds during the autumn and hatch early in the following spring. The larvae then mine within the catkins and eventually pupate in the soil, young weevils appearing in May and June. The weevils often hide during the day amongst curled leaves, dropping to the ground if disturbed. Adults are 4–5 mm long, black to brownish black, with elongate, parallel-sided elytra, an elongate rostrum and toothed femora; the elytra are also patterned with small patches of lighter hairs. At least in mainland Europe, the weevils will also attack alder (*Alnus*), birch (*Betula*) and poplar (*Populus*).

Hylesinus varius (Fabricius)
syn. *Leperisinus varius* (Fabricius)
Ash bark beetle

A common but typically secondary pest of ash (*Fraxinus excelsior*) and lilac (*Syringa*). Other hosts include common hazel (*Corylus avellana*), false acacia (*Robinia pseudoacacia*), flowering ash (*Fraxinus ornus*), hornbeam (*Carpinus betulus*), maple (*Acer*), olive (*Olea europaea*) and walnut (*Juglans*). Breeding occurs mainly in felled or fallen trees. However, the beetles will invade live trees, especially trees previously weakened or damaged by other factors. Young trees with thin bark are also favoured hosts. Adults occur from early March to May, and appear in distinct waves. Individuals (2.5–3.5 mm long) are blackish brown to brownish red, with the thorax and elytra covered in grey

scales interspersed with irregular patches of darker ones. The larvae develop beneath the bark, forming irregular galleries (each up to *c*. 20 mm long) that extend upwards and downwards from a pair of broad, more or less horizontal maternal chambers. There is a single generation annually.

Hypera arator (Linnaeus)
syn. *H. polygoni* (Linnaeus)

This widely distributed species is associated with various members of the Caryophyllaceae and is sometimes a minor pest of carnation (*Dianthus caryophyllus*) and pink (*D. plumarius*). The yellowish-grey to greenish larvae (up to 7 mm long) feed within the flowers, and damage on cultivated plants is sometimes mistaken for that caused by larvae of the carnation tortrix moth (*Cacoecimorpha pronubana*, p. 263). Pupation occurs in oval, yellowish-white cocoons formed on the foodplant. Young weevils appear about two weeks later. The adults (5–7 mm long) are yellowish brown, with pale, dark-edged longitudinal bands on the thorax and elytra and a black line along the elytral suture. They may be found from early May to the end of September.

325 Larval mines of poplar leaf-mining weevil (*Isochnus sequensi*) on *Salix*.

326 Elm leaf-mining weevil (*Orchestes alni*).

Isochnus sequensi (Stierlin) (325)

syn. *I. populi* Fabricius

Poplar leaf-mining weevil

An often common species, the larvae mining within the leaves of poplar (*Populus*) and willow (*Salix*) to form large, disfiguring, brownish-black blotches. Adults (2.0–2.5 mm long) are mainly black, with a whitish scutellum and yellowish-red antennae and legs. There is one generation a year. Larvae occur during the summer, and new adults (which eventually overwinter) appear in September.

Orchestes alni (Linnaeus) (326–328)

syn. *Rhynchaenus alni* (Linnaeus); *R. ferrugineus* (Marsham); *R. saltator* (Fourcroy)

Elm leaf-mining weevil

A widely distributed pest of elm (*Ulmus*); at least in parts of mainland Europe, also associated with alder (*Alnus*) and common hazel (*Corylus avellana*). Present throughout much of mainland Europe, from Denmark southwards; in the British Isles most numerous in the southern half of England.

327 Larva of elm leaf-mining weevil (*Orchestes alni*).

DESCRIPTION

Adult: 2.5–3.0 mm long, with yellowish-red or red, black-marked elytra and black legs and head; a yellowish form also occurs. **Larva:** up to 6 mm long; whitish to yellowish white, with a black head.

LIFE HISTORY

Adults appear in the spring to feed on the young leaves. Eggs are laid shortly after mating, each in a main vein on the underside of a leaf. The larvae mine within the leaf blade, each gallery commencing as a narrow channel that widens into a prominent brown blotch and usually terminates at the apex of the leaf. The larvae

328 Mine of elm leaf-mining weevil (*Orchestes alni*) on *Ulmus*.

feed for about two weeks. They then pupate, each in a rounded cocoon spun within the mine. New adults emerge in the summer. They feed on the leaves before eventually overwintering.

DAMAGE

Adult feeding is usually of minor importance; the larval mines are disfiguring and cause noticeable distortion of affected leaves.

Orchestes fagi (Linnaeus) (**329–332**)
syn. *Rhynchaenus fagi* (Linnaeus)
Beech leaf-mining weevil
larva = beech leaf miner
A common and widely distributed pest of beech (*Fagus sylvatica*). Present throughout Europe.

DESCRIPTION

Adult: 2.2–2.8 mm long; body black, covered with a greyish-brown pubescence; antennae brown; elytra distinctly broader than pronotum; eyes close-set; hind tibiae robust. **Larva:** up to 5 mm long; head blackish; body white, shiny and distinctly tapered posteriorly.

LIFE HISTORY

Adults overwinter under bark, in the ground and in other shelter, emerging from hibernation in late April or early May. They then invade beech trees to feed on the foliage. Eggs, usually 30–35 per female, are laid singly on the underside of the midrib of expanded leaves; less often, they are placed beneath a lateral vein. Larvae mine within the leaves from May to June, each forming a brownish-black gallery. This commences as a narrow

329 Beech leaf-mining weevil (*Orchestes fagi*).

330 Beech leaf miner (*Orchestes fagi*).

331 Mines of beech leaf miner (*Orchestes fagi*) on *Fagus.*

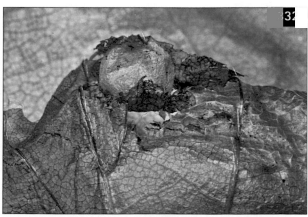

332 Pupal cocoon of beech leaf-mining weevil (*Orchestes fagi*).

333 Larval mine of oak leaf-mining weevil (*Orchestes quercus*) on *Quercus.*

334 Larva of oak leaf-mining weevil (*Orchestes quercus*).

channel, running from the midrib to the leaf margin; it then expands into a small brown blotch. There are three larval instars, fully fed larvae pupating within the feeding gallery, each in a white, spherical cocoon. The overwintered adults survive until the end of May or early June, and the new generation of adults appears in mid-June. These weevils continue feeding on the foliage for several weeks before entering their overwintering sites, often causing considerable leaf damage during June and July. Most weevils enter hibernation in late July or August, but some delay their departure until mid-September. There is just one generation annually.

DAMAGE
Adults pepper the foliage with small holes and, if the midrib is damaged, leaves may wilt. Larval galleries cause distortion, and the tips of mined leaves turn brown. Such damage is often extensive on woodland trees and is also of considerable significance on ornamental trees and hedges.

Orchestes quercus (Linnaeus) (**333–334**)
syn. *Rhynchaenus quercus* (Linnaeus)
Oak leaf-mining weevil
This common species occurs on oak (*Quercus*) and is noted occasionally on young ornamental trees. The weevils are active from April or May onwards, browsing on the surface of the leaves. Eggs are laid singly, usually in the underside of the midrib. Following egg hatch, the larva mines towards the leaf apex to form an elongate gallery which passes along the midrib and

eventually moves into the leaf blade; mining then transfers from the lower to the upper side of the leaf, the gallery eventually terminating in an irregular brownish blotch, clearly visible from above. The larvae are greenish white and translucent, with a brown head. When fully fed, usually in June or July, they pupate in rounded transparent cocoons formed within the mine. Young weevils emerge 1–2 weeks later. They feed on the leaves and then hibernate, reappearing in the spring. Adults are 2.5–3.5 mm long, and mainly reddish brown to brownish black, with a yellow pubescence.

Orchestes rusci (Herbst) (**335–336**)
syn. *Rhynchaenus rusci* (Herbst)
Birch leaf-mining weevil
Associated mainly with birch (*Betula*) and oak (*Quercus*), with adults present from March onwards. Unlike those of the previous four species, the larval gallery follows the leaf margin and then turns inwards to terminate in a more or less circular blotch. When fully grown, the larva cuts out a circular section from above and below the mine, and then forms its cocoon on the ground between these two epidermal discs. Leaves formerly containing larvae are recognized by the presence of a circular hole at the end of the deserted and abruptly terminated mine. The adult weevils are 2.0–2.5 mm long and rather oval-bodied, with a black pronotum and elytra, the latter with two bands of white or yellow scale-like hairs; the antennae are red and the legs black with brownish tarsi.

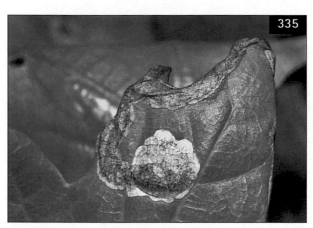

335 Larval mine of birch leaf-mining weevil (*Orchestes rusci*) on *Quercus.*

336 Remains of larval mine of birch leaf-mining weevil (*Orchestes rusci*) on *Betula.*

337 Red-legged weevil (*Otiorhynchus clavipes*).

Otiorhynchus clavipes (Bonsdorff) (337)
syn. *O. fuscipes* (Olivier); *O. lugdunensis* Boheman
Red-legged weevil

Adults attack various woody hosts, including Japanese laurel (*Aucuba japonica*), honeysuckle (*Lonicera*) and lilac (*Syringa*); the larvae are associated with the roots of soft-fruit crops, especially raspberry, and ornamentals such as lilac, maple (*Acer*), *Viburnum* and *Weigelia*. Locally common in Europe, including south-western England, Belgium, France, Germany, the Netherlands and Switzerland, particularly on light soils.

DESCRIPTION

Adult: 8–12 mm long; body blackish, oval and distinctly pointed posteriorly; sculpturing on thorax and elytra shallow; wingless; legs long and reddish. **Egg:** 0.5×0.6 mm; more or less spherical; whitish when laid but soon becoming black. **Larva:** up to 12 mm long; creamy white, with a brown head; body plump, wrinkled and strongly C-shaped.

LIFE HISTORY

Adults appear from late spring to August, the period of appearance varying according to the timing of pupation. They are active at night, feeding on leaves and other aerial parts of various woody plants. During the daytime they hide in grass tussocks, under stones and in other shelter. Eggs are laid in the soil, scattered at random close to the surface beneath host plants, each female depositing up to 300. Eggs hatch in about three weeks. The larvae then feed on plant roots, each passing through five instars. Larvae overwinter either as fully fed individuals or as young individuals which complete their development in the following spring; the former pupate in spring and the latter in the summer. Reproduction is either sexual or parthenogenetic.

DAMAGE

Adults notch the leaves. They also damage blossoms and unopened buds, and check the growth of young shoots by gnawing at the bases. Larvae are more important; they feed on roots, attacked plants wilting and sometimes dying.

Otiorhynchus ovatus (Linnaeus) (**338**)
Strawberry weevil

An often common pest of herbaceous plants. Young and seedling trees, especially Norway spruce (*Picea abies*) grown as Christmas trees, are also attacked. Eurasiatic. Present throughout Europe; introduced into Central and North America.

DESCRIPTION
Adult female: 5.0–5.5 mm long; black to reddish brown, with a slight yellow pubescence; disc of thorax distinctly furrowed. **Larva:** up to 6 mm long; creamy white, with a brown head.

LIFE HISTORY
Weevils are active from April or May onwards. They feed on the foliage and young shoots of various hosts, especially seedling conifers and strawberry plants. Eggs are laid in the soil in the spring. Larvae attack the fine roots, and cause most damage in May and June. They usually complete their development in the summer. There is a single generation annually, but details of the lifecycle vary according to local conditions.

DAMAGE
Conifers: adults destroy the young growth and needles, affecting the quality and vigour of host plants; the larvae destroy the smaller roots, weakening and often killing seedlings.

Otiorhynchus singularis (Linnaeus) (**339**)
syn. *O. picipes* (Fabricius)
Clay-coloured weevil

An often troublesome pest of ornamentals such as buddleia (*Buddleja*), *Clematis*, crab-apple (*Malus*), flowering cherry (*Prunus*), *Hydrangea*, primrose (*Primula vulgaris*), *Rhododendron*, rose (*Rosa*) *and Wisteria*; also damaging to conifers, notably western hemlock (*Tsuga heterophylla*) and yew (*Taxus*). Widespread in Europe, particularly in grassy lowland sites; an introduced pest in North America.

DESCRIPTION
Adult: 6–7 mm long; shiny black and strongly sculptured, but covered with greyish-brown scales that give the body an irregular pattern and overall matt appearance; body often encrusted with mud; wingless. **Larva:** up to 8 mm long; head brown; body creamy white, plump and wrinkled.

LIFE HISTORY
Adults occur from April to October, hiding by day in cracks in the soil, under straw mulches and in other shelter beneath host plants. At night, they feed on the foliage, buds and bark, causing most harm from April to June. Eggs are laid in the soil during the summer. The emerging larvae then feed on the roots of many kinds of plant, including various weeds, from late summer onwards. Fully grown larvae pupate in the early spring, and adults appear shortly afterwards. Some individuals survive for two or more seasons, hibernating during the winter months. Males are very rare and reproduction is mainly parthenogenetic.

DAMAGE
Larvae attack the roots, sometimes seriously weakening herbaceous plants in rockeries. Generally, however, adults are more important pests, the precise type of injury caused varying from host to host. On young trees, large irregular areas of bark are removed, stems often ring-barked and plants killed; the weevils also notch the leaves and destroy the buds of young grafts. Leaf stalks on young shoots are sometimes gnawed, causing them to fold over, wilt and drop off; buds on young bushes are also attacked, producing misshapen plants with forked shoots or numerous unwanted laterals; new growth is often girdled and killed.

Otiorhynchus sulcatus (Fabricius) (**340–343**)
Vine weevil

A major pest of greenhouse and outdoor ornamentals, including alpines (especially *Saxifraga*, *Sedum* and *Sempervivum*), *Begonia*, busy lizzie (*Impatiens*), *Calceolaria*, *Camellia*, *Cotoneaster*, *Cyclamen*, *Elaeagnus*, ferns (e.g. *Adiantum*), heather (*Erica*), lily-of-the-valley (*Convallaria majalis*), Michaelmas daisy (*Aster*), peony (*Paeonia*), *Phlox*, *Polyanthus*, primrose (*Primula vulgaris*), *Rhododendron* and various tree seedlings (coniferous and broad-leaved species). Widely distributed in Europe; an introduced pest in Australasia and North America.

DESCRIPTION
Adult: 7–10 mm long; black and shiny, the elytra parallel-sided and coated with patches of yellowish hairs; body deeply sculptured; wingless. **Egg:** 0.7 mm across; more or less spherical; white at first, soon turning brownish. **Larva:** up to 10 mm long; creamy to brownish white, with a reddish-brown head. **Pupa:** 7–10 mm long; white.

338 Strawberry weevil (*Otiorhynchus ovatus*).

339 Clay-coloured weevil (*Otiorhynchus singularis*).

340 Vine weevil (*Otiorhynchus sulcatus*).

341 Eggs of vine weevil (*Otiorhynchus sulcatus*).

342 Larva of vine weevil (*Otiorhynchus sulcatus*).

343 Pupa of vine weevil (*Otiorhynchus sulcatus*).

344 Privet weevil (*Otiorhynchus crataegi*).

345 Female of *Otiorhynchus niger*.

LIFE HISTORY
In outdoor situations, adult females of this parthenogenetic weevil emerge in May and June. They are active at night, feeding on the foliage of various plants. During the daytime, they tend to hide away beneath debris, in crevices in walls or in other shelter. The weevils mature a few weeks after commencing feeding, each being capable of depositing several hundred eggs. The eggs are laid in soil near the host plants from late July onwards. They hatch in 2–3 weeks and the larvae then attack plant roots, corms or rhizomes. Larval feeding extends over several months. Development is completed in the following spring, fully grown individuals pupating in oval subterranean cells formed in the soil; transformation to the adult stage occurs from mid-April to June. Most weevils die before the onset of winter. However, some survive for two seasons or longer, overwintering in sheltered situations and reappearing in the spring. Young adults are distinguished from older individuals by the presence of a thick spine on each mandible; this usually breaks off soon after emergence. In heated greenhouses, adults often emerge in large numbers in the autumn; also, eggs are laid over a more extended period, and all stages of the pest may occur at one and the same time.

DAMAGE
Adults notch the leaves of host plants, and often cause extensive damage to ornamentals such as camellia, lily-of-the-valley and rhododendron. They also ring-bark young plants. Most important damage, however, is caused by the larvae which destroy the finer roots, or burrow into the fleshy parts of corms and rhizomes; affected plants lack vigour and may suddenly wilt, or collapse and die. Damage is particularly important on outdoor containerized plants growing in peat composts and on greenhouse pot plants, but may also be extensive in nursery beds and rock gardens.

Otiorhynchus crataegi Germar (**344**)
Privet weevil
This small species is a pest of various ornamentals in eastern and certain other parts of mainland Europe. The weevils browse on foliage and the larvae feed on the roots of various plants, including lilac (*Syringa*) and privet (*Ligustrum vulgare*). In 1985, following an accidental introduction, extensive but localized damage was reported on privet hedges, ornamental shrubs and certain other plants, in south-central England. Adults are 5 mm long and mainly dark brown, marked irregularly with golden brown; the narrow, rounded thorax, bulbous abdomen and relatively narrow waist are characteristic.

Otiorhynchus niger (Fabricius) (**345**)
A common forestry pest in mainland Europe, particularly in mountainous areas; sometimes found on amenity trees but not an important pest of ornamentals. Adults are 9–11 mm long and mainly black, with dark orange legs and deeply sculptured elytra.

346 Lesser strawberry weevil (*Otiorhynchus rugosostriatus*).

347 Larva of lesser strawberry weevil (*Otiorhynchus rugosostriatus*).

Otiorhynchus rugosostriatus (Goeze) (**346–347**)
Lesser strawberry weevil

Although most frequently reported as a pest of strawberry, this widely distributed and locally common weevil also attacks pot plants and ornamental shrubs, the adults notching the edges of expanded leaves and the larvae feeding on the roots. Damage is most commonly reported on *Begonia*, *Cyclamen*, lilac (*Syringa*), primrose (*Primula vulgaris*) and privet (*Ligustrum vulgare*). Adults (6–7 mm long) are blackish to reddish brown, and strongly sculptured; they are most numerous from June to September. The larvae, which feed throughout the autumn and winter, are similar to those of *Otiorhynchus sulcatus* (see p. 160) but smaller, reaching a maximum length of about 8 mm.

Phloeosinus thujae (Perris)
Thuja bark beetle

A widely distributed primary and secondary pest of Cupressaceae, including coastal red wood (*Sequoia sempervirens*), common juniper (*Juniperus communis*), Italian cypress (*Cupressus sempervirens*), Sawara cypress (*Chamaecyparis pisifera*), Wellingtonia (*Sequoiadendron giganteum*) and white cedar (*Thuja occidentalis*). Most numerous in the warmer parts of central, eastern and southern Europe.

DESCRIPTION
Adult: 1.5–2.4 mm long; body stubby and egg-shaped; brownish black, with yellowish hairs; antennae yellowish red.

LIFE HISTORY
Adults occur in June. They then attack the young shoots, stems and branches of host trees. Breeding occurs in distinctive galleries formed beneath the bark of the trunks and larger branches, the beetles excavating two-branched egg chambers that run vertically upwards and downwards from the initial entry hole, and from which larval galleries eventually arise. The latter, each up to 50 mm long, emerge at right angles but soon turn vertically. There is one generation annually, but two are possible in favorable locations.

DAMAGE
Attacked established trees are weakened, but infestations on hedging plants and on young trees being raised in nurseries are most serious. When adult beetles bore into green tissue this often results in the death of shoots.

Phyllobius argentatus (Linnaeus) (**348**)
Silver-green leaf weevil

A common but minor pest of various trees and shrubs, including alder (*Alnus*), beech (*Fagus sylvatica*), birch (*Betula*), flowering cherry (*Prunus*), hawthorn (*Crataegus*) and *Sorbus*; sometimes also found on conifers. Often troublesome on young ornamentals and nursery stock. Widespread in Europe.

DESCRIPTION
Adult: 4–6 mm long; body black, but appearing bright green due to a covering of shiny, golden-green, rounded, disc-like scales; legs and antennae light brown or yellowish and partially clothed in golden-green scales; femora toothed (cf. *Polydrusus formosus*, p. 166).

LIFE HISTORY
Adults feed in the spring and early summer on the foliage and flowers of various trees and shrubs. In sunny weather, they often bask openly on the leaves; in dull conditions they tend to hide amongst folded or crinkled foliage, often sheltering in vacated habitations of leaf-rolling moth larvae and other leaf-deforming pests, but they are readily detected if a branch is tapped over a cloth or tray. Later in the season, the weevils migrate to other hosts, including herbaceous plants. Eggs are laid in the soil during the early summer. Larvae feed on the roots of various herbaceous weeds and grasses. They pupate in the spring within earthen cells, and adults emerge a few weeks later.

DAMAGE
Adults bite holes into the leaves and flower petals, sometimes causing extensive damage to young trees and shrubs; most damage is caused from April to July, but the extent of injury varies considerably from year to year and is not apparently related to the size of populations invading the trees.

Phyllobius maculicornis Germar (**349**)
Often common on young deciduous trees and shrubs, especially beech (*Fagus sylvatica*), birch (*Betula*) and hawthorn (*Crataegus*). This species is usually less numerous than *Phyllobius argentatus*, but is capable of causing extensive damage to the foliage of suitable hosts. Both species are of similar appearance and size, but *P. maculicornis* is distinguished by the black femora and stouter, partly blackish antennae (club and apex of the scape) and more prominent eyes.

Phyllobius oblongus (Linnaeus) (**350**)
Brown leaf weevil

An often abundant species, occurring throughout Europe on various trees and shrubs, including elm (*Ulmus*), lime (*Tilia*), maple (*Acer*), poplar (*Populus*) and willow (*Salix*), but most often noted on rosaceous hosts such as crab-apple (*Malus*), flowering cherry (*Prunus*) and hawthorn (*Crataegus*); frequently a problem on nursery trees. Adults (3.5–6.0 mm long) are black, with reddish-brown, slightly pubescent elytra, and brownish legs and antennae; unlike other members of the genus, the elytra are devoid of scales.

Phyllobius pyri (Linnaeus) (**351**)
Common leaf weevil

Heavy infestations of this species often occur on trees and shrubs, including alder (*Alnus*), ash (*Fraxinus excelsior*), beech (*Fagus sylvatica*), birch (*Betula*), elm (*Ulmus*), flowering cherry (*Prunus*), hawthorn (*Crataegus*), horse chestnut (*Aesculus hippocastanum*) and *Sorbus*. The weevils often occur in association with *Phyllobius argentatus*, and sometimes cause extensive damage to the foliage and blossom of ornamentals and nursery stock. In early summer, they show a particular preference for nettles (*Urtica*). Adults are 5–7 mm long and mainly black, but more or less covered with elongate, coppery, golden or greenish-bronze scales; the femora are distinctly toothed.

Phyllobius roboretanus Gredler (**352**)
syn. *P. parvulus* (Olivier)

This relatively small species is common on young trees, especially oak (*Quercus*), upon which it may be found from May onwards; also associated with nettle (*Urtica*). Adults are 3–4 mm long, with untoothed femora and the elytra coated in green scales; they are readily distinguished by the black, unscaled underside of the abdomen which bears just a few green hairs (cf. *Phyllobius viridiaeris*, p.166).

348 Silver-green leaf weevil (*Phyllobius argentatus*).

349 Adults of *Phyllobius maculicornis*.

350 Brown leaf weevil (*Phyllobius oblongus*).

351 Common leaf weevil (*Phyllobius pyri*).

352 Adult of *Phyllobius roboretanus*.

353 Adult of *Phyllobius viridiaeris.*

354 Red palm weevil (*Rhynchophorus ferrugineus*). (C. I. Carter).

Phyllobius viridiaeris (Laicharting) (**353**)
syn. *P. pomonae* (Olivier)
A generally common weevil, sometimes causing minor damage to the foliage of young ornamental trees and shrubs. Adults (3.0–4.5 mm long) are black, covered with green or yellowish-green scales both above and below; the antennae and legs are mainly reddish and the femora untoothed.

Polydrusus formosus (Mayer)
syn. *P. sericeus* (Schaller)
Green leaf weevil
A locally common but minor pest of various trees and shrubs, including alder (*Alnus*), birch (*Betula*), common hazel (*Corylus avellana*), elm (*Ulmus*), poplar (*Populus*), oak (*Quercus*) and willow (*Salix*); also occurs on conifers. Widespread in Europe; introduced into North America.

DESCRIPTION
Adult: 5–8 mm long; body black, but covered with shiny, green scales; legs brown; femora untoothed. Furrow (strobe) on each side of the rostrum, distinct and curved downwards posteriorly below the eye – typical of *Polydrusus* (in *Phyllobius* the strobes are straight and often filled with scales).

LIFE HISTORY
Adults occur during the spring and summer, feeding on the buds, flowers and foliage of trees and shrubs. Eggs are laid in the soil. Larvae feed on plant roots from summer onwards; they hibernate during the winter and pupate in the early spring.

DAMAGE
Although adults are sometimes present in large numbers, damage to ornamental trees and shrubs is usually slight.

Polydrusus pterygomalis Boheman
This local but widely distributed weevil occurs on young trees, including hawthorn (*Crataegus*), flowering cherry (*Prunus*), oak (*Quercus*) and willow (*Salix*), contributing to leaf damage caused by other more numerous leaf weevils. The adults (3–5 mm long) are black-bodied, coated in shiny green scales, with yellow legs and antennae; the femora are untoothed and the head has distinctly swollen temples.

Rhynchophorus ferrugineus Olivier (**354**)
Red palm weevil
This important Asiatic pest of date palm and other Arecaceae (including many ornamental species) has spread to various parts of the Pacific region. It has also extended it range westwards into the Mediterranean Basin, and has now reached parts of France, Greece, Italy, Portugal and Spain. The very large (*c.* 40 mm long), elongate and mainly reddish-brown adults deposit eggs singly in cavities formed within host plants, and the eggs hatch a few days later. Larvae (up to 50 mm long) are yellowish-white, plump and sac-like, with a chestnut-brown head. They feed inside the trunk and branches of host trees, and become fully grown in 1–3 months. Pupation then occurs in dense, fibrous cocoons, and new adults emerge 2–3 weeks later. Adults and larvae cause extensive damage, and attacked trees are eventually killed. As with South American palm borer (*Paysandisia archon*) (see p. 217), steps are underway to prevent further spread of red palm weevil within Europe.

Scolytus scolytus (Fabricius)
Large elm bark beetle
An important pest of elm (*Ulmus*), the beetles acting as vectors of Dutch elm disease (a disease caused by fungi in the genus *Ophiostoma*); dying or at least weakened ash (*Fraxinus excelsior*), poplar (*Populus*), oak (*Quercus*) and various other debilitated trees are also attacked. Widely distributed in Europe.

DESCRIPTION
Adult: 5–6 mm long; mainly black with reddish-brown elytra; front tibiae toothed; a median peg-like tubercle on the hind margin of the third and fourth sternites. **Larva:** white, C-shaped and legless; head dark brown.

LIFE HISTORY
Young adult beetles appear from mid-May to October, congregating at the tops of elm trees to feed on the small twigs. Beetle activity is greatest during warm weather and, if carrying Dutch elm disease spores on their mouthparts or bodies, the insects readily introduce the fungus into healthy tissue. Infections in May or June are most important, the spores usually gaining entry through the feeding channels formed by the beetles in the crotches between the twigs. Later in the summer, the beetles burrow into the trunks and branches of weak or dying hosts to form short (25 mm long) perpendicular breeding chambers immediately beneath the bark. After mating, each female lays about 50 eggs along the length of her chamber. The eggs hatch about ten days later. Larvae then burrow away from the maternal chamber, between the bark and sap wood, to produce a series of feeding galleries. These galleries form an irregular, fan-like pattern, on either side of the main chamber, which is characteristic for the species and clearly visible when the covering of dead bark is peeled away. The larvae are fully grown by the winter or following spring, their rate of development varying considerably from site to site. They then pupate, each in a slight bulb at the end of its burrow. Adults eventually emerge via small, rounded flight holes which they excavate through the bark. Host trees remain suitable as breeding sites for bark beetles and, hence, as potential reservoirs of Dutch elm disease for about two years after death.

DAMAGE
Elm: infestations cause the die-back of branches and hasten the decline of host trees. If Dutch elm disease takes a hold, the leaves of infected branches suddenly turn yellow, the fungus usually spreading and eventually killing the whole tree. **Other hosts:** attacks, which lead to the eventual death of branches or complete trees, are usually initiated only in unhealthy hosts or in those under severe root stress.

Scolytus mali (Bechstein)
Large fruit bark beetle
This widespread species is found occasionally on ornamentals such as *Chaenomeles*, *Cotoneaster*, crab-apple (*Malus*), flowering cherry (*Prunus*), hawthorn (*Crataegus*) and *Sorbus*, but is more commonly associated with rosaceous fruit trees. Breeding colonies usually occur in the trunks and larger branches of already weak or dying host plants. Each colony consists of a long maternal chamber and a series of about 50–60 more or less perpendicular larval galleries. These spread upwards and downwards, immediately beneath the bark, each terminating in a slight bulb in which pupation occurs.

Scolytus multistriatus (Marsham)
Small elm bark beetle
A relatively small species (2.0–3.5 mm long), associated mainly with elm (*Ulmus*) but also capable of damaging other trees such as oak (*Quercus*) and poplar (*Populus*). The adults are structurally similar to those of *Scolytus scolytus*, the other elm-feeding species, but have a pair of lateral teeth on the hind margin of the second, third and fourth abdominal sternites. The biology of both species is similar, and both act as vectors of Dutch elm disease; the larval galleries of *S. multistriatus* are slightly smaller than those of *S. scolytus*, and characterized by their more regular appearance.

Scolytus rugulosus (Müller)
Fruit bark beetle
Although most commonly associated with fruit trees, especially cherry and plum, infestations of this widespread species also occur on flowering cherry (*Prunus*) and some other rosaceous ornamentals. Attacks typically occur in the trunks of small trees and in branches up to 6 cm in diameter. The larval galleries, which lead away from the maternal chamber between the bark and sap wood, commonly overlap, particularly towards their extremities.

Stereonychus fraxini (Degeer) (355–357)
syn. *S. rectangulus* (Herbst)
Ash leaf weevil
A southerly-distributed pest of ash (*Fraxinus excelsior*), including ornamental trees; also associated with certain other plants, including lilac (*Syringa*) and *Phillyrea*. Widespread in south-central and southern Europe, including many parts of France and southern Germany; also found in Asia Minor and North Africa. Not present in the British Isles.

DESCRIPTION
Adult: 3 mm long; greyish brown to reddish brown, with a blackish thoracic disc and an elongate, cylindrical rostrum. **Larva:** up to 4 mm long; greenish yellow, with a black head.

LIFE HISTORY
Adults overwinter in debris on the ground, and in other shelter, emerging in the spring. They then feed on the buds and, later, on the leaves and leaf stalks. Eggs are deposited close to the veins on the underside of the expanded leaves. Following egg hatch, larvae browse on the surface of the leaves to form a series of window-like patches. When fully fed they pupate, each in an oval, yellowish-brown to brownish, parchment-like cocoon formed on the leaf surface. Young adults appear about ten days later; they feed on the foliage before entering hibernation. There is one generation annually.

DAMAGE
Weevils feeding on unopened buds in the early spring delay the appearance of the new growth. In addition, adult and larval damage to the expanded foliage is disfiguring and, if attacks are heavy, will reduce plant vigour.

Strophosoma melanogrammum (Förster) (358)
syn. *Strophosomus coryli* (Fabricius)
Nut leaf weevil
An often common pest of various trees and shrubs, including birch (*Betula*), beech (*Fagus sylvatica*), *Rhododendron* and various conifers, but most abundant on hazel (*Corylus*); sometimes an important pest of young trees and nursery stock, particularly in wooded areas. Widely distributed in Europe.

DESCRIPTION
Adult female: 4–6 mm long; body bulbous and mainly black or brown, coated in greyish-brown scales, and with a distinctive, black longitudinal line down the thorax and between the elytra; head broad, with protruding eyes.

355 Ash leaf weevil (*Stereonychus fraxini*).

356 Ash leaf weevil (*Stereonychus fraxini*) damage to leaf of

357 Pupal cocoon of ash leaf weevil (*Stereonychus fraxini*).

358 Nut leaf weevil (*Strophosoma melanogrammum*).

359 Larva of sallow leaf-mining weevil (*Tachyterges salicis*).

LIFE HISTORY

Adult females of this mainly parthenogenetic species emerge from hibernation in the spring. They browse on the leaves and attack the shoots of various trees and shrubs, feeding throughout the spring and summer months. Eggs are laid in the soil. Following egg hatch, larvae feed on the roots of various herbaceous weeds. They usually pupate in the late summer, and young adults emerge shortly afterwards.

DAMAGE

Leaf damage is of no importance but shoots are often ring-barked so that they wither and die. Infestations are of particular significance on larch (*Larix*) seedlings. Infestations in nurseries are also often of importance on western hemlock (*Tsuga heterophylla*).

360 Sallow leaf-mining weevil (*Tachyterges salicis*) damage to leaf of *Salix*.

Tachyterges salicis (Linnaeus) (359–361)
syn. *Rhynchaenus salicis* (Linnaeus)
Sallow leaf-mining weevil

A generally common species, the bright yellow larvae (up to 3 mm long) forming brown blotches in the leaves of common sallow (*Salix atrocinerea*), grey willow (*S. cinerea*) and other broad-leaved willows, and sometimes causing noticeable distortion of the foliage of nursery plants. Adults (2.0–2.5 mm long) are mainly black, with a prominent white scutellum; the elytra are patterned with white and may also be marked with yellow anteriorly.

361 Sallow leaf-mining weevil (*Tachyterges salicis*).

Order **DIPTERA** (true flies)

Family **TIPULIDAE** (crane flies)

Slow-flying insects with wings, legs and bodies long and narrow; abdomen of females distinctly pointed posteriorly; adults often called 'daddy-longlegs'. The larvae, commonly known as 'leatherjackets', are mainly soil-inhabiting, soft-bodied but tough-skinned; the hind end bears numerous fleshy papillae.

Nephrotoma appendiculata (Pierre) (362)
syn. *Pales maculata* Meigen
Spotted crane fly

A widely distributed crane fly, particularly abundant in gardens, the larvae commonly causing damage to herbaceous plants and seedlings. Adults (12–20 mm long) are yellow and black, with greyish wings; they are most abundant in May, depositing their eggs at random on the soil surface. The greyish-brown larvae (up to 30 mm long) feed on the subterranean parts of host plants, and usually complete their development in the following spring. They are distinguished from those of *Tipula paludosa, T. lateralis* and *T. oleracea* by the short, rounded anal papillae.

362 Female spotted crane fly (*Nephrotoma appendiculata*).

Tipula paludosa Meigen (363–365)
Common crane fly

An important and often common pest of protected ornamentals, hardy nursery stock and herbaceous garden plants; also damaging to lawns and sports turf. Holarctic. Present throughout Europe.

DESCRIPTION
Adult: 17–25 mm long; body grey with a brownish or yellowish-red tinge; thorax with faint longitudinal stripes; wings (13–23 mm long) shorter than body, at least in female; legs fragile, brown and very long; antennae 14-segmented. **Egg:** 1.0 × 0.4 mm; oval, black and shiny. **Larva:** up to 45 mm long; brownish grey with (unless wet) a dull, dusty appearance; body fat and slightly tapered anteriorly, with a soft but tough, leathery skin; head black, small and indistinct (cf. Bibionidae, p. 172); anal segment with a single pair of elongate anal papillae. **Pupa:** 20–30 mm long; brown and elongate, with paired respiratory horns on the head.

LIFE HISTORY
Adults emerge from June onwards but are most abundant in late summer or early autumn. Most eggs are laid just below ground level from mid-August to the end of September, each female depositing about 300 in batches of five or six. Eggs hatch about 14 days later. The larvae then attack plant roots near or at the soil surface. In mild, muggy nights they also appear on the soil surface to graze on the base of plant stems. Feeding usually ceases during the winter to be resumed in the spring. Larvae become fully grown by about June and then pupate. If adults invade and lay eggs in greenhouses in the late summer and autumn, the resulting larvae often complete their development within a few months and new adults then emerge from late March onwards.

DAMAGE
Leatherjackets graze the roots, corms, rhizomes and basal parts of stems, often causing plants to wilt; small plants may be killed, and seedlings completely destroyed. Large holes are also formed in foliage lying close to the soil or compost. On outdoor plants, leatherjacket damage is usually most extensive in the spring (but see notes on the other species) and in wet conditions; plants growing in recently ploughed or broken grassland or pasture are particularly liable to be attacked. If eggs or leatherjackets are accidentally introduced with compost, damage also occurs on plants growing in containers, pots and seed boxes.

363 Female common crane fly (*Tipula paludosa*).

364 Larva of common crane fly (*Tipula paludosa*).

365 Common crane fly (*Tipula paludosa*) damage to leaves of *Primula.*

Tipula lateralis Meigen

Infestations of this generally common crane fly also occur in association with ornamentals, the larvae causing most damage to the rooting systems of containerized plants. Adults (wings 15–20 mm long) are mainly yellowish to greyish; the wing membrane is partly yellow and has a characteristic whitish patch in the pterostigmatic area. The larvae are distinguished from those of related species by the exceptionally long anal papillae.

Tipula oleracea Linnaeus

Large common crane fly

In some regions this widespread and often common species is the most important crane fly associated with ornamental plants. It appears earlier in the year than *Tipula paludosa*, adults flying from May to August. The larvae, which possess two pairs of elongate anal papillae (cf. *T. paludosa*), often complete their development in the autumn and pupate before the winter, thereby causing most damage in the same year as eggs were laid. Adults (15–23 mm long) are distinguished from those of *T. paludosa* by the 13-segmented antennae and by the thinner legs; also, the wings (18–28 mm long) are usually noticeably longer than the body.

Family **BIBIONIDAE** (St Mark's flies)

Mainly black, hairy, robust-bodied, medium-sized to large flies; males often hover sluggishly in the air in conspicuous groups, with their noticeably long hind legs dangling downwards. Larvae are cylindrical, with a prominent head, well-developed mouthparts and fleshy processes on each body segment; they often abound in soil rich in organic matter.

Bibio spp. (**366–367**)

Several species of bibionid fly, including *Bibio johannis*, *B. marci* and *B. nigriventris*, are reported as pests of seedling trees, including conifers. Adults are slow-flying, robust, black-bodied flies that often appear in vast swarms during sunny days in the spring (e.g. *B. marci*) or early summer (e.g. *B. johannis*). Eggs are deposited in groups in the soil, particularly where there are accumulations of organic material, the egg-laying females frequently being attracted by decaying manure. The larvae occur in considerable numbers in well-manured soil, feeding indiscriminately on vegetative matter and often damaging the roots of plants. They occur throughout the summer, those of some species (e.g. *B. johannis*) overwintering and completing their development in the spring. Bibionid larvae are superficially similar in appearance to leatherjackets but have a distinct black head and do not exceed 20 mm in length; the body also bears several fleshy papillae. In the genus *Bibio*, the anal spiracles each have two pores (cf. *Dilophus*).

Dilophus febrilis (Linnaeus)
Fever fly

Larvae of this generally abundant bibionid also attack the roots of young shrubs and trees, occurring in vast numbers both in heavily manured soil and amongst decaying vegetable matter. There are two, perhaps three, generations annually. The larvae are relatively small, up to 10 mm long; they are distinguished from those of *Bibio* spp. by the presence of three pores per anal spiracle.

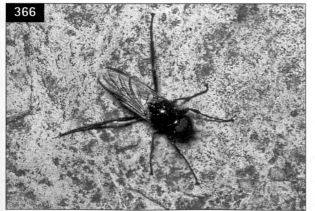

366 St Mark's fly (*Bibio marci*).

367 Larva of *Bibio johannis*.

Family **CHIRONOMIDAE** (non-biting midges)

Delicate, gnat-like flies with poorly developed mouthparts, reduced wing venation and no ocelli; male antennae are plumose.

Bryophaenocladius furcatus (Kieffer)

A widely distributed, parthenogenetic midge, the larvae sometimes damaging the roots of greenhouse plants, including ornamentals such as primrose (*Primula vulgaris*) and *Rhododendron*. Infestations are most likely to develop in damp conditions. Fully grown larvae are 5 mm long, and yellowish green with a dark brown head; the anal segment bears prominent ventral tubercles (cf. family Sciaridae).

Family **SCIARIDAE** (fungus gnats or sciarid flies) (**368**)

Small, delicate, gnat-like flies with a characteristic wing venation, 16-segmented antennae, large compound eyes (which usually meet above the antennae to form an 'eye bridge') and a somewhat humped thorax. The larvae are elongate and translucent-whitish, with a conspicuous black head.

Bradysia spp.

Various species of sciarid fly, including *Bradysia aprica*, *B. difformis* (syn. *B. paupera*), *B. ocellaris* (syn. *B. tritici*) and *B. tilocola* (syn. *B. amoena*), are associated with ornamental plants. The larvae sometimes damage cuttings, seedlings and young pot plants by tunnelling into the roots, corms or main stems. However, such attacks are usually limited to plant material already invaded by moulds or other micro-organisms. The growth of attacked cuttings and seedlings is checked; plants may also collapse and die. Adult sciarids are very active, and are often seen flitting or scurrying about the base of cuttings, seedlings or older plants. The egg-laying females are much attracted by dried-blood fertilizer and steam-sterilized soil, each depositing 100 or more eggs in the growing medium close to host plants. The eggs are small (*c.* 0.2 × 0.1 mm), oval and translucent-whitish; they hatch several days later, the incubation time varying considerably according to temperature. The larvae feed for 3–4 weeks, usually attacking the root hairs; they then construct silken cocoons within which to pupate, adults emerging about a week later. More rarely, eggs are placed directly onto plants; the larvae then feed on the leaf tissue and may eventually pupate on the foliage. Breeding is continuous under suitable conditions.

368 Larva of a sciarid fly (family Sciaridae).

Family **CECIDOMYIIDAE** (gall midges)

Minute to small, delicate flies with moniliform flagellar antennal segments, broad and often hairy wings (the venation much reduced), long, thin legs and prominent genitalia; adults are poor fliers and dispersal is often wind-assisted. The larvae are short, narrowed at both ends, and have a small, inconspicuous head; they usually possess a sternal spatula ('anchor process' or 'breast-bone'), which is often of characteristic shape for the genus or species.

Anisostephus betulinum (Kieffer) (**369–370**)
Birch leaf gall midge

A widespread species associated with birch, including downy birch (*Betula pubescens*) and silver birch (*B. pendula*). The larvae feed in distinctive galls formed on the leaves, the galls developing as yellowish-red to maroon pustules visible from above and from below.

369 Galls of birch leaf gall midge (*Anisostephus betulinum*).

370 Larva of birch leaf gall midge (*Anisostephus betulinum*).

Infestations are sometimes found on the foliage of young trees in nurseries or amenity areas. The larvae (up to 3 mm long) are whitish at first but later become bright yellow; they develop singly, from June to early August, eventually escaping through a small hole in the side of the gall and dropping to the ground to pupate in the soil. There is one generation annually.

Arnoldiola quercus (Binnie) (**371**)
Oak terminal-shoot gall midge

Associated with oak (*Quercus*), infestations causing the death of terminal shoots. The whitish larvae (up to 3 mm long) feed gregariously amongst the unfurling leaves, infested tissue failing to open and eventually turning black. There are usually two generations annually, infestations coinciding with the development of new leaf growth in the spring and summer. Attacks are of little or no importance on mature trees, but they can affect the growth of shoots on young trees and are of some significance on nursery stock.

Contarinia acerplicans (Kieffer) (**372–373**)
Sycamore leaf-roll gall midge

This widely distributed European midge infests sycamore (*Acer pseudoplatanus*), the larvae developing gregariously within reddish marginal leaf-roll galls. Galls might also develop on the leaves as elongate folds; these are typically glabrous and shiny above and hairy below, affected leaves often being considerably distorted. The galls, which eventually turn black, first appear on the young leaves in May. Fully fed larvae, which are whitish and about 2 mm long, eventually vacate the galls and drop to the ground to pupate in the soil. There are two generations annually. Although infestations sometimes occur on young ornamentals and specimen trees, they do not cause significant damage.

371 Oak terminal-shoot gall midge (*Arnoldiola quercus*) damage to young shoot of *Quercus.*

372 Young galls of sycamore leaf-roll gall midge (*Contarinia acerplicans*) on *Acer pseudoplatanus.*

373 Sycamore leaf-roll gall midge (*Contarinia acerplicans*) damage to leaf of *Acer pseudoplatanus.*

374 Mature gall of lime leaf-stalk gall midge (*Contarinia tiliarum*).

375 Section through gall of lime leaf-stalk gall midge (*Contarinia tiliarum*).

Contarinia petioli (Kieffer)

Poplar gall midge

Widely distributed in association with aspen (*Populus tremula*) and, less frequently, white poplar (*P. alba*), the orange-coloured larvae (each 3–4 mm long) developing in galls on the young twigs and leaf stalks. Galls on the leaf stalks are globular, and several are often fused together; those on the twigs develop singly, each becoming a localized lateral swelling. At maturity, each gall develops a lateral aperture through which the causal organism escapes to pupate in the soil. Although the galls attract attention, particularly when present on young trees, they are not considered harmful.

Contarinia quinquenotata (Löw, F.)

A pest of day-lily (*Hemerocallis*); widely distributed in mainland Europe, and discovered causing extensive damage in southern England in the late 1980s. The whitish larvae feed gregariously within the flower buds during the early summer. When fully grown they enter the soil and overwinter, there being just one generation annually. Infested buds become greatly swollen and fail to open, those attacked at a very early stage in their development sometimes withering and turning brown. Attacks are particularly common on yellow-flowered cultivars, infestations reaching their peak by late June; buds produced after mid-summer are unaffected.

Contarinia tiliarum (Kieffer) (**374–375**)

Lime leaf-stalk gall midge

This moderately common midge forms globular galls (*c.* 10 mm across) on the leaf stalks of lime (*Tilia*), usually just before the leaf blade; galling also occurs on the young stems. The galls are at first pale but later turn red and, finally, black. They contain several (often up to 20) yellowish-orange larvae (up to 3 mm long), each individual occupying a separate chamber. The larvae commence their development in May or June and are usually fully fed by the end of July; they then enter the soil where they eventually pupate. Adult midges appear in the spring, usually in May, there being just one generation annually. Leaves with galled stalks fail to develop properly, often becoming distorted and noticeably hairy; the underside of the midrib might also be swollen. Heavy attacks of this pest can prevent trees flowering. However, the galls are usually most common on sucker growth around the base of older trees, and this reduces the importance of infestations.

Dasineura crataegi (Winnertz) (**376–378**)
Hawthorn button-top midge
A generally common pest of hawthorn (*Crataegus*), infestations often occurring in abundance on hedges and in nurseries. Eurasiatic. Widely distributed in Europe.

DESCRIPTION
Adult: 2.0–2.5 mm long; brownish with darker crossbands on the abdomen. **Larva:** up to 3 mm long; reddish but whitish when young.

LIFE HISTORY
Adult midges appear in April or May, females depositing eggs in the tips of young shoots of hawthorn. Attacked shoots fail to elongate and, instead, develop into compact, rosette-like galls; these galls eventually turn black, remaining on the host plants long after the causal larvae have departed. The galls are often noted on nursery stock, and are most numerous on regularly trimmed hedges which provide an abundance of suitable egg-laying sites. Several larvae develop within each gall, individuals becoming fully fed in 2–3 weeks. They then drop to the ground and pupate inside silken cocoons, adults emerging shortly afterwards. There are further generations during the summer, autumn larvae overwintering in their cocoons and pupating in the spring.

DAMAGE
The galls disfigure host plants; also, by interfering with normal shoot development, they affect the shape and, hence, quality of nursery stock. Leaves immediately below the main galls often unfurl but remain disfigured by numerous red pimples. Attacks on established plants are rarely troublesome, even when infestations are heavy, but damage to nursery stock might be unacceptable.

376 Larvae of hawthorn button-top midge (*Dasineura crataegi*).

377 Gall of hawthorn button-top midge (*Dasineura crataegi*) on *Crataegus*.

378 Hawthorn button-top midge (*Dasineura crataegi*) damage to leaf of *Crataegus*.

379 Violet leaf midge (*Dasineura affinis*) galls.

Dasineura abietiperda (Henschel)

This widely distributed midge attacks the young growth of spruce (*Picea*), including Norway spruce (*P. abies*). Adults appear in April and May, the females depositing eggs near the growing points of the new shoots. The reddish larvae (up to 3 mm long) feed singly, each within a cavity formed in the bark or new wood; they will also burrow into the buds at the shoot tips. The larvae usually complete their development in the autumn. They then overwinter, pupating in the spring shortly before the emergence of the adults. Several galls usually occur close together, affected shoots remaining shorter than normal, bearing fewer needles and becoming slightly bent. Heavy infestations cause considerable damage to young trees; losses are of particular importance on spruces grown as Christmas trees.

Dasineura affinis (Kieffer) (**379**)

Violet leaf midge

A widely distributed, important and often persistent pest of wild and cultivated violet (*Viola*). Infested plants become stunted and malformed, with flower production seriously affected if not entirely prevented; in severe cases, plants might eventually die. Adult midges occur from May onwards, depositing eggs in the rolled margins of young leaves and, occasionally, in the developing flowers. Infested tissue fails to unroll and often becomes swollen into conspicuous galls, within which several whitish to whitish-orange larvae (each up to 2 mm long) develop. Larvae are fully fed in about six weeks; they then pupate in silken cocoons, still within the gall, adult midges emerging about 10–12 days later. There are usually four overlapping generations annually on outdoor plants, larvae of the final brood overwintering and completing their development in the spring. In greenhouses, there are commonly additional generations and, if conditions are suitable, breeding can continue throughout the year.

Dasineura alpestris (Kieffer)

syn. *D. arabis* (Barnes)

Arabis midge

A potentially important pest of certain kinds of *Arabis*, including rock cress (*A. alpina*), damaging infestations sometimes occurring in gardens and nurseries. Adults appear from May onwards, the females depositing reddish eggs between the leaf tissue in the developing buds. These eggs hatch within a few days, and pinkish or reddish larvae then feed in compact but open galls which develop as the leaf bases become enlarged and swollen. There are commonly 20–30 larvae in each gall, sometimes considerably more. Fully grown larvae pupate in white cocoons spun either within the gall or in the soil. Midges of the next generation emerge shortly afterwards. There are three or four generations each year, larvae of the final generation overwintering in their cocoons and pupating in the spring. Heavily infested plants produce few flowers, their central crowns are destroyed and any new growth is limited to weak, lateral shoots.

380 Galls of ash midrib pouch-gall midge (*Dasineura fraxini*) on *Fraxinus*.

381 Ash midrib pouch-gall midge (*Dasineura fraxini*) damage to leaves of *Fraxinus*.

382 Young galls of honey locust gall midge (Dasineura gleditchiae) on leaves of *Gleditsia*.

383 Galls of honey locust gall midge (*Dasineura gleditchiae*) on leaves of *Gleditsia*.

Dasineura fraxini (Bremi) (380–381)
syn. *D. fraxini* (Kieffer)
Ash midrib pouch-gall midge

This widely distributed and often common species causes the underside of the leaves of ash (*Fraxinus excelsior*) to swell into conspicuous galls (25–30 mm long) along either side of the midrib. Development of galls commences in May or June. They usually contain from four to eight orange-coloured larvae, each located in its own cell. Maturity is reached in September. The upper surface of the gall then splits open longitudinally, the fully fed (2–3 mm long) larvae dropping to the ground and overwintering in the soil. Pupation occurs in the spring, adult midges emerging shortly afterwards. Infestations occur on both young and mature trees; although affected leaves might eventually turn brown and die prematurely, damage does not affect shoot growth.

Dasineura gleditchiae (Osten-Sacken) (382–383)
Honey locust gall midge

Of North American origin but now well established on honey locust (*Gleditsia triacanthos*) in various parts of Europe, including (from the 1980s onwards) southern England. Adults emerge from late May onwards. Larvae feed gregariously on the leaves, causing them to develop into yellowish-green to purplish-red, pod-like galls that eventually turn brown. Infested shoots are also distorted, the galls often causing severe disfigurement. Pupation occurs within the galls, adults appearing shortly afterwards. There are several overlapping generations throughout the summer, development from egg to adult being completed in 3–4 weeks. Fully grown larvae are whitish and 3.0–3.5 mm long; those reared in the autumn overwinter in subterranean cocoons, pupating in the spring. The pest can also persist in the soil for more than a year before producing adults.

384 Galls of rose leaf midge (*Dasineura rosarum*) on *Rosa*.

385 Larvae of rose leaf midge (*Dasineura rosarum*).

386 Gall of *Dasineura thomasiana* on young leaf of *Tilia*.

Dasineura rosarum (Hardy) (**384–385**)
syn. *Wachtliella rosarum* (Hardy)
Rose leaf midge
A widely distributed midge, attacking both wild and cultivated rose (*Rosa*). Adult midges appear in the late spring, depositing eggs in unfurled leaves. Following egg hatch, larvae develop within reddish-tinged, pod-like galls, each formed from an expanded but still furled leaflet. Up to 50 orange-coloured larvae occur in each infested leaflet, causing the galled tissue to change from green through reddish to dark brown. Individuals are fully grown by the late summer; they then drop to the ground where they overwinter and eventually pupate; there is just one generation annually. Galled leaves often occur in groups but any effect on plant growth is slight.

Dasineura thomasiniana (Kieffer) (**386**)
Minor infestations of this midge are sometimes noticed on lime (*Tilia*), the larvae feeding within distorted leaves at the tips of young shoots. Affected leaves remain folded, the veins becoming thickened and deformed; buds are also affected. Each gall contains several (2–3 mm long) whitish to orange or reddish larvae. There are two or more generations annually.

387 Galls of lime leaf-roll gall midge (*Dasineura tiliamvolvens*) on *Tilia.*

388 Mature galls of lime leaf-roll gall midge (*Dasineura tiliamvolvens*) on *Tilia.*

Dasineura tiliamvolvens (Rübsaamen) (**387–388**)
Lime leaf-roll gall midge

The striking galls of this midge occur on lime (*Tilia*), developing on the young leaves from April or May onwards. Each gall appears as an upward rolling of the leaf edge, affected tissue turning dark red; galling might extend around most if not all of the leaf margin, producing considerable distortion; the leaf blade may also become mottled with red (cf. galls formed by the mite *Phytoptus tetratrichus*, p. 405). The galls, which eventually turn black, enclose several whitish to orange larvae (each up to 3 mm long). These larvae complete their development in June; they then enter the soil where they eventually overwinter. Adult midges appear in late April and May, there being just one generation annually.

389 Young galls of lime leaf gall midge (*Didymomyia tiliacea*) on *Tilia.*

Didymomyia tiliacea (Bremi) (**389–390**)
syn. *D. reaumuriana* (Löw, F.)
Lime leaf gall midge

A local species, associated with lime (*Tilia*), forming conspicuous, greenish-white, pustule-like galls on the leaves, visible from above and from below. The galls, which occur from May onwards, are about 4 mm in diameter. Larvae, which are up to 3 mm long and white to pale yellow, develop singly within the galls and become fully grown by the late summer. At maturity, a characteristic capsule containing the fully fed larva is ejected through the underside of the gall. Individuals overwinter on the ground within these protective structures, adult midges appearing in the spring from late April onwards. There are commonly up to 40 galls on an infested leaf but, apart from possible distortion of the leaf blade, growth is not affected.

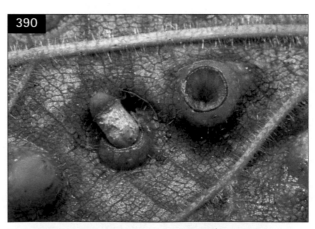

390 Pupal capsule of lime leaf gall midge (*Didymomyia tiliacea*).

391 Galls of *Harmandia cavernosa* on leaf of *Populus tremula.*

392 Galls of aspen leaf gall midge (*Harmandia tremulae*).

Harmandia cavernosa (Rübsaamen) (**391**)

Galls formed by this species commonly occur in pairs at the base of the leaves of aspen (*Populus tremula*) and white poplar (*P. alba*). The galls develop as light green swellings (up to 5 mm across) on the underside of the leaf blade, with a distinct opening through the upper surface. Each gall contains a single yellowish-red larva, which eventually escapes to pupate in the soil. Mature galls are distinctly blackened; however, although very noticeable, they cause little or no distortion of leaves.

Harmandia tremulae (Winnertz) (**392**)
Aspen leaf gall midge

Leaves of aspen (*Populus tremula*) are often disfigured by spherical, red to purplish-red galls induced by this pest; they arise on the upper surface, each from a major vein, but do not cause distortion. The galls (up to 4 mm in diameter) develop from June onwards, reaching maturity in August or September. There are usually several galls per infested leaf, each containing a single larva. Fully fed larvae vacate the galls in the autumn and eventually pupate in the soil, adults appearing in the spring. There is one generation annually.

Hartigiola annulipes (Hartig) (**393**)
syn. *Hormomyia piligera* Loew, H.; *Pegobia tornatella* (Bremi)
Beech hairy-pouch-gall midge

Beech (*Fagus sylvatica*) is commonly attacked by this widespread gall midge. The adults, which occur from mid-May to early June, insert eggs through the upper epidermis of expanded leaves, generally close to the midrib. In July, inner tissue of the leaf bursts through the upper surface at each oviposition point to form yellowish, hairy, cylindrical galls (*c.* 5 × 3 mm), each

393 Gall of beech hairy-pouch-gall midge (*Hartigiola annulipes*).

enclosing a small (up to 3 mm long) white larva. The galls eventually turn reddish brown and, in the late summer, when mature, they break away from the leaf and drop to the ground. Larvae overwinter within the galls and pupate in the spring. An infested leaf might bear several galls, sometimes as many as 50, but plant growth is not affected.

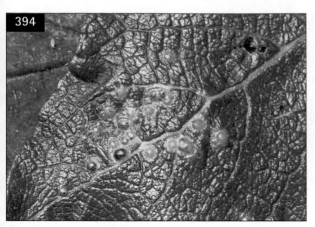

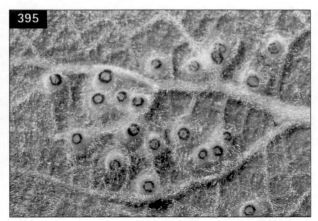

394 Galls of *Iteomyia capreae* on leaf of *Salix*, viewed from above.

395 Galls of *Iteomyia capreae* on leaf of *Salix*, viewed from below.

Helicomyia saliciperda (Dufour)
Willow shot-hole midge

A widely distributed pest of willow (*Salix*), especially crack willow (*S. fragilis*), weeping willow (*S. vitellina* var. *pendula*) and white willow (*S. alba*); most numerous in central and southern Europe, where attacks often occur on ornamentals. The larvae feed beneath the bark, attacking down to the sap wood. Infested parts of the tree become gnarled, the bark splitting open; also, affected branches and stems are sometimes ring-barked so that tissues above the galled area dies. In strong winds or following heavy snowfall, damaged young branches often break off. Adult midges occur in May and June, sometimes earlier or later, eggs being laid in long chains on the bark of host plants. Larvae (up to 3 mm long) are orange-yellow to reddish. They feed from June or July onwards, and eventually pupate in the feeding galleries. There is one generation annually.

Iteomyia capreae (Winnertz) (**394–395**)

Larvae of this widely distributed and often common species occur in galls formed on the leaves of eared willow (*Salix aurita*), pussy willow (*S. caprea*) and certain other broad-leaved willows. The galls occur from May or June onwards. From above, they appear as polished, yellowish-green, red-flushed swellings, each 2–3 mm in diameter; below, they are light green and volcano-like, each with a purplish-red rim surrounding a prominent central opening. Each gall contains a tiny (up to 2 mm long) translucent-whitish, and finally red, larva which feeds throughout the summer, eventually dropping to the ground to pupate. There is just one generation annually. The galls are sometimes invaded by mites, including the eriophyid *Aculops tetanothrix* (p. 418).

Janetiella lemei (Kieffer) (**396–397**)

A widely distributed but local species, associated with elm (*Ulmus*). The yellow-coloured larvae develop during the summer in pale, flask-like galls formed as swellings of the veins on the underside of leaves. The galls also occur on the leaf stalks and young shoot. When fully grown, the larvae drop to the ground and eventually pupate in the soil, adult midges appearing in the spring from April or May onwards.

Macrodiplosis dryobia (Löw, F.) (**398–400**)
An oak fold-gall midge

Adults of this widely distributed and often common midge occur in the late spring, depositing eggs close to the tips of lateral veins on the leaves of oak (*Quercus*). Subsequently, the leaf tissue folds over to meet the underside of the leaf blade; up to four yellowish-white, translucent larvae (each up to 3 mm long) feed within this protective envelope. The flap covering the larvae soon becomes mottled and, eventually, the upper surface of the leaf immediately above the gall also becomes discoloured. Fully grown larvae escape from the galls from July onwards, individuals eventually pupating in the soil. There is just one generation annually. Although often abundant on young trees, the galls do not affect plant growth.

Macrodiplosis volvens Kieffer (**401**)
An oak fold-gall midge

This species is essentially similar to *Macrodiplosis dryobia* but larval development takes place in somewhat broader and shallower folds which arise between the lateral veins and overlap the upper surface of the leaves. The larvae (up to 3 mm long) are orange-red, and up to five are present per gall.

396 Galls of *Janetiella lemei* on leaf of *Ulmus*.

397 Galls of *Janetiella lemei* on shoot of *Ulmus*.

398 Galls of oak fold-gall midge (*Macrodiplosis dryobia*).

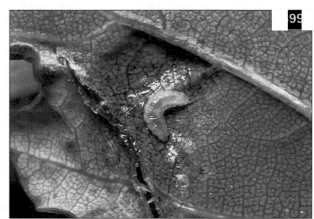

399 Larva of oak fold-gall midge (*Macrodiplosis dryobia*).

400 Oak fold-gall midge (*Macrodiplosis dryobia*) damage to leaf of *Quercus*.

401 Galls of *Macrodiplosis volvens* on leaf of *Quercus*.

Mikiola fagi (Hartig) (402–403)
Beech smooth-pouch-gall midge
This midge also occurs on beech (*Fagus sylvatica*), the larvae developing in bulbous leaf galls, each 4–10 mm long. The galls, which have a smooth, often plum-like surface, occur on the leaves from May to late September or October and, if numerous, can cause the death of young plants. The larvae are whitish and plump (*c.* 4 mm long when fully grown), and feed within the galls throughout the summer. They then overwinter within the fallen galls, eventually pupating and producing adults in March or April. Although present in England, this pest appears more common and of greater significance in mainland Europe.

Monarthropalpus buxi (Laboulbène) (404–405)
Box leaf mining midge
An often important pest of common box (*Buxus sempervirens*), particularly in nurseries. Widely distributed in Europe; also present in North America.

DESCRIPTION
Adult: 2–3 mm long; yellowish-orange; female with abdomen terminating in a long, curved spine. **Larva:** up to 2.5 mm long; white, later orange; flattened and narrowing posteriorly.

402 Gall of beech smooth-pouch-gall midge (*Mikiola fagi*).

403 Larva of beech smooth-pouch-gall midge (*Mikiola fagi*).

404 Galls of box leaf mining midge (*Monarthropalpus buxi*), viewed from below.

405 Galls of box leaf mining midge (*Monarthropalpus buxi*), viewed from above.

LIFE HISTORY

Adults are active in the spring, depositing eggs in the young leaves by inserting the ovipositor from below. Eggs hatch in about three weeks; the larvae then mine within the leaf tissue to form distinctive yellowish or brownish blister-like mines on the underside of the leaves. Where several larvae are present in the same leaf, the blisters often coalesce and may eventually occupy the complete leaf blade. Larvae overwinter within the mines, completing their development in the spring. They then pupate, midges appearing 2–3 weeks later.

DAMAGE

Larval mines are visible from above as discoloured swellings, disfiguring the foliage and spoiling the appearance of both specimen ornamentals and hedges. Heavy infestations weaken host plants, infested leaves dropping prematurely. Persistent attacks gradually reduce the overall density of the foliage.

Oblodiplosis robiniae (Haldeman) (**406**)

False acacia leaf midge

This invasive North American species first appeared in Europe in 2003, in northern Italy. It has since spread rapidly and is now firmly established on false acacia (*Robinia pseudoacacia*) in many parts of mainland Europe. The pest has also been found in Japan and Korea. Larvae (up to 4 mm long) are at first whitish and translucent, but later turn yellow. They feed in tightly rolled leaf edges, on average each gall containing about two individuals. Infestations affect the appearance of ornamental plants, and also cause premature leaf fall. There are two or more generations annually.

Parallelodiplosis cattleyae (Molliard)

Cattleya gall midge

This tropical midge sometimes occurs in European greenhouses, initiating elongate or pea-like galls on the aerial roots of orchids, including various species of *Cattleya* and *Laelia*. The pest is usually of little significance but the growth of heavily infested plants is retarded. There are several generations annually, with numerous orange-coloured larvae developing within individual cells in each gall. Fully grown larvae pupate in their cells. Pupae eventually protrude from the surface of the gall following emergence of the adults.

Resseliella oculiperda (Rübsaamen)

larva = red bud borer

A local and usually sporadic pest of budded stock and grafts, attacking rose (*Rosa*) and rosaceous fruit trees such as apple. Widely distributed in Europe.

DESCRIPTION

Adult: 1.4–2.0 mm long; dark reddish brown. **Larva:** up to 3.5 mm long; salmon-pink to red, with a bilobed spatula.

LIFE HISTORY

Adults occur in three generations, in late May to late June, July to early August and late August to early September, depositing eggs in graft slits or cuts in the bark of newly budded stock. Eggs hatch about a week later. Larvae then feed in small groups on sap in the cambium between the scion and stock. They are fully fed in 2–3 weeks, each dropping to the ground to pupate in a small cocoon a few centimetres below the surface. Larvae of the autumn generation overwinter in their cocoons and pupate in the spring.

DAMAGE

Infestations prevent grafts or buds from taking, so that the scions or buds wither and die. Most damage is caused by larvae which feed from August onwards, when losses on unprotected nursery stock may be considerable.

406

406 Galls of false acacia leaf midge (*Oblodiplosis robiniae*).

407 Larva of willow rosette-gall midge (*Rabdophaga rosaria*).

408 Gall of willow rosette-gall midge (*Rabdophaga rosaria*).

409 Section through gall of willow rosette-gall midge (*Rabdophaga rosaria*).

410 Old gall of willow rosette-gall midge (*Rabdophaga rosaria*).

Rabdophaga rosaria (Loew, H.) (**407–410**)
syn. *Rhabdophaga cinerearum* (Hardy)
Willow rosette-gall midge
A generally common pest of willow (*Salix*), especially crack willow (*S. fragilis*), pussy willow (*S. caprea*) and white willow (*S. alba*). Present throughout Europe.

DESCRIPTION
Adult female: 2.2 mm long; head black; thorax blackish to reddish; abdomen red; wings clear, 2.8 mm long. **Larva:** up to 4 mm long; whitish to pinkish-orange.

LIFE HISTORY
Adults occur in late April or early May, depositing eggs in association with vegetative buds. Infested buds develop during the summer into leafy, cabbage-like rosette galls, each containing a single larva (cf. *Rabdophaga heterobia*). The galls eventually turn brown and remain on the shoots throughout the winter. They are particularly obvious in the early spring, just before the emerging new green growth forces them to drop off; at this stage they are sometimes referred to as 'camellia galls'. Larvae overwinter within the galls. They then pupate in the spring, shortly before adults emerge.

DAMAGE
The galls disfigure host plants and cause the death of leading shoots, but tend to be of significance only on small bushes.

411 Gall of willow button-top midge (*Rabdophaga heterobia*).

412 Section through gall of willow button-top midge (*Rabdophaga heterobia*).

Rabdophaga clausilia (Bremi)
syn. *Rhabdophaga inchbaldiana* (Mik)
Willow leaf-folding midge
This widely distributed midge causes a discontinuous marginal rolling of willow (*Salix*) leaves. Each roll-gall, unlike those of *Rabdophaga marginemtorquens* (p. 188), consists of a short length of folded tissue which encloses a single yellowish-orange larva. The galls, which mature in the autumn, occur mainly on white willow (*S. alba*). They are noted occasionally on cultivated plants but damage is not important.

Rabdophaga heterobia (Löw, F.) (**411–412**)
syn. *Rhabdophaga saligna* (Hardy)
Willow button-top midge
A widespread species, breeding in galled terminal buds, lateral buds and male catkins of certain kinds of willow, especially almond willow (*Salix triandra*). Adults occur in two main generations, from late April to September, depositing eggs on or close to the growing points. Infested terminal buds develop into loose rosette galls, each enclosing up to 40 tiny (*c*. 2 mm long), whitish to orange-red larvae. When fully fed, the larvae pupate in white cocoons formed within the gall. Adults emerge shortly afterwards. Larvae of the second generation induce further terminal and lateral galls. The larvae eventually overwinter. They then pupate in the spring, usually still within the protective remnants of the dead rosette leaves which remain attached to host plants long after leaf fall.

413 Galls of osier leaf-folding midge (*Rabdophaga marginemtorquens*).

414 Larva of yew gall midge (*Taxomyia taxi*).

415 Gall of yew gall midge (*Taxomyia taxi*).

416 Old gall of yew gall midge (*Taxomyia taxi*).

Rabdophaga marginemtorquens (Bremi) (**413**)
syn. *Rhabdophaga marginemtorquens* (Winnertz)
Osier leaf-folding midge

Leaves of osier (*Salix viminalis*) are frequently infested by larvae of this common and widespread species. Adults occur in the spring, and deposit eggs along the margins of young leaves. Attacked leaves develop long, continuous marginal folds which become yellowish green and tinged with red; the galls enclose several orange larvae (up to 2 mm long) (cf. *Rabdophaga clausilia*), the position of each being disclosed by a local swelling. Fully fed larvae vacate the galls in August or September; they then overwinter in the soil and pupate in the spring. The galls do not affect plant growth, although severe infestations in nursery beds give an impression of poor plant quality. Similar galls are formed on osier by the mite *Aculus truncatus* (p. 418).

Taxomyia taxi (Inchbald) (**414–416**)
Yew gall midge

A common pest of yew (*Taxus*); also associated with cowtail pine (*Cephalotaxus harringtonia*). Present throughout Europe.

DESCRIPTION
Adult: 3–4 mm long; yellowish orange. **Larva:** up to 3 mm long; yellowish orange, lacking a sternal spatula.

LIFE HISTORY
This species usually has a two-year lifecycle. Adult midges occur in May and June. Eggs are deposited singly in the terminal buds of yew trees, and hatch about one to three weeks later. The young larvae then burrow into the immature buds to begin feeding. Each infested bud develops into a distinctive gall, formed from a tight

417 Galls of hornbeam leaf gall midge (*Zygobia carpini*).

418 Drone fly (*Eristalis tenax*).

cluster of terminal leaves. The galled buds, within which larval development takes place, remain relatively small in the first year. However, in the next year they become larger and less compact, and may then exceed 15–20 mm in diameter; these are commonly described as 'artichoke' galls. Fully fed larvae eventually pupate within their galls, and adults emerge in the following spring. Vacated galls eventually turn brown, and often remain *in situ* long after the midges have flown. Although the vast majority of larvae develop slowly over two years, some individuals complete their development within just one year. These, however, do not initiate artichoke galls.

DAMAGE
The galls abort shoot growth and are disfiguring, host plants being stunted by persistently heavy infestations. Attacks are most damaging in nurseries and on plants already infested by yew gall mite (*Cecidophyopsis psilapsis*) (p. 420).

Zygobia carpini (Löw, F.) (**417**)
Hornbeam leaf gall midge
Galls of this widely distributed midge develop on the leaves of hornbeam (*Carpinus betulus*), either occurring singly or as a series of swellings along the length of the midrib. Infestations are sometimes noticed on cultivated plants but, although disfiguring, they have no apparent effect on plant growth. The whitish to yellowish-white larvae (up to 3 mm long), one per gall, develop throughout the summer; they vacate the galls in September and eventually pupate in the soil. Infestations also occur on European hop-hornbeam (*Ostrya carpinifolia*).

Family **SYRPHIDAE** (drone flies or hover flies)

Small to large, often brightly patterned flies, some of which hover in the air and emit a bee-like hum; body sometimes very hairy. Larvae of many species are important predators of aphids, whereas others feed on living plant tissue, including bulbs and corms, or are associated with rotting wood or boggy (aquatic or semi-aquatic) habitats.

Eristalis tenax (Linnaeus) (**418**)
Drone fly
Adults of this widespread and generally common syrphid are often attracted in large numbers to flowers, especially Asteraceae (e.g. *Aster* and *Chrysanthemum*) and Apiaceae, where they feed avidly on nectar. They also bask in sunshine, darting into the air and hovering nearby if disturbed. The flies are normally harmless but sometimes enter greenhouses, particularly in the autumn, and might then contaminate the petals of chrysanthemums and other flowers with droplets of excrement. The insects breed in wet, decaying organic matter. Drone fly larvae (commonly known as 'rat-tailed maggots') possess a very long, extensible tube which allows them to breathe whilst submerged well below the surface of mud or stagnant water. The bee-like adults (12–15 mm long) are mainly brownish black, with the thorax clothed in yellowish to brownish-yellow hairs and the abdomen variably marked with yellowish or yellowish brown. Species of *Eristalis*, including *E. tenax*, are distinguished from other closely related syrphids by the dark bands of hairs across the eyes and by the simple, unbranched arista.

Eumerus tuberculatus Rondani (**419–422**)
A small narcissus fly

A generally common pest of *Narcissus* bulbs but usually confining its attacks to bulbs that are mechanically damaged or diseased – e.g. infected with basal rot (*Fusarium oxysporum* f. sp. *narcissi*). Holarctic. Present throughout Europe.

DESCRIPTION
Adult: 5–6 mm long; robust-bodied and mainly black; head and thorax with a golden sheen, the thorax also with a pair of whitish longitudinal lines; abdomen with three pairs of white, crescent-shaped marks, and with golden hairs along the sides and towards the tip; pterostigma of fore wings dark brown or blackish; hind femora (particularly of male) with a basal projection. **Egg:** 0.7 mm long; white, elongate-oval and tapered at one end. **Larva:** up to 9 mm long; whitish grey to pale yellowish white, with an elongate reddish-brown spiracular cone at the hind end, that protrudes beyond two long and two short, fleshy papillae. **Puparium:** 6–7 mm long; yellowish white.

LIFE HISTORY
Adults appear in the early spring, and deposit batches of eggs in association with unhealthy narcissus bulbs. Larvae emerge a few days later; they then enter the bulbs to feed on the inner tissue in groups of five or six but often in much larger numbers. Pupation takes place in or around the neck of the host bulb in June or early July. Adults emerge about two weeks later. Larvae of a second generation become fully fed by the autumn. They vacate the remains of the bulb, overwinter in the soil and then pupate.

DAMAGE
Narcissus bulbs are often totally destroyed by the larvae. However, attacks rarely occur on healthy stock and are, therefore, of secondary importance.

419 Small narcissus fly (*Eumerus tuberculatus*).

420 Larvae of small narcissus fly (*Eumerus tuberculatus*).

421 Puparium of small narcissus fly (*Eumerus tuberculatus*).

422 Small narcissus fly (*Eumerus tuberculatus*) damage to bulb of *Narcissus*.

Eumerus strigatus (Fallén)
A small narcissus fly

This fly is essentially similar to *Eumerus tuberculatus* but has a wider host range, including *Colchicum*, hyacinth (*Hyacinthus orientalis*), *Iris*, *Lilium* and *Narcissus*; the larvae also invade vegetable crops such as carrot, onion, parsnip and potato, increasing damage to previously injured tissue. Adults are distinguished from those of *E. tuberculatus* by the yellow or light brown pterostigma on the fore wings and by the absence of a basal projection on the hind femora.

Merodon equestris (Fabricius) (**423–424**)
Large narcissus fly

A locally important pest of *Narcissus*; various other cultivated bulbs, including belladonna lily (*Amaryllis belladonna*), *Galtonia*, hyacinth (*Hyacinthus orientalis*), *Iris*, snowflake (*Leucojum*), snowdrop (*Galanthus nivalis*), squill (*Scilla*) and *Vallota* are also attacked. Widely distributed in mainland Europe; in the

423 Large narcissus fly (*Merodon equestris*).

424 Larva of large narcissus fly (*Merodon equestris*).

British Isles most numerous in south-western England. Also present in parts of Australasia, Japan and North America.

DESCRIPTION

Adult: 12–14 mm long; stout-bodied and very hairy, the body hairs ranging in colour from black to greyish, yellowish, orange or red and often forming a bumblebee-like pattern; each hind leg bears a characteristic tooth-like projection. **Egg:** 1.6 mm long; elongate-oval and pearly white. **Larva:** up to 18 mm long; dirty yellowish-white and plump, with a short, dark brown respiratory cone at the hind end, bordered on either side by small inconspicuous tubercles. **Puparium:** 10–12 mm long; brownish.

LIFE HISTORY

Adults occur mainly from May to July but individuals may appear as early as February in forcing houses. In cool, inclement weather, they hide in hedgerows and in other suitable situations, but in favourable conditions they visit various flowers in search of nectar. They may also be found sunning themselves on nearby banks, posts, tree trunks and leaves. In flight they emit a characteristic, bee-like buzz, and often occur in large numbers in narcissus fields during warm, sunny afternoons. Eggs are deposited singly, either on foliage in the neck region of host bulbs, or directly on the bulb, or in the soil, the female usually crawling into the hole left by the withering foliage and flower stem. Most eggs are laid in June and early July. Eggs hatch in about two weeks. Each larva then crawls to the base plate of a bulb before burrowing in to begin feeding. The tissue immediately around the larval entry hole soon turns rusty red. This becomes visible if the dead tissue around the base plate is scraped away. At first, the larva (typically one per infested bulb) quickly hollows out a large cavity which becomes filled with blackish frass and rotting tissue. Larvae normally complete their development in the original bulb but will move from bulb to bulb if necessary; most individuals become fully grown by the winter. Pupation occurs in the following spring, either in the neck region of the bulb or in the soil. Adults emerge 5–6 weeks later.

DAMAGE

Infested bulbs become soft, particularly in the neck region, and much of the inner tissue is destroyed. Small bulbs are often completely destroyed and, if planted, fail to grow. Larger ones may appear sound, but will produce weak, distorted, yellowish foliage or merely a ring of small, grass-like leaves.

Family **TEPHRITIDAE**

Small flies, often with large, colourful eyes, patterned wings and, in females, a rigid sheath around the ovipositor. Larvae are maggot-like; they feed mainly within plant tissue, some living as leaf miners.

Euleia heraclei (Linnaeus) (**425–426**)
Celery fly

Although mainly regarded as a pest of celery, this generally common species also attacks other cultivated umbelliferous hosts (Apiaceae), including giant hogweed (*Heracleum mantegazzianum*). Adults (wings 5 mm long) are mainly brown to black, with a yellow scutellum, yellow legs and smoky-patterned wings.

They appear in late April and early May, depositing eggs on the underside of leaves. The eggs hatch in about a week, larvae then mine within the leaf blades to form expansive brownish blotches in which black deposits of frass accumulate. The larvae (up to 7 mm long) are greenish white and translucent, with a pointed head end and prominent black mouth-hooks. They often feed gregariously within the same blotch, and development takes about three weeks. Fully fed larvae pupate within the mine or in the soil, each in a yellowish (5 mm long) puparium. Adults emerge about four weeks later, usually in late July and August. Larvae of the second generation complete their development in the autumn, individuals overwintering within puparia; in favourable conditions a third generation is possible.

425 Mine of celery fly (*Euleia heraclei*) in leaf of *Heracleum*.

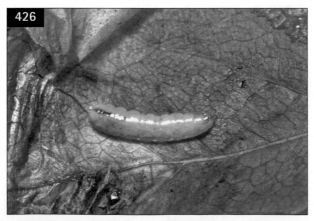

426 Larva of celery fly (*Euleia heraclei*).

Trypeta zoe (Meigen) (427)

larva = chrysanthemum blotch miner

A minor pest of *Chrysanthemum*. Attacks of significance tend to occur most frequently on autumn-flowering or winter-flowering plants. Certain other cultivated members of the family Asteraceae are also attacked. Widely distributed in Europe.

DESCRIPTION

Adult: 4–5 mm long; yellow-bodied, with iridescent wings distinctly marked with brown. **Larva:** up to 7 mm long; yellowish and rather stout-bodied, pointed anteriorly and truncated posteriorly.

LIFE HISTORY

Adults occur in the spring, depositing eggs on the leaves of chrysanthemum and certain other Asteraceae, including *Aster*, common ragwort (*Senecio jacobaea*), groundsel (*S. vulgaris*), mugwort (*Artemisia vulgaris*) and tansy (*Chrysanthemum vulgare*). The larvae then feed within the leaves during May and June, each forming a characteristic mine. This feeding gallery commences as an irregularly rounded blotch but later develops into an expanded, somewhat linear, mine which frequently follows the midrib and major veins. When fully fed, usually in early July, the larvae enter the soil and pupate. A second generation of adults appears about two weeks later. Larvae of the second generation feed during the autumn; they eventually pupate to produce adults in the following spring.

DAMAGE

When numerous, the mines seriously disfigure the foliage of chrysanthemum plants, affected tissue eventually turning brown.

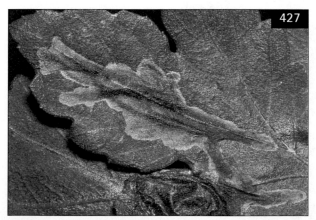

427 Mine of chrysanthemum blotch miner (*Trypeta zoe*).

Family **PSILIDAE**

Small to relatively large, colourful flies; wings with a pale streak or fold extending from the broken costa. Larvae typically mine within plant tissue.

Psila nigricornis Meigen

larva = chrysanthemum stool miner

An infrequent but formerly common pest of *Chrysanthemum*, occurring mainly where clean stock plants are not used. Also attacks other herbaceous plants. Widely distributed in Europe; also found in Canada.

DESCRIPTION

Adult: 4–5 mm long; shiny bluish black, with head and legs brownish yellow; antennae mainly black. **Larva:** up to 6 mm long; creamy white, tough-skinned and slender-bodied. **Puparium:** 4 mm long; dark brown.

LIFE HISTORY

Adults of the first generation occur in May and June, laying eggs in the soil close to chrysanthemum plants. The eggs hatch in about two weeks. Larvae then invade the chrysanthemum stools, and are fully fed in 1–2 months. They return to the soil to pupate. Adult flies emerge from late August to early October. Larvae resulting from this second generation feed throughout the winter, usually taking 3–4 months to reach maturity; under greenhouse conditions, however, the rate of development is increased and the first new adults often appear in February or March.

DAMAGE

Larvae form long, superficial tunnels in chrysanthemum roots and stools, the outer tissue drying out and then splitting open; shoots, especially of cuttings, are also attacked. Infested plants are weakened and shoot production is reduced. In addition, attacked cuttings may be killed. Damage is most significant in stool beds during the autumn, particularly where early-flowering cultivars (e.g. cv. Favourite) are used for producing cuttings in heated greenhouses. Damage may also be caused to the roots of lettuces planted into infested chrysanthemum beds.

Family **EPHYDRIDAE**

Very small to small flies, found mainly near water or in damp situations.

Scatella spp.
Glasshouse wing-spot flies

Two species, *Scatella stagnalis* and *S. tenuicosta*, of these small (*c.* 3 mm long), black-bodied flies are commonly present in greenhouses and other protected sites, where they are associated with algal growths in nutrient-film troughs and on potting composts, rockwool growing-media and so forth. They are often mistaken for sciarid flies (p. 173) and, when present in large numbers, may cause concern. Although of little significance, they contaminate plants with specks of faecal material and also aid the spread of fungal diseases. Unlike the larvae of sciarids, those of *Scatella* are harmless; they do not attack plant roots, cuttings or seedlings. The adults are characterized by their twice-broken costal veins, and by the slightly darkened wings which have several small, clear patches in the membrane; differences between species are slight.

Family **DROSOPHILIDAE**

Very small to small flies, much attracted to fermenting juices and often called 'vinegar flies'. Compound eyes bright red; arista of the antennae generally plumose, with a bifid tip. Larvae of a few species are leaf miners, but most feed in rotting plant tissue, or act as predators or parasitoids.

Scaptomyza flava (Fallén) (**428**)
syn. *S. apicalis* Hardy

A widely distributed, often abundant leaf miner which attacks various brassicaceous plants, including ornamentals; nasturtium (*Tropaeolum*) is also a host. In severe cases infested leaves are killed, but any effect on plant growth is unimportant. There are several overlapping generations each year, and adults occur from April to September. The larvae feed within the leaf blades, forming conspicuous whitish mines that vary from simple blotches to irregular, branched galleries which often follow the major veins. Fully fed larvae are 3–4 mm long, white and translucent, with four prominent tubercles at the hind end. They pupate externally on the ground, each in an elongate (3.0–3.5 mm long) reddish-brown puparium. Adults (wings 2.5 mm long) are pale yellow, with grey markings and red eyes.

428 Mines of *Scaptomyza flava* in leaves of *Tropaeolum*.

Family **AGROMYZIDAE**

Very small to small flies. The larvae of most species are host-specific, usually mining leaves to form serpentine or blotch-like mines characterized by double lines of frass (cf. lepidopterous and hymenopterous leaf miners).

Agromyza demeijerei Hendel (**429**)

A locally abundant pest of *Laburnum*, particularly in England, Germany, the Netherlands and Sweden. Widely distributed in Europe.

DESCRIPTION
Adult: wings 2.4–3.0 mm long; black with mainly yellow legs. **Larva:** up to 3 mm long; whitish.

LIFE HISTORY
This species has two generations annually, and adults appear in May and in August. Larvae feed from June to July and from September to October. Fully grown larvae vacate their mines to pupate on the ground. Each mine, which is restricted to the upper leaf surface, commences as a narrow gallery along the leaf margin, but then widens into a substantial blotch (cf. *Phytomyza cytisi*, p. 204; *Leucoptera laburnella*, p. 220); completed mines eventually turn brown.

DAMAGE
The foliage of severely infested trees looks distinctly scorched, and attacks lead to premature leaf fall.

Agromyza alnibetulae Hendel (**430**)
syn. *A. albitarsis* Hendel
This species occurs widely in central and northern Europe and is often common on young birch (*Betula*) trees in parks, gardens and nurseries. The larvae (up to 3 mm long) are whitish to orange-coloured. They feed in very long, serpentine galleries formed on the upper side of the leaves. The mines, unlike those formed on birch by lepidopterous pests such as *Lyonetia clerkella* (p. 221), are distinctly widened towards their end and contain a double row of black frass; the underside of the gallery is also noticeably swollen. The larvae occur in two main broods, from June to July and from August to September, fully grown individuals pupating on the ground. Adults (wings 2.2–2.5 mm long) are mainly greyish black; they occur in May or June and from late July to early August.

429 Mine of *Agromyza demeijerei* in leaf of *Laburnum*.

430 Mine of *Agromyza alnibetulae* in leaf of *Betula*.

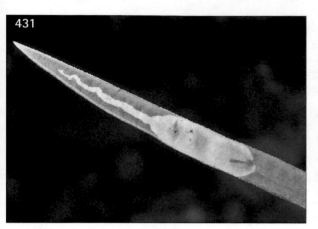

431 Mine of *Amauromyza flavifrons* in leaf of *Dianthus*.

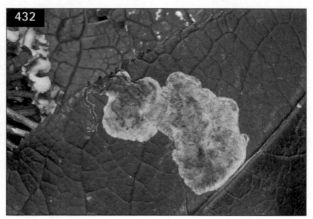

432 Mine of *Amauromyza verbasci* in leaf of *Buddleja*.

Agromyza idaeina (Hardy)
syn. *A. potentillae* (Kaltenbach); *A. spiraeae* Kaltenbach
Irregular, light brown blotch mines formed by this widely distributed and locally common species sometimes occur on the leaves of ornamental *Geum* and *Potentilla*, but are more frequently noted on raspberry and strawberry. Attacks check the growth of young plants but are rarely important. Although mainly a European pest, this leaf miner is also present in Canada.

Agromyza johannae de Meijere
Widespread and locally common in association with broom (*Cytisus*) and Spanish broom (*Spartium junceum*). The larvae mine the leaves, and sometimes cause significant damage to cultivated bushes. Each mine begins as a thin, linear gallery, usually following the leaf margin towards the apex; it then turns abruptly and develops into an elongate, central blotch. Fresh mines are inconspicuous, and infestations often pass unnoticed until mined leaves dry up prematurely and die. Adults (wings 2.2–2.9 mm long) are greyish black with a reddish frons.

Amauromyza flavifrons (Meigen) (**431**)
Larvae of this widespread but local species mine the leaves of various plants, including carnation (*Dianthus caryophyllus*) and pink (*D. plumarius*), *Gypsophila* and *Silene*; outbreaks on carnations and pinks are particularly serious, and heavily infested plants may be killed. The yellow larvae (up to 3 mm long) feed during the summer, each forming a characteristic white blotch mine, preceded by a narrow gallery (cf. *Delia cardui*, p. 206). Pupation occurs on the ground in a reddish-brown puparium. Adults (wings 2 mm long) are black and yellow, with bright yellow halteres and black legs.

Amauromyza maculosa (Malloch)
A polyphagous, mainly tropical or subtropical American species found occasionally in northern Europe on imported *Chrysanthemum* cuttings. The larvae feed gregariously within expansive leaf blotches, fully fed individuals pupating externally in reddish puparia. The posterior spiracles of larvae each have three pores. There are several generations each year, but infestations are unlikely to become established in northern Europe, except in heated greenhouses.

Amauromyza verbasci (Bouché) (**432**)
Large, conspicuous blotches formed by larvae of this species are sometimes noted on the leaves of cultivated buddleia (*Buddleja*). Each mine commences as a narrow, contorted gallery but soon widens into a broad blotch. Although disfiguring, particularly if present in large numbers, damage is not important. Cape figwort (*Phygelius capensis*), figwort (*Scrophularia*) and mullein (*Verbascum*) are also hosts. Adults (wings 2.5 mm long) are mainly greyish black, with a yellow frons and bright yellow knees.

Aulagromyza hendeliana (Hering) (**433**)
syn. *Paraphytomyza hendeliana* (Hering)
Infestations of this locally common species occur on honeysuckle (*Lonicera*), pheasant berry (*Leycesteria crocothyrosos*) and snowberry (*Symphoricarpos rivularis*), the yellowish larvae (up to 3 mm long) forming long, brownish to whitish leaf mines. Occupied mines occur from late April or May onwards, the larvae pupating externally. There are several generations annually. Adults (wings 2.5–2.75 mm long) are mainly greyish to brownish, with yellow knees.

433 Mine of *Aulagromyza hendeliana* in leaf of *Symphoricarpos.*

434 Mines of *Cerodontha iraeos* in leaves of *Iris pseudacorus.*

435 Mines of *Cerodontha iridis* in leaves of *Iris foetidissima.*

436 Puparia of *Cerodontha iridis.*

Cerodontha iraeos (Robineau-Desvoidy) (**434**)
syn. *C. ireos* (Goureau)
larva = iris leaf miner
Conspicuous leaf mines formed by this widely distributed species are often common on both wild and cultivated yellow flag (*Iris pseudacorus*). Each mine is broadly elongate with dark discrete patches of frass clearly visible; the larva pupates in a black puparium orientated lengthwise at the end of the mine. Adults (wings 2.0–2.7 mm long) are mainly black, with bright yellow knees.

Cerodontha iridis (Hendel) (**435–436**)
larva = gregarious iris leaf miner
An abundant, southerly distributed species, associated with *Iris foetidissima*; the larvae feed gregariously, forming relatively large, opaque, greenish to yellowish leaf mines. Damage also occurs on other cultivated irises, including *Iris ochroleuca* and *I. spuria*. Pupation takes place at the end of the mine in a stack of reddish-brown or dark brown puparia, each puparium orientated crosswise. Adults (wings 2.3–3.2 mm long) are mainly black with just the fore knees yellow.

Chromatomyia syngenesiae Hardy (437–439)

syn. *Phytomyza syngenesiae* (Hardy)

larva = a chrysanthemum leaf miner

A generally abundant pest of greenhouse-grown ornamentals, including *Chrysanthemum*, *Cineraria*, pot marigold (*Calendula officinalis*) and sunflower (*Helianthus annuus*). Widespread in Europe; serious infestations have also occurred in Australasia and in North America.

DESCRIPTION

Adult: wings 2.2–2.6 mm long; greyish black, with pale yellow markings on the head and sides; legs mainly black, with yellow knees; costa reaching vein R4 + 5, at apex of wing (cf. *Liriomyza trifolii*, p. 201). **Egg:** 0.35 × 0.15 mm; oval, white and shiny. **Larva:** up to 3.5 mm long; greenish white. **Puparium:** yellowish brown to dark brown; oval but rather flattened.

437 Adult of *Chromatomyia syngenesiae.*

LIFE HISTORY

Eggs are laid mainly on the upper surface of the leaves of host plants, particularly where foliage is shaded. They hatch in 3–6 days and larvae then mine the leaves, each forming a long white to brownish, serpentine gallery on the upper side, within which grains of black frass are distributed at irregular intervals. The larvae are fully fed in about 7–10 days; they then burrow through the leaf to the lower surface before pupating, with the anterior spiracles of the puparium protruding through the lower epidermis: cf. *Chromatomyia horticola* (p. 199) and *Liriomyza trifolii* (p. 201). Adults emerge in 9–12 days at normal greenhouse temperatures. In common with other agromyzids, the adults feed on host leaves by inserting their ovipositor into the leaf tissue and then imbibing the exuded sap; the flies are also relatively inactive, usually making only short, jerky flights. Under protection there are several generations each year, and breeding is continuous whilst conditions remain favourable. Outdoors, where development may be protracted, there are normally two generations annually, adults occurring in May or June and again in late July and August; the winter is passed in the pupal stage. During the summer, greenhouses may be invaded by flies which emerge from wild hosts, such as groundsel (*Senecio vulgaris*) and sow-thistles (*Sonchus*), or from cultivated garden plants such as outdoor chrysanthemums.

DAMAGE

Foliage is disfigured by the relatively large, wavy-edged (cf. *Liriomyza trifolii*, p. 201) adult feeding and egg-laying punctures; leaves of some hosts also develop wart-like wounds. Larval mines, which may be extensive, spoil the appearance of ornamental plants;

438 Puparium of *Chromatomyia syngenesiae* projecting from underside of leaf of *Chrysanthemum.*

439 Mines of *Chromatomyia syngenesiae* in leaf of *Chrysanthemum.*

440 Mine of *Chromatomyia aprilina* in leaf of *Lonicera*.

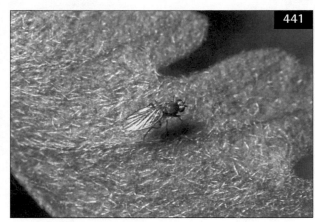

441 Adult of *Chromatomyia horticola*.

heavily infested leaves shrivel and turn brown, weakening host plants and sometimes causing their death. Although infestations on greenhouse-grown plants are often severe, attacks in the open are usually of little or no consequence.

Chromatomyia aprilina (Goureau) (**440**)
syn. *Phytomyza aprilina* Goureau; *P. lonicerella* (Hendel)

The characteristic mines of this widely distributed species occur on honeysuckle (*Lonicera*), causing considerable disfigurement of affected leaves. Each mine commences as a stellate gallery but later becomes distinctly linear, with frass deposited in black lines. Infestations are noted most often on wild plants but also occur on cultivated bushes.

Chromatomyia horticola (Goureau) (**441–443**)
syn. *Phytomyza horticola* Goureau
larva = a chrysanthemum leaf miner

A widely distributed and generally common leaf miner. Adults are similar in appearance to those of *Chromatomyia syngenesiae*. The larvae are very polyphagous, forming galleries in the leaves of various ornamentals, including *Chrysanthemum*, *Petunia*, *Phlox*, poppy (*Papaver*), sweet pea (*Lathyrus odoratus*), tobacco plant (*Nicotiana*) and wallflower (*Cheiranthus cheiri*). In common with *Chromatomyia syngenesiae* pupation takes place at the end of the larval mine, but the puparium protrudes through the epidermis immediately adjacent to the gallery.

442 Mines of *Chromatomyia horticola* in leaf of *Cheiranthus*.

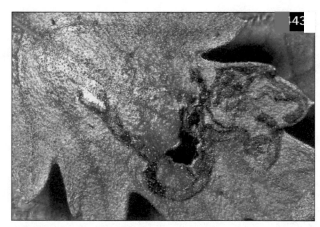

443 Puparium of *Chromatomyia horticola* on upper surface of leaf of *Chrysanthemum*.

444 Mines of *Chromatomyia lonicerae* in leaf of *Lonicera*.

Chromatomyia lonicerae (Robineau-Desvoidy) (**444**)

syn. *Paraphytomyza lonicerae* Robineau-Desvoidy

A widely distributed pest of honeysuckle (*Lonicera*) and snowberry (*Symphoricarpos rivularis*), heavy infestations sometimes developing on cultivated bushes in parks and gardens. The mines are relatively small but distinctly broadened in their later stages. Several are often present in the same leaf. Unlike the previous species, there is just one generation annually and occupied mines occur in the early summer. Adults are distinguished from those of *Aulagromyza hendeliana* by the entirely dark knees on the middle and hind legs.

Chromatomyia primulae (Robineau-Desvoidy) (**445**)

syn. *Phytomyza primulae* Robineau-Desvoidy

Leaves of wild primrose (*Primula vulgaris*) and cultivated species of both *Polyanthus* and *Primula* are attacked by this common and widespread species. The larvae form long, silver-white linear mines, in which widely scattered lumps of black frass are clearly visible. Pupation occurs within the mine in a whitish puparium. Adults (wings 2 mm long) have black legs with yellow knees.

Chromatomyia scolopendri (Robineau-Desvoidy) (**446**)

syn. *Phytomyza scolopendri* Robineau-Desvoidy

Associated with ferns, especially common polypody (*Polypodium vulgare*). Cultivated ferns may be attacked but damage is limited to the presence of mines and, therefore, not important. On hart's-tongue fern (*Asplenium scolopendrium*) the yellowish-green mines are extremely long, sometimes exceeding 10 cm, and are very distinctive. Pupation occurs within or outside the mine. Adults (wings 2.1–2.6 mm long) are black and yellow.

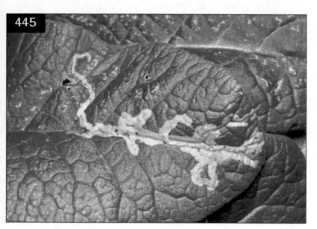

445 Mine of *Chromatomyia primulae* in leaf of *Primula*.

446 Mine of *Chromatomyia scolopendri* in leaf of *Asplenium*.

Liriomyza trifolii (Burgess) (**447–449**)

larva = American serpentine leaf miner

A mainly North American species but frequently introduced into Europe on *Chrysanthemum* cuttings imported from America, the Canary Islands, Kenya and Malta; now established in greenhouses or outdoors in various parts of mainland Europe. In addition to chrysanthemum, infestations also occur on many other hosts, including vegetable crops. This insect is a notifiable pest in several European countries.

447 Fully-fed larva of American serpentine leaf miner (*Liriomyza trifolii*).

DESCRIPTION

Adult: wings 1.2–1.5 mm long; greyish black, with a mainly yellow head and a bright yellow scutellum; antennae bright yellow; legs with yellow coxae and femora, and brown tibiae and tarsi; costa reaching vein M1 + 2, short of apex of wing (cf. *Chromatomyia syngenesiae*, p. 198). **Egg:** 0.2×0.1 mm; oval, smooth and translucent. **Larva:** up to 2 mm long; yellowish orange; posterior spiracles with three pores, the outer ones elongate. **Puparium:** 1.5 mm long; yellowish brown.

LIFE HISTORY

Eggs are deposited singly in the leaves of host plants, each female being capable of depositing several hundred over a period of about a month. Eggs hatch within a few days. The larvae then mine the leaves, each forming a contorted, whitish gallery which contains a meandering double line of dark frass along its length; fully fed larvae vacate the mines about 10–14 days later and pupate in the soil (cf. *Chromatomyia syngenesiae*, p. 198). Adults emerge a week or two later. Breeding is continuous under favourable conditions, the lifecycle being greatly shortened by high temperatures (but protracted by low temperatures and poor lighting conditions), and populations build up rapidly following an original infestation. Once established, infestations may also spread to outdoor plants, including common weeds such as bittersweet (*Solanum dulcamara*), common ragwort (*Senecio jacobaea*) and groundsel (*Senecio vulgaris*). In most instances, populations in northern Europe are unlikely to survive the winter outdoors.

448 Mines of American serpentine leaf miner (*Liriomyza trifolii*) on *Chrysanthemum*.

449 Adult feeding punctures of American serpentine leaf miner (*Liriomyza trifolii*) in leaf of *Chrysanthemum*.

DAMAGE

Leaves are disfigured by small, rounded adult feeding punctures (cf. those of *Chromatomyia syngenesiae*, p. 198); the leaf mines, when numerous, cause considerable damage to the foliage, checking growth and affecting crop quality and marketability.

Liriomyza congesta (Becker) (**450**)

larva = pea leaf miner

A generally common, polyphagous leaf miner; sometimes a minor pest of sweet pea (*Lathyrus odoratus*). The larvae (up to 2 mm long) are greenish white. They form narrow linear mines on the upper side of the leaves, with black frass dispersed characteristically along either side of a green central band; the mines, usually no more than one per infested leaf, terminate in an expanded blotch. Fully fed larvae pupate externally on the ground (cf. *Chromatomyia horticola*, p. 199). There are two or more generations annually, populations often building up throughout the season; however, the mines cause little or no distortion and are usually insufficiently numerous to affect plant growth. Adults (wings 1.3–1.8 mm long) are greyish black, with the frons, scutellum and sides of both thorax and abdomen, antennae and legs yellow.

Liriomyza huidobrensis (Blanchard) (**451**)

larva = South American leaf miner

A very polyphagous, mainly South American species found occasionally in Europe on imported plants, especially *Chrysanthemum*; also well established in California. The yellowish-white larvae (up to 3.25 mm long) feed in distinctive linear leaf mines that often commence their development in association with the midrib and other major veins. Later, the mines frequently turn back upon themselves to give the appearance of a broad, blotch-like gallery; the larvae may also burrow alongside mines formed in the same leaf by other individuals. Fully grown larvae pupate in the soil or within the feeding gallery, each forming a yellowish-brown or reddish-brown puparium; characteristically, each posterior spiracle has about 6–9 small pores arranged in an arc. Adults are similar in appearance to those of *Liriomyza trifolii* but larger (wings 1.7–2.2 mm long) and slightly darker.

Liriomyza sativae Blanchard

A mainly American species but sometimes introduced into Europe on imported plants, especially *Chrysanthemum* cuttings. The larval mines are sinuous but relatively small; however, large numbers can occur in the same leaf, so that damage becomes extensive and host plants severely weakened or killed. Fully fed larvae pupate externally in pale orange-yellow puparia; the posterior spiracles each bear three stout bulbs. In common with other alien species, infestations may occur on various ornamentals and protected vegetables; larvae also invade weeds such as bittersweet (*Solanum dulcamara*), common ragwort (*Senecio jacobaea*) and groundsel (*Senecio vulgaris*). Breeding is continuous under suitable conditions and host plant availability.

Liriomyza strigata (Meigen) (**452**)

syn. *L. pumila* (Meigen); *L. violae* Curtis

This common and widespread species attacks a wide variety of herbaceous plants, including ornamentals, but is usually present in only small numbers. The leaf mines, which usually occur only on older plants, tend to follow the midrib, with distinctive lateral branches extending into the leaf blade. Although the presence of mines on ornamentals may be disfiguring, they rarely, if ever, cause actual harm to host plants.

Nemorimyza posticata (Meigen)

Widely distributed and locally common on golden-rod (*Solidago virgaurea*), the larvae forming characteristic blotch mines on the leaves; the mines are extensive and adorned by wavy lines of frass. Pupation takes place externally. Although usually an unimportant species, significant damage to cultivated hosts has occurred in southern England. Adults (wings 3.0–3.3 mm long) are mainly black, with the knees of the fore legs and (in males) the anterior abdominal tergites yellow.

Phytomyza ilicis Curtis (**453**)

larva = holly leaf miner

An often abundant pest of holly (*Ilex*), occurring on wild and cultivated plants. Widely distributed in Europe; also now well established in parts of North America, having been introduced from Europe along with its foodplant.

DESCRIPTION

Adult: wings 2.5–3.0 mm long; body mainly dark brown or black; legs black. **Egg:** 0.4 × 0.2 mm; white. **Larva:** up to 3 mm long; whitish to yellowish white. **Puparium:** 2.5 mm long; brown.

450 Mines of *Liriomyza congesta* in leaf of *Lathyrus.*

451 Mine of South American leaf miner (*Liriomyza huidobrensis*) on *Chrysanthemum.*

452 Mine of *Liriomyza strigata* in leaf of *Valeriana.*

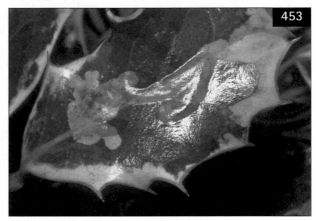

453 Mine of holly leaf miner (*Phytomyza ilicis*).

LIFE HISTORY

Adults occur in May and June, depositing eggs near the base of the midrib on the underside of young holly leaves. Eggs soon hatch and the larvae enter the midrib to begin their development. They feed slowly until the end of the summer, and then move into the leaf blade. The larvae usually remain undetected until December, when the first signs of a yellowish to purplish-brown blotch appear on the upper surface of the leaf. The often somewhat linear mine continues to develop throughout the winter, and is completed by the spring, the larva becoming fully fed in March or April. Pupation occurs within the mine, with the anterior tip of the puparium protruding through the upper surface of the leaf. The adult eventually emerges through a pin-head-sized hole. Galled leaves remain on bushes throughout the year and, within each, the remains of the puparium may be found close to the adult emergence hole.

DAMAGE

Infested leaves look unsightly and heavy attacks on nursery stock may cause concern; however, plant growth is rarely affected.

Phytomyza aconiti Hendel

Unlike most other phytomyzids, larvae of this species feed gregariously. They form large, irregular, brown blotch mines in the leaves of *Delphinium* and monkshood (*Aconitum angelicum*); significant damage often occurs on such plants in gardens, as in southern England. Up to six larvae occur within each blotch, individuals eventually pupating in dark brown puparia. Adults (wings 2.4–2.7 mm long) are black with the sides of the thorax yellowish.

454 Larva of *Phytomyza aquilegiae.*

455 Mines of *Phytomyza aquilegiae* in stipules of *Aquilegia.*

Phytomyza aquilegiae Hardy (**454–455**)

Severe attacks by this leaf miner commonly occur on wild and cultivated columbine (*Aquilegia vulgaris*). The larvae form large, distinctive, greenish-white blotch mines in the leaves, foliage often being destroyed and plants seriously weakened; galleries are also formed in the stipules. Pupation takes place on the ground in reddish-brown puparia (2 mm long). Adults are mainly black with a pale frons, and are larger (wings 2.1–2.5 mm long) than the other columbine-infesting species *Phytomyza minuscula.*

Phytomyza calthophila Hendel

The long, dark, snake-like mines of this widespread but local species occur in the leaves of wild and cultivated marsh marigold (*Caltha palustris*). Although disfiguring, they are not harmful.

Phytomyza cytisi Brischke (**456–457**)

Leaf mines of this species are sometimes abundant during the summer and again in the autumn on *Laburnum*. They are, however, less destructive than those of *Agromyza demeijerei* (p. 195) and *Leucoptera laburnella* (p. 220), two other leaf miners on laburnum. The mines are formed mainly towards the upper leaf surface. They are irregular, white and linear, with obvious black frass scattered along their length. Larvae pupate on the ground in brown puparia. Adults wings (1.8–2.2 mm long) are mainly black with a pale frons. Adult feeding punctures commonly disfigure the foliage of host trees.

Phytomyza minuscula Goureau (**458–459**)

This species forms short, conspicuous, whitish, irregular, linear leaf mines on columbine (*Aquilegia vulgaris*) and meadow-rue (*Thalictrum*). Attacks are common on columbines in gardens, often occurring in company with the more destructive pest *Phytomyza aquilegiae.* Pupation occurs in an orange puparium, frequently attached to the lower surface of a leaf, close to the end of the mine. Adults are relatively small (wings 1.7–2.0 mm long) and mainly black with a pale frons. Adult feeding punctures on the leaves of host plants are sometimes very conspicuous.

Phytomyza spondylii Robineau-Desvoidy (**460**)

A common species on wild hogweed (*Heracleum sphondylium*); also well established on certain introduced ornamental umbelliferous plants (Apiaceae), including giant hogweed (*Heracleum mantegazzianum*). The larvae are yellowish, and form broadly linear mines. Larvae eventually pupate on the ground after escaping through a slit made in the lower surface of the leaf. Adults (wings 2.1–2.4 mm long) are blackish, with the sides of the thorax pale; the legs are black, with just the knees of the fore legs yellow.

456 Mine of *Phytomyza cytisi* in leaf of *Laburnum.*

457 Adult feeding punctures of *Phytomyza cytisi* in leaf of *Laburnum.*

458 Mines of *Phytomyza minuscula* in leaf of *Aquilegia.*

459 Puparium of *Phytomyza minuscula* on leaf of *Aquilegia.*

460 Mine of *Phytomyza spondylii* in leaf of *Heracleum.*

Phytomyza vitalbae Kaltenbach (**461**)

This widespread leaf miner is associated with wild traveller's joy (*Clematis vitalba*) and will also attack various kinds of *Clematis* grown in cultivation. The larvae form long, irregular linear mines on the upper surface of leaves, causing noticeable distortion. Larvae eventually pupate externally in dark brown puparia. Adults (wings 2.2 mm long) are mainly black with a yellow scutellum.

Pseudonapomyza dianthicola Venturi

syn. *Paraphytomyza dianthicola* (Venturi)
Mediterranean carnation leaf miner
The primarily Mediterranean leaf miner is a potentially serious pest of carnation (*Dianthus caryophyllus*). The larvae form linear mines, which extend down the leaf blade, usually on the underside, each gallery gradually widening into an elongate blotch (cf. mines of *Amauromyza flavifrons*, p. 196); feeding may also occur in the stems. There are several generations each year. Serious attacks have occurred in Belgium, but reports of this pest in northern Europe are usually limited to the discovery of mines on plants imported from Mediterranean regions, including Crete, Greece, Italy and the South of France.

Family **ANTHOMYIIDAE**

Adults are 'house fly'-like. The maggot-like larvae possess distinctive mouthparts; several species attack roots and stems of cultivated plants.

Delia cardui (Meigen) (**462**)

Carnation fly
A common pest of greenhouse and outdoor carnation (*Dianthus caryophyllus*), pink (*D. plumarius*) and sweet william (*D. barbatus*). Widespread in Europe.

DESCRIPTION
Adult: 6 mm long; mainly greyish brown. **Larva:** up to 10 mm long; creamy white.

LIFE HISTORY
Adults occur throughout the summer but do not become sexually mature until the autumn. Eggs are then laid in the leaf axils of host plants or on the soil close by. The eggs hatch about two weeks later. The larvae then burrow into the leaves to form elongate blotch mines (cf. *Amauromyza flavifrons*, p. 196). Larvae may also bore into the pith of the shoots. When fully grown, either in late autumn or late winter, the larvae enter the soil. They eventually pupate in the spring, and adults emerge about two months later.

DAMAGE
Affected leaves are extensively discoloured by the mines, and may eventually wither and die. Infested shoots may also be killed.

461 Mine of *Phytomyza vitalbae* in leaf of *Clematis*.

462 Mine of carnation fly (*Delia cardui*) in leaf of *Dianthus*.

Delia radicum (Linnaeus)
syn. *D. brassicae* (Bouché); *D. brassicae* (Wiedemann)
Cabbage root fly
A notorious and generally common pest of vegetable brassicas, but also sometimes a problem on brassicaceous ornamentals such as *Alyssum*, stock (*Matthiola*) and wallflower (*Cheiranthus cheiri*). Holarctic. Present throughout Europe.

DESCRIPTION
Adult: 6–7 mm long; grey to blackish. **Egg:** 1 mm long; elongate-oval, white and ribbed longitudinally. **Larva:** up to 10 mm long; creamy white, with prominent papillae on the anal segment. **Puparium:** 6–7 mm long; elongate-oval, reddish brown.

LIFE HISTORY
Individuals overwinter as pupae within puparia. Adults emerge in the spring from mid-April onwards, the precise timing of their appearance depending on temperature. Eggs are deposited in the soil close to the stems of host plants, the period of egg laying often coinciding with the flowering of cow parsley (*Anthriscus sylvestris*). Eggs hatch after 3–7 days, the larvae immediately attacking the roots of adjacent host plants. They feed for 3–4 weeks and then, when fully grown, move away through the soil for a few centimetres before pupating. Adults of the second generation appear in late June and July, and those of the third from mid-August onwards, the two generations tending to overlap so that subsequent egg laying can occur at virtually any time from July to September.

DAMAGE
Seedlings or recent transplants collapse and die, the fibrous roots and much of the tap root being destroyed. Older or less heavily infested plants wilt in warm, dry weather and make poor growth; damaged root systems are also liable to subsequent attack by fungal pathogens.

Delia platura (Meigen)
syn. *D. cilicrura* (Rondani)
Bean seed fly
Although associated mainly with vegetable crops, infestations of this world-wide pest also occur on ornamentals such as *Anemone*, *Freesia* and hollyhock (*Alcea rosea*); damage is also reported on conifer seedlings. Attacked plants lack vigour and may be killed; seedlings arising from attacked freesia corms sometimes turn bluish. Adults are active from May onwards. Eggs are then deposited in the soil, particularly in the presence of decaying organic matter. They hatch within a few days. The larvae then tunnel inside germinating bean seeds, young stems and other suitable plant tissue. They feed for 1–3 weeks before pupating in the surrounding soil, each in an oval (4–5 mm long), reddish-brown puparium. New adults appear 2–3 weeks later; after mating, females initiate a further generation, each depositing about 50 eggs. There are usually 3–5 generations annually. Adults (6 mm long) are greyish brown; the larvae (up to 8 mm long) are white and relatively robust, with distinct, curved mouth-hooks and 12 posterior tubercles. The posterior respiratory stigmata each have 8–10 projections.

Order **LEPIDOPTERA** (butterflies and moths)

Family **ERIOCRANIIDAE**

A small group of small or very small, primitive, metallic-looking, day-flying moths with reduced mouthparts. The larvae are apodous leaf-miners, with a very small head partly shielded by a large prothorax.

Eriocrania semipurpurella (Stephens) (**463–464**)

Generally common and often abundant on birch (*Betula*), mainly on trees no more than a few metres in height. Present throughout central and northern Europe.

DESCRIPTION

Adult: 10–16 mm wingspan; fore wings with a purplish-golden sheen; hind wings mainly bronzy grey. **Larva:** up to 8 mm long; whitish, with a light brown head; legless.

463 Larva of *Eriocrania semipurpurella*.

464 Mature mine of *Eriocrania semipurpurella* in leaf of *Betula*.

LIFE HISTORY

Adults occur in March and April. They fly in sunshine, and often fly around birch trees in noticeable swarms. Eggs are laid in the leaf buds and hatch shortly afterwards. The leaf-mining larvae feed from late March to early May, each forming an expansive, brownish-white blotch. Characteristically, the mines contain quantities of black frass distributed throughout in criss-crossing threads. Fully grown larvae drop to the ground and enter the soil to pupate in tough, silken cocoons. There is one generation annually.

DAMAGE

Larval mines occupy a large proportion of the leaf blade, disfiguring and sometimes totally destroying some of the earliest-expanded leaves. Growth of plants, however, is not affected.

Eriocrania subpurpurella (Haworth) (**465–467**)

Generally common on oak (*Quercus*) and locally abundant; occasionally troublesome on trees in parks and gardens. Present throughout most of Europe.

DESCRIPTION

Adult: 9–14 mm wingspan; fore wings pale gold and shiny, partly suffused and speckled with purple; hind wings yellowish grey but purple apically. **Larva:** up to 8 mm long; whitish, with a light brown head; legless.

LIFE HISTORY

Adults are active in sunny weather during April and May. In dull weather they rest on the bark of oak trees. Eggs are laid singly in the buds. Later, from late May to July, larvae mine within the leaves. Each mine is an expansive, brownish blotch that contains numerous intertwining threads of blackish frass. When fully fed, larvae vacate their mines and enter the soil to pupate in tough, silken cocoons.

DAMAGE

Mined leaves are distorted but infestations have little or no effect on tree growth. However, heavy infestations on ornamental trees are unsightly and sometimes cause concern, the foliage appearing extensively scorched.

Eriocrania sangii (Wood) (**468–469**)

A locally common species on birch (*Betula*), the characteristically grey-coloured larvae (up to 9 mm long) feeding from late March to May. Large, brownish blotch mines are formed in the leaves, and heavily infested trees appear scorched.

465 Adult of *Eriocrania subpurpurella*.

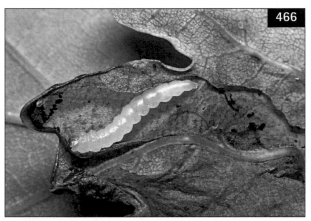

466 Larva of *Eriocrania subpurpurella*.

467 Mine of *Eriocrania subpurpurella* in leaf of *Quercus*.

468 Larva of *Eriocrania sangii*.

469 Mine of *Eriocrania sangii* in leaf of *Betula*.

Eriocrania sparrmannella (Bosc) (**470–471**)

Mines are formed in the leaves of birch (*Betula*) in June and July, much later than those of other birch-feeding eriocraniids. The mines are also characterized by the presence of an initial narrow, linear gallery, which then opens abruptly into an expansive, brown blotch. Adults (10–13 mm wingspan) occur in May, and have golden, purple-marked fore wings. The larvae (up to 10 mm long) are whitish, with a small brown head.

Family **HEPIALIDAE** (swift moths)

Medium-sized moths, with vestigial mouthparts and short antennae. The larvae are elongate, with well-developed legs and crotchets on the abdominal prolegs arranged into several concentric circles. Larvae are soil-inhabiting and feed on the roots of various wild and cultivated plants.

Hepialus lupulinus (Linnaeus) (**472–473**)
syn. *Korscheltellus lupulinus* (Linnaeus)
Garden swift moth

A common horticultural pest, particularly in grassy sites or in the vicinity of grassland. Larvae attack the roots of various annual and perennial herbaceous plants, and also damage bulbs, corms, rhizomes and tubers. Infestations are most important on ornamentals such as *Anemone*, bell flower (*Campanula*), *Chrysanthemum*, daffodil (*Narcissus*), *Dahlia*, *Delphinium*, *Gladiolus*, *Iris*, lily-of-the-valley (*Convallaria majalis*), lupin (*Lupinus*), Michaelmas daisy (*Aster*), peony (*Paeonia*) and *Phlox*. Widely distributed in Europe.

DESCRIPTION

Adult: 25–40 mm wingspan; fore wings yellowish brown, variably marked with white (notably in male); hind wings yellowish grey, darker in the male. **Egg:** 0.5 mm diameter, almost spherical; whitish when laid but soon becoming black. **Larva:** up to 35 mm long; white, shiny and translucent, the dark gut contents often clearly visible; head and prothoracic plate light brown. **Pupa:** 20 mm long; reddish brown; abdominal segments with ventral projections and dentate dorsal ridges.

LIFE HISTORY

Adults fly at dusk in May and June, and occasionally also in August and September, often skimming over grassland, lawns and pastures in considerable numbers.

470 Mine of *Eriocrania sparrmannella* in leaf of *Betula*.

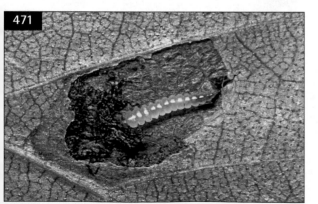

471 Larva of *Eriocrania sparrmannella*.

472 Male garden swift moth (*Hepialus lupulinus*).

Each female lays up to 300 eggs which she broadcasts at random whilst in flight. Larvae feed in the soil on the roots of grasses and many other plants. If disturbed, the larvae retreat rapidly backwards down narrow, feeding burrows, but they are readily unearthed during soil cultivation. Larval development continues throughout the winter. Larvae usually pupate in the following April, each in a loosely woven silken tunnel formed among the root system of the host. Adults emerge about six weeks later, and often appear in distinct flushes after rainfall.

DAMAGE

Larvae bite off the roots and tunnel into bulbs, corms, rhizomes and tubers of hosts with fleshy underground root systems. This retards growth and often causes plants to wilt; badly damaged plants may be killed. If unchecked, infestations on perennial hosts may persist and increase in importance from year to year. Most serious damage is caused in autumn, winter and early spring, and is often experienced when plants are grown in recently ploughed pasture or grassland.

Hepialus humuli (Linnaeus) (474–476)

Ghost swift moth

A widely distributed and generally common species, infesting a similar range of hosts to the previous species but usually more damaging to grassland, pastures and lawns. The larvae (up to 50 mm long) are whitish, robust and relatively opaque, with a reddish-brown head and prothoracic plate, and prominent, dark brown pinacula (cf. *Hepialus lupulinus*). When young, they feed on plant rootlets but older individuals attack the larger roots and also bite into stolons and the lowermost parts of stems. In common with the previous species, individuals construct silk-lined feeding tunnels in the soil, retreating into them or curling up if disturbed. The larvae consume large amounts of food but growth is slow, the period of development usually extending over two, and occasionally three, years. Damage to plants is particularly severe in the second summer of larval growth, individuals then pupating in the following April or May. Adults occur mainly in June and July, and are active at dusk. The females are relatively large (50–70 mm wingspan), with yellowish-ochreous, orange-marked fore wings; males are much smaller (46–50 mm wingspan), with silvery-white wings.

473 Larva of garden swift moth (*Hepialus lupulinus*).

474 Female ghost swift moth (*Hepialus humuli*).

475 Male ghost swift moth (*Hepialus humuli*).

476 Larva of ghost swift moth (*Hepialus humuli*).

Family **NEPTICULIDAE**

Minute moths with the first antennal segment forming an 'eye-cap'. The larvae, which feed in leaves and form sinuous mines or blotches, have a wedge-shaped head and are virtually apodous, the thoracic legs being reduced to short, extendible lobes and the abdominal prolegs to fleshy humps without crotchets. Pupation usually takes place outside the mine in small, parchment-like cocoon.

Stigmella anomalella (Goeze) (**477**)

larva = rose leaf miner

A generally common pest of rose (*Rosa*), and often abundant on both wild and cultivated bushes. Present throughout Europe.

DESCRIPTION

Adult: 5–6 mm wingspan; head orange, often suffused with dark brown; fore wings mainly greenish bronze to golden, with a partly coppery tinge, the apical region purple; hind wings brownish grey. **Larva:** up to 5 mm long; yellow, with a brown head.

LIFE HISTORY

Adults occur in May and in August, eggs being deposited on the underside of leaves, usually close to the midrib. The larvae form long, contorted mines which become filled with greenish-grey to blackish frass; each gallery widens considerably in its later stages to leave a clear marginal line along both sides of the central band of frass. Occupied mines occur mainly in July and October. Fully grown larvae pupate in brownish or reddish-brown cocoons spun at the base of a leaf stalk, in the angle between two shoots or on the surface of a leaf.

DAMAGE

Infested leaves are unsightly but attacks have little or no effect on plant growth.

Stigmella hemargyrella (Kollar) (**478**)

Widely distributed and locally common on beech (*Fagus sylvatica*), the larvae feeding in elongate galleries formed in the leaves. Infestations are often discovered on beech hedges, the mines disfiguring the leaves but not causing significant damage. There are two generations annually, larvae feeding in June and from August to September; adults (5–6 mm wingspan, the fore wings bronzy brown and with a distinct silver or golden crossband) occur from April to May and from late July to early August.

Stigmella lapponica (Wocke) (**479**)

Generally common on birch (*Betula*), including young amenity and nursery trees. Adults occur in May, eggs being laid on the underside of the leaves, usually close to a major vein. The larvae feed from mid- or late June onwards, forming long, angular galleries which often follow but may also cross the leaf veins. Development is completed in early July, there being normally just one generation annually. The first quarter of the mine is filled with greenish frass; there is then an abrupt change to a central black line of frass which continues throughout the rest of the gallery. There may be several mines per leaf, but the leaves are not distorted and shoot growth is unaffected.

Stigmella obliquella (Heinemann) (**480**)

Widely distributed on narrow-leaved willows, including ornamentals such as weeping willow (*Salix vitellina* var. *pendula*). Larval mines appear as narrow, frass-filled galleries; these eventually end in a blotch, with frass

477 Mine of rose leaf miner (*Stigmella anomalella*) on *Rosa*.

478 Mine of *Stigmella hemargyrella* in leaf of *Fagus*.

accumulated in the centre. Adults appear in May and in August, and occupied mines occur from June to July, and from September to October. Larvae (up to 5 mm long) are orange-yellow, with a brown head and, on the ventral surface, a line of dark pear-shaped spots. When fully grown, they pupate externally in brownish-orange cocoons surrounded by strands of silk. Adults (4–5 mm wingspan) have mainly dark brown fore wings, each marked by a narrow, yellowish crossband.

Stigmella roborella (Haworth) (**481**)

A widely distributed and generally common leaf miner on oak (*Quercus*) but its true status uncertain owing to confusion with closely related species, especially *Stigmella atricapitella* and *S. ruficapitella*. The larvae form long galleries on the leaves, characterized by the clearly defined central line of black frass. Adults (5.0–6.5 mm wingspan, with dark bronzy-brown,

purplish-tinged fore wings) are present in May and June, with a second generation appearing in August and September; occupied mines occur mainly in late June, July, September and October.

Stigmella suberivora (Stainton) (**482**)

This local species is reported in southern England in association with holm oak (*Quercus ilex*), and is gradually extending its range; in Mediterranean areas, including France and Italy, it is found on cork oak (*Q. suber*). Adults occur in May and in September, depositing eggs on the upper side of the leaves; the yellow-bodied larvae feed in July and August, and from November onwards, those of the second generation pupating in the late winter or early spring. Although not affecting plant growth, the elongate, serpentine mines disfigure the foliage and become very conspicuous as leaves age and turn brown.

479 Mine of *Stigmella lapponica* in leaf of *Betula*.

480 Mine of *Stigmella obliquella* in leaf of *Salix vitellina*.

481 Mine of *Stigmella roborella* in leaf of *Quercus*.

482 Mine of *Stigmella suberivora* in leaf of *Quercus ilex*.

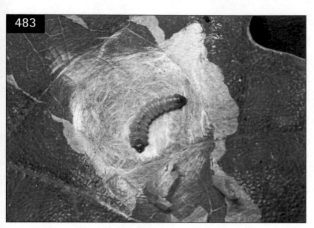

483 Larva of *Tischeria ekebladella*.

484 Young mines of *Tischeria ekebladella* in leaf of *Castanea*.

485 Young mines of *Tischeria angusticollella* in leaf of *Rosa*.

Family TISCHERIIDAE

A small family of very small moths with narrow, pointed wings. The leaf-mining larvae are very flat-bodied, with vestigial abdominal prolegs.

Tischeria ekebladella (Bjerkander) (483–484)

A locally common leaf miner on oak (*Quercus*) and sweet chestnut (*Castanea sativa*). Present throughout Europe.

DESCRIPTION
Adult: 8–11 mm wingspan; fore wings deep ochreous yellow, speckled with blackish scales apically; hind wings grey. **Larva:** up to 7 mm long; pale yellow, with the gut partly visible; head light brown; prothoracic plate brown.

LIFE HISTORY
Adults occur in June, depositing eggs on the leaves of oak and sweet chestnut. The larvae form whitish, initially shell-shaped mines on the upper side of leaves, and feed from about September to November; unlike many leaf miners, a slit is made at the edge of the mine, through which frass is ejected. When fully fed, each larva forms a circular chamber in the middle of its mine, within which the winter is passed and pupation eventually takes place.

DAMAGE
Although several mines may be present on a leaf, the tissue is not distorted; also, the lateness of attacks lessens their importance.

Tischeria angusticollella (Duponchel) (485)

Minor infestations of this widely distributed but local species occur on rose (*Rosa*), including cultivated bushes. The small, light green larvae form characteristic blotch mines on the upper side of the leaves. Although causing distortion, and affecting the appearance of foliage, the mines have no effect on plant growth. Occupied mines occur from June to July and again in September and October. The insignificant, dark purplish-brown adults (8–9 mm wingspan) appear in the spring, with a second generation emerging in the late summer.

Family **INCURVARIIDAE**

Small, metallic-looking, mainly day-flying moths with well-developed antennae. The larvae commence feeding as leaf miners but later become case-dwellers; they have a single transverse band of crotchets on each abdominal proleg.

Incurvaria pectinea Haworth (**486**)

syn. *I. pectinella* (Fabricius); *I. zinckenii* (Zeller)

A minor pest of birch (*Betula*) and hazel (*Corylus*). Widely distributed and locally common in central and northern Europe; in mainland Europe, alder (*Alnus*), dogwood (*Cornus*), hornbeam (*Carpinus betulus*), maple (*Acer*) and rowan (*Sorbus aucuparia*) are also attacked.

DESCRIPTION

Adult: 12–16 mm wingspan; head yellow (male with antennae strongly bipectinate); fore wings light brownish bronze, each marked on the dorsal margin with two whitish spots; hind wings greyish bronze. **Larva:** up to 7 mm long; whitish, with a brown head and yellowish-grey thoracic plates. **Case:** 8 × 5 mm; oval and flattened.

LIFE HISTORY

Adults occur in April and May, depositing eggs on the leaves of host plants. Larvae commence feeding in June, at first mining the leaves and forming small, circular blotches. Later, they construct portable cases by cutting out and spinning together oval pieces from each leaf surface. The larvae then wander or fall away to feed on dead leaves and other vegetable debris, completing their development in the autumn. The cases are then attached to upright surfaces, including fences, posts or tree trunks. Larvae pupate in the winter and adult moths emerge in the spring.

DAMAGE

Larval mines disfigure the foliage but are less noticeable than the series of holes left by the case-forming larvae. Damage caused in not significant.

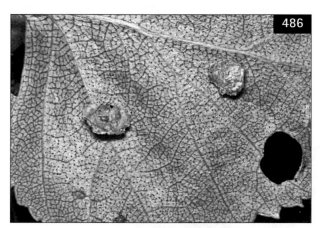

486 Larval cases of *Incurvaria pectinea* on leaf of *Betula*.

Family **COSSIDAE**

Large or very large moths. The larvae are wood-borers, feeding in the trunks and branches of trees and shrubs.

Cossus cossus (Linnaeus) (**487**)
Goat moth

Infestations of this widely distributed but generally uncommon, wood-boring species occur in various mature trees, including ash (*Fraxinus excelsior*), birch (*Betula*), elm (*Ulmus*), flowering cherry (*Prunus*), oak (*Quercus*) and willow (*Salix*). The large (70–100 mm wingspan), dull, greyish-brown moths occur in June and July, and are sometimes found resting on tree trunks during the daytime. Eggs are deposited in crevices in the bark, usually in groups of about fifty. After egg hatch, the pinkish naked-looking larvae immediately burrow into the trunks to feed within the sap and heart wood. Larval development is greatly protracted, lasting up to three or four years. Fully grown individuals (80–100 mm long) eventually pupate just below the bark, or in the ground, in strong, silken cocoons. Larval feeding galleries are very extensive and the larvae may eventually cause the death of host trees. Infested trees are characterized by the presence of large emergence holes in the bark, from which sap is sometimes exuded. An unpleasant, goat-like smell is also emitted from the larval galleries.

Zeuzera pyrina (Linnaeus) (**488–490**)
Leopard moth

A sporadically important pest of various trees and shrubs, including ash (*Fraxinus excelsior*), birch (*Betula*), *Cotoneaster*, crab-apple (*Malus*), flowering cherry (*Prunus*), hawthorn (*Crataegus*), honeysuckle (*Lonicera*), horse chestnut (*Aesculus hippocastanum*), lilac (*Syringa*), maple (*Acer*), *Rhododendron* and sycamore (*Acer pseudoplatanus*). Widespread in mainland Europe; in Britain restricted mainly to the southern half of England and eastern Wales. Also introduced into North America.

DESCRIPTION

Adult: 45–65 mm wingspan; wings white and translucent, with black or blue-black spots; body similarly coloured and rather velvety; male considerably smaller than female and with the antennae strongly bipectinate basally. **Egg:** 1 mm long; oval and pinkish-orange. **Larva:** up to 60 mm long; yellowish, with prominent black pinacula; head and prothoracic plate brownish black, the latter with a characteristically scalloped hind margin; young larvae at first pinkish, readily distinguished from those of other wood-boring species by the characteristic prothoracic plate. **Pupa:** 25–35 mm long; reddish brown.

LIFE HISTORY

Adults occur in June or July, depositing eggs in groups in wounds or cracks in the bark of host plants. The newly emerged larvae initially attack the leaf stalks and major leaf veins, buds and shoots of host plants. Later, they then enter the larger twigs and branches to feed in the heart wood. Larval galleries extend for 40 cm or more, a larva taking two or three years to complete its development. Pupation occurs in the feeding gallery in a slight cocoon into which particles of wood are incorporated. In early summer the pupa wriggles out of the cocoon and breaks through the surface of the branch where it remains protruding after emergence of the adult.

DAMAGE

Attacks usually occur in branches, stems or trunks less than 10 cm in diameter. The presence of a larva is indicated by the accumulation of frass and particles of wood which are forced out of the entry holes and, later, by the withering and die-back of the leaves and shoots. Infested branches are weakened and may snap off in a strong wind.

487 Goat moth (*Cossus cossus*) damage to trunk of *Quercus*.

488 Male leopard moth (*Zeuzera pyrina*).

489 Larva of leopard moth (*Zeuzera pyrina*).

490 Pupa of leopard moth (*Zeuzera pyrina*).

Family **CASTNIIDAE**

A small family of medium-sized to very large day-flying moths with dull-coloured fore wings, brightly coloured hind wings and clubbed antennae. Larvae feed within the roots, crowns or stems of host plants.

Paysandisia archon (Burmeister)
Palm borer moth

A South American pest of palm trees, which (probably following the importation of infested palm trees from Argentina) has recently become established in southern Europe (France, Italy and Spain).

DESCRIPTION
Adult: 90–110 mm wingspan; fore wings olive-brown; hind wings orange-red, with black and white markings.
Larva: up to 70 mm long; robust, brownish white and glossy.

LIFE HISTORY
Day-flying adults of this attractive moth occur from June to September. Eggs are laid on the stems of palm trees, close to the growing points. Following egg hatch, larvae attack the young leaves. Later, they tunnel within the crowns and stems, expelling large quantities of sawdust-like debris. Larvae probably take up to two years to complete their development, each pupating in a tough cocoon formed within the feeding gallery.

DAMAGE
Attacked leaves are partly shredded and stems hollowed, retarding growth and resulting in the distortion or death of crowns. As with red palm weevil (*Rhynchophorus ferrugineus*) (see p. 166), steps are underway to prevent further spread of South American palm borer within Europe.

Family **ZYGAENIDAE**

A family of small to medium-sized, often brightly-coloured, day-flying moths, with either bipectinate or clubbed antennae. Larvae have a small, retractile head and are often extremely colourful, with clusters of body hairs arising from large verrucae.

Aglaope infausta Linnaeus (**491–493**)
Almond leaf skeletonizer moth

A southerly-distributed pest of hawthorn (*Crataegus*) and other Rosaceae such as blackthorn (*Prunus spinosa*) and *Cotoneaster*, as well as almond (*Prunus dulcis*) and rosaceous fruit trees. Most often reported in southern France, north-western Italy, Portugal and Spain.

DESCRIPTION
Adult: 20 mm wingspan; head and thorax black, the latter with a red crossband anteriorly; fore wings mainly greyish black, suffused with red basally, particularly along the wing margins; hind wings greyish black, suffused with red, particularly on the apical half; antennae bipectinate in both sexes. **Larva:** up to 15 mm long; body mainly purple, black and yellow, but creamy white below, with clusters of hairs arising from distinct verrucae; prothoracic plate relatively large and fleshy,

mainly black, marked with purple and yellow. **Pupa:** 10 mm long; light brown.

LIFE HISTORY
Adults occur in July, depositing eggs on the bark of host plants. Eggs hatch in about 10 days. Larvae then feed for a short time before spinning silken, scale-like hibernacula, also on the bark of host trees. They then remain in diapause from mid-summer onwards and do not recommence feeding until the following spring. In wetter years, however, when host trees are able to produce late new growth, feeding is protracted and the larvae do not form their hibernacula until late summer. In spring, larvae feed fully exposed on the leaves. If disturbed, they either drop to the ground or remain temporarily suspended on a thread of silk. Larvae complete their development in about three weeks. They then pupate, each in a broad (8 × 4 mm), pinkish or yellowish-white cocoon spun either singly or in batches on the foliage and developing fruits, or amongst withered leaves. New adults appear about a month later.

DAMAGE
Feeding by overwintered larvae has an adverse effect on the development of host trees, with leaves grazed extensively, and often reduced to a skeletal collection of midribs. In wetter summers, damage caused by young larvae is also of some significance.

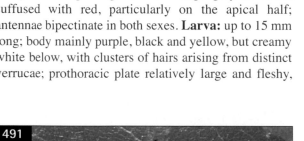

491 Adult female of almond leaf skeletonizer moth (*Aglaope infausta*).

492 Larva of almond leaf skeletonizer moth (*Aglaope infausta*).

493 Pupal cocoons of almond leaf skeletonizer moth (*Aglaope infausta*).

Family **LYONETIIDAE**

Small moths with narrow wings fringed by long cilia. The larvae have a complete circle of crotchets on each abdominal proleg.

Bucculatrix thoracella (Thunberg) (**494–496**)
syn. *B. hippocastanella* (Duponchel)
Locally common on lime (*Tilia*), and of increasing importance as a pest of amenity or shade trees in towns and cities, particularly in Germany and the Netherlands. Widely distributed in Europe.

DESCRIPTION
Adult: 8 mm wingspan; fore wings pale ochreous yellow, with blackish markings; hind wings grey. **Larva:** up to 7 mm long; pale creamy white, with an orange tinge.

LIFE HISTORY
Adults of the first generation occur in May, depositing eggs on the leaves of lime. At first the larvae mine within the leaves. Later they feed externally on the undersurface, the change between instars occurring beneath opaque, silken webs. When fully grown the larvae descend on silken threads to pupate in ribbed cocoons (*c*. 5 mm long) formed on the trunks of trees or on fallen leaves or amongst other debris on the ground. A second flight of adults occurs in about July. Larvae of this second generation complete their development in late August or September. Pupal cocoons persist on the trunks of infested trees for several seasons, some with a split through which the pupa has burst on emergence of the adult moth, and others with a rounded hole through which a parasitoid wasp would have emerged.

DAMAGE
Infested leaves are extensively disfigured by whitish or brownish patches, visible from above and from below. In public places the larvae descending on their silken threads are a nuisance, particularly on trees growing close to market stalls and open-fronted food shops.

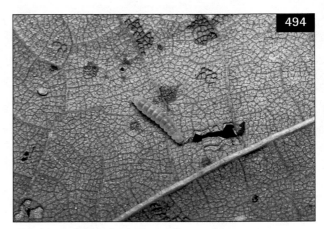

494 Larva of *Bucculatrix thoracella*.

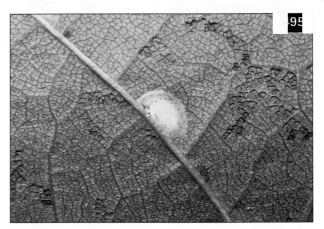

495 Larval cocoon of *Bucculatrix thoracella* on underside of leaf of *Tilia*.

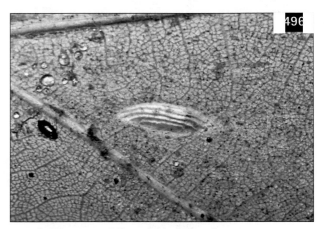

496 Pupal cocoon of *Bucculatrix thoracella*.

497 Adult of *Leucoptera laburnella.*

498 Laburnum leaf miner (*Leucoptera laburnella*).

499 Mine of laburnum leaf miner (*Leucoptera laburnella*) on *Laburnum.*

500 Pupal cocoon of laburnum leaf miner (*Leucoptera laburnella*) on *Laburnum.*

Leucoptera laburnella (Stainton) (**497–500**)

larva = laburnum leaf miner

An often common pest of *Laburnum*; also found on lupin (*Lupinus*). Present throughout Europe; also present in North America.

DESCRIPTION

Adult: 7–9 mm wingspan; fore wings shiny white, marked towards the apex with brownish and yellowish orange; hind wings white; body mainly white. **Larva:** up to 6 mm long; greyish white, with a distinct greenish gut; head light brownish; prothoracic plate broad and greyish brown; abdominal prolegs reduced but with crotchets forming a complete circle. **Pupa:** 4 mm long; yellowish brown.

LIFE HISTORY

Adults occur in May and in August; in favourable conditions, a third generation is possible in the autumn. Eggs are deposited on the underside of leaves of laburnum and, less frequently, on lupin. Larvae feed singly within the leaves, each burrowing to the upper side and then forming a large, pale blotch; blackish frass within the mine forms a distinctive pattern: cf. mines of the dipterous leaf miners *Agromyza demeijerei* (p. 195) and *Phytomyza cytisi* (p. 204). Occupied mines are most common from June to July and in August. When fully grown, each larva bites its way through the upper surface of the mine and spins a white web on the underside of the same or an adjacent leaf. Pupation then takes place. Pupae occur from July to August and from September to May.

DAMAGE

Mines distort the foliage and disfigure host plants; heavy infestations reduce the vigour of young hosts.

501 Adult of apple leaf miner (*Lyonetia clerkella*).

502 Apple leaf miner (*Lyonetia clerkella*).

503 Mine of apple leaf miner (*Lyonetia clerkella*) on *Amelanchier*.

504 Pupal cocoon of apple leaf miner (*Lyonetia clerkella*).

505 Apple leaf miner (*Lyonetia clerkella*) damage to leaf of *Amelanchier*.

506 Mines of apple leaf miner (*Lyonetia clerkella*) on *Betula*.

Lyonetia clerkella (Linnaeus) (501–506)

larva = apple leaf miner

A generally common pest of trees and shrubs, including birch (*Betula*), cherry laurel (*Prunus laurocerasus*), *Cotoneaster*, crab-apple (*Malus*), flowering cherry (*Prunus*), hawthorn (*Crataegus*), rowan (*Sorbus aucuparia*) and snowy mespilus (*Amelanchier laevis*); often present on nursery stock, amenity trees and other ornamentals. Palaearctic. Widely distributed throughout Europe.

DESCRIPTION

Adult: 8–9 mm wingspan; fore wings shiny white (often partly or entirely suffused with brown), marked apically with a dark spot and by several black streaks which extend through the fringe of cilia; hind wings dark grey. **Larva:** up to 8 mm long; body green, translucent and moniliform; head and legs brown. **Pupa:** 3.5 mm long; light green, with yellowish-brown wing cases.

LIFE HISTORY

Adults occur in June and August, and from October onwards, individuals of the autumn generation hibernating under loose bark, amongst thatch and in out-buildings. They reappear in the following April. Eggs are usually laid singly on the underside of a leaf, and hatch about two weeks later. Each larva then commences to mine towards the upper surface and eventually forms a very long, pale-coloured gallery. This widens gradually throughout its length and terminates in a distinct, elongate chamber. Feeding is completed in three or four weeks. The larva then emerges and wanders about on the foliage for a few hours before beginning to spin a slight, 6–7 mm long, hammock-like cocoon, attached by strands of silk to a leaf or rough bark. Pupation then takes place, and the adult moth appears about two weeks later. There are usually three generations each year.

DAMAGE

The mines disfigure the foliage and, if numerous, cause distortion and premature death of infested leaves. On some hosts, notably snowy mespilus, infested leaves become greatly discoloured, particularly where the gallery isolates a section of the leaf blade. On cherry laurel the leaf tissue often splits along the length of the mine; portions of the leaf blade may also fall away to leave rounded or irregular holes.

Family **HIEROXESTIDAE**

Opogona sacchari (Bojer) (**507**)

larva = sugar cane borer

This tropical species is a native of various African offshore islands, including the Canary Islands, Mauritius, St. Helena and the Seychelles, where it is a pest of banana and sugar cane. Infestations also occur on various plants imported into Europe as ornamentals, including blue-stem yucca (*Yucca guatamalensis*), *Dracaena*, *Hibiscus* and rubber plant (*Ficus elastica*).

DESCRIPTION

Adult: 20–28 mm wingspan; fore wings lanceolate, pale ochreous suffused with brown and blackish brown; hind wings brownish and shiny. **Larva:** up to 25 mm long; whitish, with small yellowish pinacula; head chestnut-brown; prothoracic and anal plates brown.

LIFE HISTORY

Larvae burrow into the stems of host plants, forming extensive, silk-lined, frass-filled galleries. Fully grown individuals pupate close to the surface, and the pupae protrude from the galleries following emergence of the adult moths. Infested plants imported into greenhouses in northern Europe usually produce moths in the winter months, but such insects are unable to survive the winter outdoors.

DAMAGE

The stems of infested plants are riddled with galleries, spoiling the appearance of ornamentals, reducing vigour and, sometimes, causing death of plants.

507 Sugar cane borer (*Opogona sacchari*) damage to stem of *Dracaena*.

Family **GRACILLARIIDAE**

Very small moths with narrow wings fringed by long cilia; adults adopt a typical resting posture, with the head end raised up by the long, widely splayed front legs. The larvae are 14-legged (prolegs absent on the sixth abdominal segment); they are mainly leaf miners and, in their young stages, are apodous and have a peculiar flattened head. The pupa typically protrudes from the cocoon following emergence of the adult.

Acrocercops brongniardella (Fabricius) (508–510)

An often abundant pest of oak (*Quercus*), including holm oak (*Q. ilex*). Widely distributed in mainland Europe; very local in the British Isles, and found mainly in the southern half of England, south-east Wales and southern Ireland. Also present in Asia and North Africa.

DESCRIPTION

Adult: 8–10 mm wingspan; fore wings ochreous brown to brown, marked with white; hind wings grey. **Larva:** up to 7 mm long; translucent, pale yellowish brown, with a blackish gut; pinacula and a series of oval dorsal plates, shiny greyish brown; head and prothoracic plates pale yellowish brown.

LIFE HISTORY

Eggs are laid in the spring on the upper side of oak leaves. Larvae then mine the leaves, several individuals commonly feeding in the same leaf. The mines commence as silvery galleries and, later, develop into blotches; these eventually join to form an expansive mine that becomes distinctly bloated and may occupy most of the leaf. The blotches often become partly filled with moisture but this does not appear to harm the occupants. Larvae are fully fed by late June or early July. They then pupate in cocoons formed amongst debris on the ground. Adults appear in late July or early August and then overwinter; they reappear in April and May. There is usually just one generation annually but, in favourable situations, there may be at least a partial second.

DAMAGE

The larval mines disfigure host plants, and are most obvious and important on hedges of holm oak.

508 Adult of *Acrocercops brongniardella*.

509 Larva of *Acrocercops brongniardella*.

510 Mine of *Acrocercops brongniardella* in leaf of *Quercus ilex*.

Caloptilia azaleella (Brants) (**511–514**)

syn. *Caloptilia azaleae* (Busck)

larva = azalea leaf miner

A native of eastern Asia but now a well-established pest of indoor azalea (*Rhododendron*) in various parts of Europe, including Belgium, southern England, France, Germany and the Netherlands; survives outdoors in favourable areas. Also introduced into New Zealand and North America.

DESCRIPTION

Adult: 10–12 mm wingspan; fore wings violet brown, with a large, elongate, golden blotch and fine brownish-black markings; hind wings grey. **Larva:** up to 12 mm long; pale greenish yellow; head pale brownish yellow; early instars legless.

LIFE HISTORY

Eggs are laid singly, close to the midrib on the underside of the leaves of azaleas. Immediately after egg-hatch each larva forms a small, narrow mine which is then extended into an expansive leaf blotch, the latter often occurring without a preliminary mine. After completion of the blotch, the larva moves to the leaf tip; this is then rolled back and secured with silk to form a distinctive cone-like shelter, within which further feeding occurs; a second such shelter is constructed before the larva is fully grown. Pupation takes place within a white cocoon on the underside of a longitudinally folded leaf. Outdoors, this species completes two generations (with adults present in May and August, and larvae in June and September). The winter is spent in the pupal stage. Under glass, however, there are several generations annually and, if conditions remain favourable, breeding (and, hence, damage) may be continuous.

DAMAGE

Heavily infested shrubs are seriously disfigured, both by the larval shelters and by the blotch mines; the latter commonly turn brown and infested leaves often shrivel and fall prematurely.

511 Adult of azalea leaf miner (*Caloptilia azaleella*).

512 Azalea leaf miner (*Caloptilia azaleella*).

513 Mine of azalea leaf miner (*Caloptilia azaleella*).

514 Habitation of azalea leaf miner (*Caloptilia azaleella*).

Caloptilia stigmatella (Fabricius) (515–518)

Generally common on *Populus*, including aspen (*P. tremula*); also reported on birch (*Betula*) (a rare host) and willow (*Salix*). Holarctic. Present throughout Europe.

DESCRIPTION

Adult: 12–14 mm wingspan; fore wings chestnut-brown, with a whitish, triangular, costal blotch. **Larva:** up to 8 mm long; head pale yellowish brown; body whitish yellow to whitish green.

LIFE HISTORY

Adults overwinter, emerging in the spring and eventually depositing eggs on the underside of leaves of willows or poplars. Eggs hatch in late June or July. Each larva forms an elongate gallery which eventually terminates in a small blotch. The larva then feeds externally, forming a characteristic cone-like shelter at the leaf tip. When the surface tissue within the feeding habitation is consumed, the larva moves to another leaf and forms a second shelter; rarely, feeding may continue on a further leaf. Fully fed individuals pupate in silken cocoons formed on the underside of an adjacent leaf. Larvae develop from July to September, and adults appear from September onwards.

DAMAGE

Damage is of no significance but the presence of larval cones on young ornamental trees is disfiguring; affected tissue often turns black.

515 Adult of *Caloptilia stigmatella*.

516 Larva of *Caloptilia stigmatella*.

517 Larval habitation of *Caloptilia stigmatella* on leaf of *Salix*.

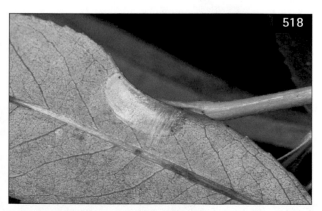

518 Pupal cocoon of *Caloptilia stigmatella*.

519 Adult of *Caloptilia syringella*.

520 Lilac leaf miner (*Caloptilia syringella*).

521 Habitation of lilac leaf miner (*Caloptilia syringella*) on *Syringa*.

522 Habitation of lilac leaf miner (*Caloptilia syringella*) on *Fraxinus*.

Caloptilia syringella (Fabricius) (519–522)

syn. *Gracillaria syringella* (Fabricius)

larva = lilac leaf miner

An often abundant pest of lilac (*Syringa*) and privet (*Ligustrum vulgare*); ash (*Fraxinus excelsior*) and, less frequently, mock privet (*Phillyrea latifolia*) and white jasmine (*Jasminum officinale*) are also attacked. Eurasiatic. Widespread in Europe; also present in North America.

DESCRIPTION

Adult: 10–13 mm wingspan; fore wings golden brown, with various dark-edged, whitish-yellow markings; hind wings dark brown. **Larva:** up to 7 mm long; greenish white to yellowish white and translucent, with a darker green gut; head pale brownish yellow. **Pupa:** 4 mm long; brownish yellow.

LIFE HISTORY

Adults occur in two distinct generations, from April to May and in July. They may be seen at rest on host plants, adopting a typical position with the wings and body held at an angle to the leaf surface. Eggs are laid singly or in rows on the midrib of host plants. Young larvae mine in the leaves, the galleries commencing as narrow channels but soon widening into expansive blotches; several larvae usually feed together. Later in their development the larvae feed within distinctive, frass-filled shelters formed from webbed-down leaves. Occupied larval mines or tents are most common in June and from August to September. Fully grown individuals pupate within greyish-white cocoons constructed on the underside of a leaf or amongst litter on the ground. Individuals of the first generation usually pupate in June or July, and adults emerge shortly afterwards. Those of the second pupate in October, and adults appear in the following spring.

DAMAGE

Feeding by young larvae leads to distinctive leaf discoloration and distortion. The prominent larval tents, formed either from a single leaf (as on lilac) or from several webbed-down leaves or leaflets (as on ash), disfigure host plants; badly affected sections of privet hedges appear scorched.

523 Horse chestnut leaf miner moth (*Cameraria ohridella*).

524 Larva of horse chestnut leaf miner moth (*Cameraria ohridella*).

525 Pupa of horse chestnut leaf miner moth (*Cameraria ohridella*).

526 Horse chestnut leaf miner moth (*Cameraria ohridella*) damage to leaf of *Aesculus*.

Cameraria ohridella Deschka & Dimic (523–526)

Horse chestnut leaf miner moth

A major pest of horse chestnut (*Aesculus hippocastanum*) in both urban and rural areas; Japanese horse chestnut (*A. turbinata*) is also susceptible. Larvae are unable to survive on the hybrid red horse chestnut (*Aesculus carnea*); Indian horse chestnut (*A. indica*), amongst other Asian species, is also resistant. Discovered as recently as 1985, near Lake Ohrid in Macedonia. Since then the pest has spread very rapidly, and is now firmly established in most of Europe, including Britain and Scandinavia.

DESCRIPTION

Adult: 8 mm wingspan; fore wings brownish orange, with several whitish (partly black-bordered) streaks and crosslines; hind wings dark grey. **Larva:** up to 5 mm long; pale yellowish green, with a yellowish-brown head. **Pupa:** 4 mm long; dark brown, and characteristically (unlike those of *Phyllonorycter* spp., pp. 228–33) lacking a cremaster.

LIFE HISTORY

This pest overwinters in the pupal stage, within fallen leaves. Adults emerge in spring, and eggs are eventually deposited on the upper surface of expanded leaves. The eggs hatch 2–3 weeks later. Larvae then feed within the leaves, each forming a pale blotch. Larvae are fully fed in about a month. They then pupate within their mines, and young adults emerge about two weeks later. On average, in western Europe, there are three overlapping generations annually, but up to five are possible in hot, dry regions.

DAMAGE

Infested leaves become peppered with mines. This diminishes photosynthetic activity, resulting in premature leaf fall, and has an adverse impact on host vigour. Infested trees also become very unsightly. Affected leaves are often mistaken for those infected by the fungal pathogen *Guignardia aesculi*, the cause of guignardia leaf blotch; pest mines, however, lack the yellow band that usually surrounds diseased tissue.

Parectopa robiniella (Clemens) (527)

This North American leaf miner is associated with false acacia (*Robinia pseudoacacia*). In Europe, infestations were first reported in 1970 in Italy. Since then the pest has spread to various other parts of central and southern Europe, including Austria, Croatia, the Czech Republic, France, Germany, Hungary, Romania, Switzerland and the former Yugoslavia, where it is now firmly established. Moths appear in May or June, eggs being deposited singly on the underside of the leaves of host trees. The larvae form large, irregular blotch mines on the upper side of leaves (cf. *Phyllonorycter robiniella*, p. 232), which become very obvious and spoil the appearance of infested trees. Unlike many related leaf miners, larval frass is ejected through a small hole in the lower surface of the feeding gallery. The outline of the blotch mine is also, typically, digitate. Fully grown larvae are 4.0–4.5 mm long and mainly green. They

527 Mines of *Parectopa robiniella* on *Robinia*, viewed from above.

528 Mine of *Phyllonorycter maestingella* in leaf of *Fagus.*

pupate in white cocoons spun at the edges of leaves, and moths emerge shortly afterwards. The adults (8–9 mm wingspan) are mainly dark brown, the fore wings being ornamented with several narrow, white, wedge-shaped marks. In favourable areas there are up to three generations annually. Fully-grown larvae of the final brood overwinter within cocoons on fallen leaves.

Phyllonorycter maestingella (Müller) (528)
syn. *P. faginella* (Zeller)
Generally common on beech (*Fagus sylvatica*), and sometimes a minor pest of ornamental and nursery plants. Present throughout central and northern Europe.

DESCRIPTION
Adult: 7.5–9.0 mm wingspan; fore wings dark brown, with white, wedge-shaped markings; hind wings greyish brown. **Larva:** up to 5 mm long; pale greenish yellow, with a darker green gut; head light brown. **Pupa:** 3 mm long; light brown.

LIFE HISTORY
Adults occur in two distinct generations, from May to early June and in August. Eggs are deposited on the underside of beech leaves. The larvae then develop in brownish blotch mines formed on the underside of the leaves between two lateral veins. The mines extend from the midrib to, or almost to, the leaf edge, their elongate form distinguishing them from those formed on beech leaves by *Phyllonorycter messaniella*. Larvae may be found in July and again from September to October. Pupation takes place in the mine within a white, silken cocoon; pupae occur in July and August, and from November to May.

DAMAGE
Mines cause slight distortion of the leaves but infestations, unless heavy, have little effect on the general appearance of hosts; they do not affect plant growth.

Phyllonorycter messaniella (Zeller) (529)
Zeller's midget moth
Generally common on holm oak (*Quercus ilex*) and occasionally an important pest; deciduous oak (*Quercus*) trees, as well as beech (*Fagus sylvatica*), hornbeam (*Carpinus betulus*) and sweet chestnut (*Castanea sativa*), are also attacked. Widespread throughout much of Europe, except in more northerly areas; an introduced pest of ornamentals in Australia and New Zealand.

DESCRIPTION
Adult: 7–9 mm wingspan; fore wings shiny pale golden ochreous, with whitish, slightly brownish-tinged markings partly edged with dark brown; hind wings greyish brown. **Larva:** up to 4.5 mm long; yellow to whitish yellow; head brown. **Pupa:** 3.0–3.5 mm long; dark brown; a pair of large spines on each of the first three abdominal segments.

LIFE HISTORY
Adults occur mainly from April to May, in August and from late October to late November, females depositing eggs on the underside of oak leaves. Larvae mine the underside of the leaves, each forming a light brown, oval or elongate blotch. The fully grown larva pupates in a slight cocoon spun to one side of the gallery. The adult emerges 2–3 weeks later. Larvae are most frequent in July and October. On holm oak they also occur from December to March.

DAMAGE
On holm oak, mines develop a strong elongate crease on their lower surface, distorting the leaf so the blade folds downwards. The upper surface over the mine becomes slightly mottled and eventually turns brown. Heavy infestations significantly disfigure host plants. On deciduous oak, mines are relatively small. Nevertheless, they cause some distortion of the leaf, a distinct crease developing on the upper surface of the leaf blade above the mine. Mines on beech are also tightly folded (cf. *Phyllonorycter maestingella*).

Phyllonorycter platani (Staudinger) (**530**)

A pest of plane (*Platanus*), originating in Asia Minor but now widely distributed in Mediterranean regions. The pest has also spread much further north, and now occurs throughout France and in many other countries, including Belgium, England, Germany and the Netherlands.

DESCRIPTION
Adult: 8–10 mm wingspan; fore wings golden brown, with white, dark-edged markings. **Larva:** up to 6 mm long; whitish, with a yellowish-brown head.

LIFE HISTORY
Adults appear in May or June, eggs being laid on the underside of the leaves, mainly on those on the lower branches. Larvae feed within the leaves, forming distinctive blotch mines visible from below. Occasionally, mines also occur on the upper side of leaves. There are often several mines per leaf, the blotches varying considerably in size and sometimes exceeding 20 mm in length. Fully fed larvae pupate within the feeding gallery, adults of a second generation emerging in August. Larvae of the second brood complete their development in the autumn. They overwinter as pupae in fallen leaves.

DAMAGE
Mines disfigure the foliage and, if numerous, cause considerable deformation of the leaf blade. Attacks on mature trees are of minor importance but heavy infestations on nursery trees are troublesome.

529 Mine of *Phyllonorycter messaniella* in leaf of *Quercus ilex.*

530 Mines of *Phyllonorycter platani* in leaf of *Platanus.*

531 Mine of *Phyllonorycter comparella* in leaf of *Populus tremula*.

532 Adult of *Phyllonorycter comparella*.

533 Mine of *Phyllonorycter corylifoliella* in leaf of *Sorbus aucuparia*.

Phyllonorycter comparella (Duponchel) (531–532)

A local, southerly distributed and usually uncommon species, associated with *Populus*, including aspen (*P. tremula*). Not of importance as a pest but sometimes noted on young amenity trees, particularly in mainland Europe. The larvae mine the underside of the leaves, forming relatively small blotches that cause noticeable discoloration of the upper surface. Occupied mines occur in mid-summer and in the autumn. Unlike most species of *Phyllonorycter* the adults are mainly white and grey, and the insect overwinters in the adult stage.

Phyllonorycter corylifoliella (Hübner) (533)

Widely distributed and often common on rosaceous trees, including crab-apple (*Malus*), hawthorn (*Crataegus*), rowan (*Sorbus aucuparia*) and other kinds of *Sorbus*; less frequently, birch (*Betula*) is also attacked. The larvae feed throughout July and again in September and October, each forming a large, silvery-white, russet-flecked blotch on the upper side of leaves; the mines commonly overlap the midrib or a major vein and eventually cause infested leaves to crinkle. Fully grown larvae are 5–6 mm long, with a brown head and the body dirty whitish to pale yellowish but appearing greenish inter-segmentally. Adults occur in May and June, and in August. They are relatively large (8–9 mm wingspan); the fore wings are chestnut-brown, suffused with blackish scales and marked with two or three narrow, whitish striae and a long, white basal streak.

Phyllonorycter issikii (Kumata)

This invasive Asiatic species has recently become established in Europe, having spread rapidly westwards into various countries, including Austria, Bulgaria, the Czech Republic, Germany, Hungary, Italy, Poland and Romania. The pest is associated with lime (*Tilia*), upon which the larvae develop in small blotch mines formed on the underside of leaves. The mines are relatively small but clearly visible from above. There are two generations annually; adults of the summer generation are paler in appearance than those of the overwintering (autumn) form, which have the fore wings heavily marked with black.

Phyllonorycter leucographella (Zeller) (**534–535**)

Firethorn leaf miner moth

A common pest of firethorn (*Pyracantha*), but native to Mediterranean areas where it is associated with spiny broom (*Calicotome spinosa*). In recent years this insect has greatly extended its range into more northern parts of Europe, including Austria, Britain, France, Germany, the Netherlands and Switzerland. Adults (8–9 mm wingspan) are mainly yellowish orange, the fore wings each marked with distinctive white striae. They occur from May to June and from August to mid-September. The larvae mine within the leaves, each forming a large, silvery blotch on the upper surface along the midrib. On firethorn the leaf mine often extends over the whole of the upper surface, affected leaves eventually folding longitudinally at the completion of larval development. Mines also occur on other rosaceous plants, including *Chaenomeles*, *Cotoneaster*, crab-apple (*Malus*), flowering cherry (*Prunus*), hawthorn (*Crataegus*) and *Sorbus* growing in the immediate vicinity of infested firethorn bushes; however, mines on such hosts tend to split open and larval development is then aborted. The larvae occur during the summer and from September onwards; they continue to feed throughout the winter, except in very cold weather, but are unable to survive in fallen leaves. Badly affected leaves tend to drop during the spring and summer, and persistent attacks on bushes lead to noticeable defoliation.

Phyllonorycter oxyacanthae (Frey) (**536**)

Generally common on hawthorn (*Crataegus*), the larvae forming blotch mines on the underside of the leaves. The mines occur most frequently on the leaf lobes closest to the stalk, and cause considerable distortion of the leaf blade. Infestations are often present on nursery stock but do not affect plant growth. The larvae are pale yellowish green; they feed in July and from September to October. Adults (6–8 mm wingspan) occur in May and August; the fore wings are yellowish brown, marked with dark brown scales and white striae.

534 Mine of *firethorn* leaf miner moth (*Phyllonorycter leucographella*) on *Pyracantha*.

535 *Firethorn* leaf miner moth (*Phyllonorycter leucographella*) damage on *Pyracantha*.

536 Mine of *Phyllonorycter oxyacanthae* in leaf of *Crataegus*.

Phyllonorycter quercifoliella (Zeller) (537)

Generally common on oak (*Quercus*), including young trees, the larvae forming mainly oval mines on the underside of the leaves. The mines usually occur between the major veins, their upper surface being characterized by a central patch of unconsumed tissue. The larvae feed in July and from September to November. The golden-brown, white-marked adults (7–9 mm wingspan) emerge in late April or May and in August.

Phyllonorycter rajella (Linnaeus) (538)

syn. *Phyllonorycter alnifoliella* (Hübner)

A widespread and often abundant leaf miner on alder (*Alnus*), including amenity and nursery trees. There are two generations annually, larvae mining the underside of the leaves during the summer and again in the autumn. There are often several mines in each infested leaf but they cause only slight distortion. Adults (7–9 mm wingspan) are brown, marked with white; they occur in May and again in August.

Phyllonorycter robiniella (Clemens) (539–540)

Following its accidental introduction to Switzerland in 1983, this North American species has become widely distributed in mainland Europe on false acacia (*Robinia pseudoacacia*). The larvae mine within large blotch mines, formed on the underside of the leaves (cf. *Parectopa robiniella*, p. 228). Trees are often heavily infested, and the mines frequently coalesce; the larvae then feed gregariously. The pest completes two or three generations annually.

537 Mines of *Phyllonorycter quercifoliella* in leaf of *Quercus*.

538 Adult of *Phyllonorycter rajella*.

539 Leaf mines of *Phyllonorycter robiniella* on *Robinia*, viewed from above.

540 Leaf mines of *Phyllonorycter robiniella* on *Robinia*, viewed from below.

Phyllonorycter salicicolella (Sircom) (541)

Blotch mines of this generally common species occur on the underside of leaves of common sallow (*Salix atrocinerea*), grey willow (*S. cinerea*) and other broad-leaved willows. They are often present on young shrubs and, if formed close to the leaf margin, cause noticeable distortion of the leaf blade. There are two generations annually, larvae feeding in July and from September to October. Larvae are whitish green, with an orange spot on the sixth abdominal segment. Adults (7.0–8.5 mm wingspan) have light brown, white-striated fore wings. They occur in May and from late July to August.

Phyllonorycter trifasciella (Haworth) (542)

Generally common on honeysuckle (*Lonicera*) and snowberry (*Symphoricarpos rivularis*), and sometimes a minor pest of cultivated plants. The larvae feed in distinctive blotches formed on the underside of the leaves, causing considerable distortion of the leaf blade. There are usually three generations, larvae feeding in March to April, July and October. The mainly brownish-orange adults (7–8 mm wingspan) appear in May, August and November.

Phyllonorycter ulmifoliella (Hübner) (543)

A generally common species on birch (*Betula*), the larvae developing in blotch mines formed on the underside of leaves. The mines are characterized by several longitudinal folds which coalesce across the middle; the upper surface appears slightly mottled. Although relatively small, the mines cause noticeable distortion of leaves and often disfigure young trees, including nursery stock. There are two generations annually, occupied mines occurring in July and again in the autumn. The larvae (up to 5 mm long) are pale yellowish green, with a yellow spot on the sixth abdominal segment. Adults (7–9 mm wingspan) occur in May and August; the fore wings are golden brown, marked with several white, partly blackish-bordered striae and a white basal streak.

541 Mine of *Phyllonorycter salicicolella* in leaf of *Salix*.

542 Mine of *Phyllonorycter trifasciella* in leaf of *Symphoricarpos*.

543 Mines of *Phyllonorycter ulmifoliella* in leaf of *Betula*.

Family **PHYLLOCNISTIDAE**

A small family of very small moths with an 'eye-cap'. The larvae are apodous, sap-feeding leaf miners.

Phyllocnistis citrella (Stainton) (**544**)
Citrus leaf miner moth

In recent years this Asiatic leaf-mining species has become well established as a pest in citrus-fruit orchards in southern Europe. *Citrus* plants being raised as ornamentals are also attacked. The mines are very long and sinuous, and occur on either side of the leaves. Occasionally, young shoots are also attacked. Heavy infestations lead to premature leaf fall and dieback of shoots. Adults (5–8 mm wingspan) are mainly white; the fore wings are partly suffused with yellowish orange, and marked with black striae and a black pre-apical spot. Larvae (up to 3.5 mm long) are translucent and partly yellowish. The pest overwinters as adults, and under suitable conditions completes several generations annually.

Phyllocnistis unipunctella (Stephens) (**545–547**)
syn. *P. suffusella* (Zeller)

larva = poplar leaf miner

An often abundant leaf miner on black poplar (*Populus nigra*) and Lombardy poplar (*P. nigra* 'Italica'), including trees growing in gardens and nurseries. Although present throughout much of mainland Europe, in Britain it is found mainly in England and south-east Wales. Adults occur in July and from September to April, those of the second generation hibernating and eventually depositing eggs in the spring. The larvae feed in June and in August, each forming a very long, slightly raised but inconspicuous, gallery on the upper side of the leaves. The mines are most obvious when a leaf is viewed at an angle; they then appear as silvery, slug-like trails. The larvae (up to 5 mm long) are pale greenish white, with a transparent head and a blackish prothoracic plate; unlike members of the Gracillariidae they are sap feeders throughout their development, excreting a clear liquid instead of faecal pellets. Fully grown larvae pupate in whitish cocoons, each usually spun externally under a fold of the leaf margin at the end of the larval gallery. The adults (7–8 mm wingspan) are superficially similar in appearance to those of various species of *Phyllonorycter* (p. 228–233). Damage caused by this insect is unimportant and usually noticed only when foliage of trees is examined closely for other disorders; however, in some situations mined leaves are reduced in size, suggesting that persistent attacks on young hosts might have an adverse effect on tree growth.

544 Mine of citrus leaf miner moth (*Phyllocnistis citrella*) on *Citrus*.

545 Mines of poplar leaf miner (*Phyllocnistis unipunctella*) on *Populus*.

546 Poplar leaf miner (*Phyllocnistis unipunctella*).

Family **SESIIDAE** (clearwing moths)

Unusual, wasp-like moths with partly clear (scale-less) wings and a distinct fan-like anal tuft of scales. The larvae are stem-borers in shrubs and trees; crotchets on the abdominal prolegs are arranged into two transverse bands.

Sesia apiformis (Clerck) (**548**)
Hornet moth

Associated mainly with black poplar (*Populus nigra*) but also found on other kinds of *Populus*, including aspen (*P. tremula*). Sometimes damaging in plantations, particularly on drought-stressed trees; specimen trees are also attacked. Palaearctic. Widely distributed in mainland Europe; present in the British Isles, but rarely found nowadays and of little or no pest status. An introduced pest in North America.

DESCRIPTION

Adult: 35–45 mm wingspan; wings mainly clear, with brownish scales along the costal margin, brownish veins and cilia; head yellow; thorax mainly brown, with tegulae distinctively yellow anteriorly; abdomen mainly yellow, ringed with brown or black. **Larva:** up to 50 mm long; yellowish white, with a large reddish-brown head and a yellowish prothoracic plate. **Pupa:** 20–30 mm long; dark reddish brown; very plump, with rows of backwardly directed spines on the abdominal tergites.

LIFE HISTORY

Adults occur mainly in June and July, and freshly emerged individuals often bask in morning sunshine on the trunks of host trees. Eggs are laid in bark crevices or holes, close to the base of the trunks of poplar trees, and hatch in about two weeks. Larvae then burrow into the trees to feed, forming extensive, frass-filled galleries between the bark and heart wood; the feeding galleries may also extend into the roots. Larvae continue to feed for at least two years. Fully fed individuals overwinter in tough cocoons, formed just below the bark at the base of infested trees, and pupate in the spring. Adults emerge several weeks later.

DAMAGE

Infested or formerly infested trees are recognized by the presence at the base of the trunks of adult emergence holes, each about 8 mm in diameter.

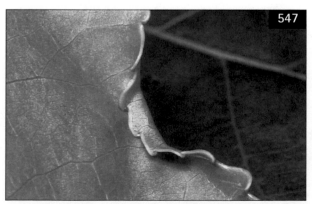

547 Pupal habitation of poplar leaf miner (*Phyllocnistis unipunctella*).

548 Hornet moth (*Sesia apiformis*) emergence holes at base of *Populus* tree trunk.

549 Fig-tree skeletonizer moth (*Choreutis nemorana*).

550 Fig-tree skeletonizer (*Choreutis nemorana*).

551 Apple leaf skeletonizer moth (*Eutromula pariana*).

552 Apple leaf skeletonizer (*Eutromula pariana*).

Family **CHOREUTIDAE**

Small moths, the wings held flat over the body when in repose. The larvae have long, pencil-like abdominal prolegs.

Choreutis nemorana (Hübner) (**549–550**)

Fig-tree skeletonizer moth
larva = fig-tree skeletonizer

A generally common but minor pest of common fig (*Ficus carica*). Widely distributed in the Mediterranean area, including southern Europe and North Africa; also present in parts of Asia, and in the Canary Islands and Madeira.

DESCRIPTION

Adult: 16–20 mm wingspan; fore wings mainly reddish brown to ochreous brown, suffused with black, and marked with white to grey scales; hind wings brownish, each with a pair of pale spots towards the margin.

Larva: up to 20 mm long; body light green, shiny and semitransparent, with a pale dorsal line and numerous large, black verrucae; head ochreous yellow and shiny, marked with black.

LIFE HISTORY

Overwintered adults appear in the early spring, and eventually deposit eggs in groups on the leaves of fig. Larvae feed from mid-May onwards, each protected by a thick web of silken threads. Larvae are fully grown a few weeks later. They then pupate, each in a dense, white, boat-shaped cocoon spun on the leaves or elsewhere on the foodplant. Adults of the summer generation appear in July, and second-brood larvae feed from the end of July to early October. Adults emerge in the autumn and then hibernate.

DAMAGE

Larvae cause noticeable distortion, discoloration, scarification and tattering of leaves.

553 Juniper shoot moth (*Argyresthia dilectella*).

554 Juniper shoot borer (*Argyresthia dilectella*).

555 Juniper shoot borer (*Argyresthia dilectella*) damage to shoot of *Juniperus*.

Family **YPONOMEUTIDAE**

A very large family of small moths with elongate, relatively broad wings and well-developed, projecting maxillary palps. The larvae are of variable form; crotchets on the abdominal prolegs are often arranged into several concentric circles.

Argyresthia dilectella Zeller (**553–555**)
Juniper shoot moth
larva = juniper shoot borer
A locally common pest of juniper (*Juniperus*), and sometimes troublesome in gardens and nurseries; also reported on other members of the Cupressaceae. Restricted to central Europe and to the more southerly parts of northern Europe.

DESCRIPTION
Adult: 8–10 mm wingspan; fore wings white, suffused and ornamented with golden brown; hind wings light grey. **Larva:** up to 5 mm long: yellowish green, sometimes reddish intersegmentally, with a brownish-black head and blackish prothoracic and anal plates.

LIFE HISTORY
Moths occur in July and deposit eggs on the shoots of juniper. After egg hatch the larvae tunnel within the shoots, feeding throughout the winter months and completing their development by the end of May. They then vacate their feeding galleries, to pupate externally in white cocoons.

DAMAGE
Affected shoots turn purplish and then brown. This spoils the appearance of infested plants and also affects their development.

Eutromula pariana (Clerck) (**551–552**)
Apple leaf skeletonizer moth
larva = apple leaf skeletonizer
Although associated mainly with apple, including crab-apple (*Malus*), this species is occasionally damaging to the foliage of other ornamental rosaceous plants, including flowering cherry (*Prunus*), hawthorn (*Crataegus*) and ornamental pear (*Pyrus calleryana* 'Chanticleer'). The light green to yellowish, black-spotted larvae (up to 14 mm long) cause extensive damage to the leaves, typically removing tissue from the upper surface but leaving the lower one intact. They feed beneath silken webs from May onwards, and individuals eventually pupate in white, boat-shaped cocoons spun beneath a leaf or among plant debris. The greyish-brown adults (11–13 mm wingspan) appear in July and August. Larvae of a second generation complete their development in the autumn. They then pupate and adults, that then overwinter, appear shorty afterwards.

Argyresthia thuiella (Packard) (556–557)

American thuja shoot moth

larva = American thuja leaf miner

A North American pest of Cupressaceae, especially Lawson cypress (*Chamaecyparis lawsoniana*) and white cedar (*Thuja occidentalis*); accidentally introduced in the late 1970s into parts of mainland Europe, including Austria, Germany and the Netherlands, where it is now well established. Reports of this species attacking juniper (*Juniperus*) are due to confusion with *Argyresthia trifasciata*.

DESCRIPTION

Adult: 8 mm wingspan; head and thorax whitish; fore wings white to greyish white, with silver-grey markings. **Egg:** greenish. **Larva:** up to 5 mm long; body green or brownish, often with a reddish tinge at the posterior margin of each segment; head and prothoracic plate shiny black or brown. **Pupa:** 3.5 mm long; green to brown.

LIFE HISTORY

Adults occur mainly in late June and July, when they may be found settled on the foliage of host plants; if disturbed, they fly a short distance before resettling. Eggs are laid between scales, and hatch in 2–3 weeks. The larvae then bore into the shoots to begin feeding, each mining away from the tip. Larval development continues from July onwards, and fully grown individuals eventually pupate in the following May or June.

DAMAGE

Tips of mined shoots become discoloured, and eventually turn brown and die.

Argyresthia trifasciata (Staudinger) (558)

A common pest of juniper (*Juniperus*) in southern Europe. In recent years it has greatly extended its range and is now found in many countries, including Austria, Belgium, Britain, the Czech Republic, Denmark, France, Germany, Italy, the Netherlands, Poland, Sweden and Switzerland. The pest also occurs on *Chamaecyparis*, *Cupressocyparis*, *Cupressus* and *Thuja*.

DESCRIPTION

Adult: 10 mm wingspan; fore wings blackish, each marked with whitish crosslines; hind wings dark grey. **Larva:** up to 5 mm long; greenish, with a blackish head.

556 American thuja shoot moth (*Argyresthia thuiella*).

557 American thuja leaf miner (*Argyresthia thuiella*) damage to shoot of *Thuja*.

558 *Argyresthia trifasciata* damage to shoot of *Juniperus*.

LIFE HISTORY

Adults occur from mid-May to mid-June, eggs being deposited on the shoots of juniper. Larvae mine within the shoots from late June onwards, and complete their development in the following spring. They then pupate on a main shoot, each in a small cocoon hidden beneath a flake of bark.

DAMAGE

Infested shoots turn brown, and damage to hosts is often extensive. Particularly severe infestations have occurred on pencil cedar (*Juniperus virginiana* 'Sky Rocket').

Argyresthia sorbiella (Treitschke) (**559–560**)

A locally common species associated with rowan (*Sorbus aucuparia*). The larvae tunnel within the young shoots, and cause the tips to wilt and turn black; infestations are found occasionally on young amenity trees but are rarely extensive. Larvae are whitish (up to 6 mm long), with a brownish-black head, prothoracic plate and anal plate; they feed mainly in May, and may be found if an infested shoot is split open. Adults (*c.* 12 mm wingspan) are mainly whitish, with golden-brown markings on the fore wings; they occur in June and July.

Plutella xylostella (Linnaeus) (**561**)

Diamond-back moth

This notorious worldwide crop pest occasionally attacks brassicaceous ornamentals such as *Alyssum*, candytuft (*Iberis*) and wallflower (*Cheiranthus cheiri*). However, it is of far greater significance on vegetable brassicas. The narrow-winged adults (11–16 mm wingspan) are habitual migrants, and in some years arrive in northern Europe in considerable numbers. Larvae (up to 12 mm long) are mainly pale yellowish green, with a light brown head; the hind set of abdominal prolegs (anal claspers) are elongate and clearly visible from above. The larvae feed on the underside of leaves, beneath flimsy silken webs, and are capable of causing extensive defoliation. When fully fed they pupate on the foodplant, each in a net-like cocoon. In particularly favourable areas, such as southern Europe, there are several overlapping generations annually, but in northern Europe there may be no more than two.

559 *Argyresthia sorbiella* damage to shoot of *Sorbus.*

560 Larva of *Argyresthia sorbiella.*

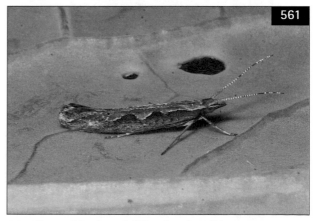

561 Diamond-back moth (*Plutella xylostella*).

562 Ash bud moth (*Prays fraxinella*).

563 Ash bud moth (*Prays fraxinella*) damage to shoot of *Fraxinus*.

Prays fraxinella (Bjerkander) (562–563)
syn. *P. curtisella* (Donovan)
Ash bud moth

A generally common pest of ash (*Fraxinus excelsior*); damage to young trees is sometimes extensive. Present throughout much of Europe.

DESCRIPTION

Adult: 15 mm wingspan; fore wings white, with brownish-grey cilia and irregular blackish markings, including a distinct triangular patch; head white; antennae black; melanic forms, largely suffused with black, also occur. **Larva:** up to 12 mm long; dirty reddish green, with paler sides, more or less marked with grey; head, prothoracic and anal plates black.

LIFE HISTORY

Adults occur in June and July, and deposit eggs on the shoots of ash. Larvae commence feeding in the autumn. They mine within the buds or leaves before overwintering beneath the bark close to a terminal bud. In the following spring they attack the terminal shoots, boring into the pith. In May, growth from infested shoots wilts, and affected tissue soon turns black. If examined, a distinct hole may be seen at the base of each wilted shoots and a larva found feeding inside. Larvae complete their development by the end of May; they then pupate, and adults emerge a few weeks later. There is just one generation annually in northern Europe, but two are completed in southern Europe.

DAMAGE

Growth from affected buds is aborted, and shoot development then becomes forked as paired laterals become dominant.

Scythropia crataegella (Linnaeus) (564–567)
Hawthorn webber moth

A locally important pest of hawthorn (*Crataegus*); also associated with *Cotoneaster* and *Prunus*. Widespread in central and southern Europe; also found in the more southerly parts of northern Europe.

DESCRIPTION

Adult: 13–14 mm wingspan; fore wings silvery white, mottled with ochreous and brown. **Larva:** up to 15 mm long; reddish brown, the thoracic segments marked dorsally with yellowish orange; body hairs whitish and relatively long; head black. **Pupa:** 7–8 mm long; mainly black, suffused with dirty creamy white dorsally.

564 Hawthorn webber moth (*Scythropia crataegella*).

LIFE HISTORY

Infestations are usually first noticed during the spring, when the overwintered larvae feed gregariously on the foliage of host plants, protected by a flimsy but expansive web. Larvae are fully grown from mid-June onwards. They then pupate within the web, and adults emerge in late June and July. Eggs deposited during the summer hatch in the late summer. The young larvae then mine briefly within the leaves before hibernating, the mined leaves remaining attached to host plants throughout the winter.

565 Larvae of hawthorn webber moth (*Scythropia crataegella*).

DAMAGE

Larvae cause significant defoliation. Also, their webs disfigure host plants, affecting the development of the new shoots.

566 Pupa of hawthorn webber moth (*Scythropia crataegella*).

567 Hawthorn webber moth (*Scythropia crataegella*) damage

568 Common small ermine moth (*Yponomeuta padella*).

569 Larvae of common small ermine moth (*Yponomeuta padella*).

Yponomeuta padella (Linnaeus) (**568–571**)
Common small ermine moth

Often abundant on hawthorn (*Crataegus*) and *Prunus*, including blackthorn (*P. spinosa*) and wild cherry (*P. avium*) but not bird-cherry (*P. padus*) (cf. *Yponomeuta evonymella*), and sometimes injurious to hedges and ornamentals. A locally distributed race attacks rowan (*Sorbus aucuparia*) and is sometimes a pest of such trees on alpine camp sites. Widely distributed in Europe.

DESCRIPTION

Adult: 20–22 mm wingspan; fore wings whitish to whitish grey, with light grey cilia and four longitudinal rows of black dots (including a row of 4–7 dots towards the lower margin); hind wings grey. **Egg:** flat, yellowish to dark reddish. **Larva:** up to 20 mm long; dirty yellowish grey to greenish grey, marked with black spots and a blackish dorsal stripe; head, prothoracic plate and anal plate black. **Pupa:** 10 mm long; yellowish, with head, thorax, wing cases and tip blackish; cremaster with six filaments.

LIFE HISTORY

Adults occur in July and August, and commonly rest in numbers on the original foodplant or on nearby hosts. Eggs are laid in overlapping tile-like rows in flat, scale-like batches on the bark of shoots and branches of host plants. They are then coated with a gelatinous secretion which soon hardens to form a protective cap. Eggs hatch 2–3 weeks later, the larvae biting their way through the lower surface of the egg shells but leaving the upper surface of the egg cluster intact. The young larvae remain *in situ* throughout the winter; they are very resistant to low temperatures, and survive without apparent difficulty even in exposed mountainous regions. Activity is resumed in the following May, the

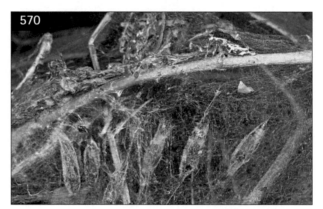

570 Pupae of common small ermine moth (*Yponomeuta padella*).

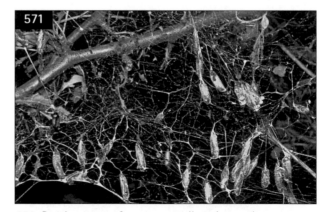

571 Pupal cocoons of common small ermine moth (*Yponomeuta padella*).

572 Web of *Yponomeuta cagnagella* on *Euonymus*.

573 Adult of *Yponomeuta cagnagella*.

larvae attacking the foliage to feed in communal webs which eventually cover the shoots and branches of host trees. Larvae are fully grown in June. They then pupate in pendulous, semi-transparent, silken cocoons dispersed throughout the communal web (cf. *Yponomeuta evonymella*). Adults emerge a few weeks later.

DAMAGE
Host plants are disfigured by the presence of webbing and loss or death of foliage; they are also weakened following severe attacks. Large numbers of larvae and pupae are often killed by parasitoids, but natural control is often insufficient to prevent infestations from reaching damaging levels.

574 Larva of *Yponomeuta cagnagella*.

Yponomeuta cagnagella (Hübner) (572–575)
syn. *Y. cognatella* Treitschke; *Y. evonymi* Zeller
Infestations of this relatively common species occur on wild European spindle (*Euonymus europaeus*) and are also noted occasionally on cultivated spindle bushes. The lifecycle is similar to that of *Yponomeuta padella*, but larval activity commences slightly earlier in the spring. At first, the larvae attack the unfurling leaves. Later, they form conspicuous silken webs on the shoots or branches. Adults (24–26 mm wingspan) are mainly white, the fore wings with white cilia and four longitudinal rows of black dots (including a row of 4–7 dots towards the lower margin). The larvae (18–22 mm long when fully grown) are yellowish grey to greenish grey, marked with black spots; head, prothoracic and anal plates black; younger larvae are darker bodied. Fully fed larvae pupate in groups in white, opaque cocoons (cf. *Yponomeuta plumbella*). Adults emerge in July.

575 Pupal cocoons of *Yponomeuta cagnagella*.

Yponomeuta evonymella (Linnaeus) (576–577)
syn. *Y. padi* Zeller

A widely distributed and often abundant species on bird-cherry (*Prunus padus*), the greenish-grey, black-spotted larvae commonly causing complete defoliation of host trees and coating the trunks and branches in dense, polythene-like sheets of webbing. Adults (23–25 mm wingspan) are larger than those of *Yponomeuta padella*; also, the fore wings are pure white, with white cilia and five or six longitudinal rows of black dots (including a row of 9–11 dots towards the lower margin); the hind wings are dark grey. The biology of both species is similar but individuals of *Y. evonymella* pupae in white, opaque cocoons formed in distinct clusters within the larval web.

Yponomeuta plumbella (Denis & Schiffermüller)
This uncommon, relatively small species occurs on spindle (*Euonymus*). At first, the larvae feed within the shoots, causing them to wilt, but they later inhabit webs on the foliage. Adults (16–18 mm wingspan) have greyish-white fore wings (each with four longitudinal

576 Web of *Yponomeuta evonymella* on *Prunus padus*.

577 Adult and pupal cocoons of *Yponomeuta evonymella*.

rows of black dots) and dark grey hind wings. The larvae are yellowish grey, marked with black spots. Pupae are light brown, and formed within the larval web in spatially separated silken cocoons.

Yponomeuta rorrella (Hübner) (578–581)
syn. *Y. rorella* (Hübner)

This uncommon, local, southerly-distributed species is essentially similar to *Yponomeuta padella*, but the larvae feed on white willow (*Salix alba*) and sometimes on other species of *Salix*, including grey willow (*S. cinerea*), osier (*S. viminalis*) and pussy willow (*S. caprea*). The larvae inhabit compact webs on the branches and cause noticeable defoliation. Individuals (18–20 mm long when fully grown) are greyish green, marked with black or dark grey; the head, prothoracic plate and anal plate are dark brown. The adults (22–25 mm wingspan) are greyish white, with four longitudinal rows of black dots on each fore wing and dark grey hind wings. The pupae occur amongst the webbing, but with little protection.

Yponomeuta sedella Treitschke
syn. *Y. vigintipunctata* (Retzius)
Sedum small ermine moth

This moth is a local, mainly southerly-distributed pest of cultivated stonecrop (*Sedum*), occurring most commonly on orpine (*S. telephium*). The yellowish-grey, black-spotted larvae (up to 25 mm long) feed in small groups in June to July and in September to October, forming silken webs on the foliage. Fully grown larvae pupate in opaque, silvery-white cocoons spun amongst withered leaves at the base of the host plant. Adults (22–25 mm wingspan) are mainly grey, with three longitudinal rows of black spots on each fore wing. They occur in April and May, and again in August.

Ypsolopha dentella (Fabricius) (582–583)
syn. *Y. harpella* (Denis & Schiffermüller)

Generally common on honeysuckle (*Lonicera*) and occasionally a minor pest of cultivated plants. Widely distributed in Europe.

DESCRIPTION
Adult: 17–21 mm wingspan; fore wings dentate at tip, chocolate-brown, marked with yellow and white; hind wings grey. **Larva:** up to 18 mm long; yellowish green, with a broad, reddish, black-marked stripe down the back; head yellowish, marked with brown.

LIFE HISTORY

Larvae feed on honeysuckle during May and June, each sheltering in a rolled leaf. They are extremely active if disturbed, and wriggle violently backwards out of their habitation and drop to the ground. When fully fed, the larvae pupate in light brown, boat-shaped cocoons formed either on the foodplant or amongst debris on the ground. The adult moths appear in July and August.

DAMAGE

Larvae cause loss of leaf tissue and slight distortion of new growth but damage is not important.

578 Web of *Yponomeuta rorrella* on *Salix fragilis*.

579 Larvae of *Yponomeuta rorrella*.

580 Adult of *Yponomeuta rorrella*.

581 Pupae of *Yponomeuta rorrella*.

582 Adult of *Ypsolopha dentella*.

583 Larva of *Ypsolopha dentella*.

Family **COLEOPHORIDAE** (casebearer moths)

A distinctive family of small moths with long, narrow, pointed wings. The larvae commence feeding as leaf miners but then inhabit characteristic portable cases formed from silk and a cut-out portion of leaf or other plant tissue; crotchets on the abdominal prolegs are arranged into transverse bands.

Coleophora laricella (Hübner) (**584–585**)

Larch casebearer moth
larva = larch casebearer
An often common pest of larch (*Larix*), including ornamentals, but mainly of importance in forestry. Widespread in central and northern Europe; introduced into North America.

DESCRIPTION

Adult: 9–11mm wingspan; fore wings shiny grey; hind wings grey; head grey, with light grey antennae. **Larva:** up to 4.5 mm long; dark brown, with a black head and prothoracic plate, and a pair of small, black plates on the second thoracic segment; 16-legged. **Case:** 4.0–4.5 mm long; pale straw-coloured to greyish white.

LIFE HISTORY

Adults occur in June and July, depositing eggs on the shoots of larch. Larvae feed on the needles from September onwards, and hibernate during the winter. In spring, the larvae attack the young needles, entering a short distance from the tip of a needle and eating out the contents. Individuals are fully fed by the end of May or early June. They then pupate, and adults emerge a few weeks later. There is just one generation annually.

DAMAGE

The tips of damaged needles appear whitish and, when examined under a lens, the rounded entry hole (characteristic of casebearer feeding blotches) may be seen; the tips of damaged needles often break off.

Coleophora potentillae Elisha (**586–587**)

Small blotch mines formed by this locally common species are sometimes found on the leaves of cultivated rose (*Rosa*), disfiguring the foliage but usually not causing distortion. Attacks also occur on ornamental cinquefoil (*Potentilla*), goat's beard (*Aruncus dioicus*), *Rubus* and strawberry (*Fragaria*). The larval cases are cone-shaped, and reminiscent of tiny snail shells. They may be found on the underside of leaves from August to September. Pupation takes place in the larval case, attached to fallen leaves or other debris on the ground. Adults (10 mm wingspan) are mainly greyish-bronze; they occur in June.

Coleophora serratella (Linnaeus) (**588–589**)

syn. *C. fuscedinella* Zeller
Hazel casebearer moth
An abundant pest of trees, including alder (*Alnus*), birch (*Betula*), crab-apple (*Malus*), elm (*Ulmus*), hazel (*Corylus*) and hornbeam (*Carpinus betulus*); often associated with ornamentals and young amenity trees. The larvae inhabit short, cylindrical cases, attached to the underside of leaves. When feeding, they form pale blotches, which sometimes cause noticeable distortion of the leaf blades. Cases occur on host plants from September onwards, but are most obvious in spring and early summer. The brownish, ochreous-tinged adults (12 mm wingspan) occur in July and early August.

584 Case of larch casebearer (*Coleophora laricella*).

585 Larch casebearer (*Coleophora laricella*) damage to needles of *Larix*.

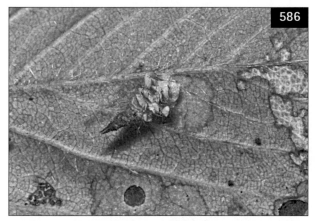

586 Larval case of *Coleophora potentillae*.

587 Mines of *Coleophora potentillae* in leaf of *Rosa*.

588 Case of *Coleophora serratella* on leaf of *Ulmus*.

589 Mines of *Coleophora serratella* in leaf of *Betula*.

Family **OECOPHORIDAE**

A group of relatively small moths with prominent labial palps and broadly elongate fore wings. The larvae have a well-developed prothoracic plate and are very active.

Agonopterix conterminella (Zeller) (**590–591**)
Willow shoot moth
A generally common but minor pest of willow (*Salix*). Present throughout central Europe; also found in the more southerly parts of northern Europe.

DESCRIPTION
Adult: 18–19 mm wingspan; fore wings reddish brown, suffused with blackish and yellowish-white scales; hind wings brownish white. **Larva:** up to 18 mm long: light green, with black pinacula; head yellowish brown; young larva pale, with the head, prothoracic plate and pinacula black.

LIFE HISTORY
Larvae feed on willow trees in May and June, spinning the leaves of the new shoots together to form a shelter. Fully fed larvae drop to the ground and eventually pupate, each in a subterranean, silken cocoon. Adults occur from July to September.

DAMAGE
The larvae cause noticeable distortion of the shoot tips but infestations are rarely extensive.

590 Willow shoot moth (*Agonopterix conterminella*).

591 Larva of willow shoot moth (*Agonopterix conterminella*).

Agonopterix ocellana (Fabricius) (**592–593**)

This widely distributed species is also associated with willow (*Salix*). The larvae feed within spun leaves or shoots, and sometimes cause slight damage to cultivated plants. Adults (23 mm wingspan) are characterized by the elongate, pale greyish-buff fore wings, each suffused with black and marked with a prominent red streak; they occur from August to April, hibernating throughout the winter. The larvae (up to 20 mm long) are green, with yellowish intersegmental divisions and small, but distinct, black pinacula.

Carcina quercana (Fabricius) (**594–595**)

A minor pest of deciduous ornamental trees and shrubs, including beech (*Fagus sylvatica*), crab-apple (*Malus*), flowering cherry (*Prunus*) and oak (*Quercus*). The bright, apple-green, somewhat flattened larvae (up to 15 mm long) feed on the underside of leaves during May and June, each sheltered beneath a slight web. Pupation occurs beneath an opaque web from mid-June onwards, and the bright yellow to greyish-pink adults (15–20 mm wingspan) appear in July and August.

592 Adult of *Agonopterix ocellana.*

593 Larva of *Agonopterix ocellana.*

594 Adult of *Carcina quercana.*

595 Larva of *Carcina quercana.*

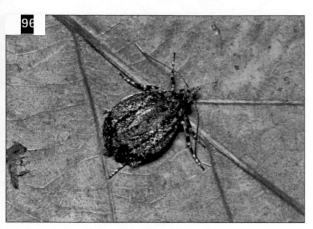

596 Female of *Diurnea fagella*.

597 Male of *Diurnea fagella*.

598 Larva of *Diurnea fagella*.

Diurnea fagella (Denis & Schiffermüller) (596–598)

A generally abundant woodland species, attacking various ornamental trees and shrubs, including beech (*Fagus sylvatica*), birch (*Betula*), crab-apple (*Malus*), hazel (*Corylus*), oak (*Quercus*) and willow (*Salix*). Widely distributed in Europe; also present in Asia Minor.

DESCRIPTION

Adult female: 18–20 mm wingspan, but wings much reduced and pointed; fore wings whitish, suffused to a greater or lesser degree with brownish black and bearing blackish and buff scale tufts; hind wings greyish to brownish black; palps long. **Adult male:** 25–28 mm wingspan; coloration similar to female. **Larva:** up to 18 mm long; pale, dull yellowish green, with yellowish intersegmental markings; head light brown; prothoracic plate mainly yellowish, with a pair of brown lateral markings; third pair of thoracic legs fleshy and projecting well beyond width of body. **Pupa:** 10 mm long; brown; cremaster with a cluster of long, hooked setae.

LIFE HISTORY

Adults appear in March and April. The females, although incapable of flight, are very active and crawl with considerable speed if disturbed. Eggs are laid on various hosts, the larvae then feeding from June to October in spun or, occasionally, rolled leaves. Fully fed larvae spin cocoons amongst debris on the ground. The winter is passed in the pupal stage.

DAMAGE

Damage is rarely extensive and usually limited to loss of a few leaves.

Family **GELECHIIDAE**

Anacampsis blattariella (Hübner) (**599–600**)
syn. *A. betulinella* Vári

A common but minor pest of birch (*Betula*) including, occasionally, nursery and ornamental trees. Widely distributed in Europe.

DESCRIPTION
Adult: 16–18 mm wingspan; fore wings light grey, marked extensively with black or dark grey, and with several black scale tufts and black terminal spots; hind wings dark grey. **Larva:** up to 14 mm long; greyish green, with large black pinacula; head and prothoracic plate shiny black.

LIFE HISTORY
Larvae of this single-brooded species feed in webbed leaves of birch during May and June, and are often mistaken for those of the family Tortricidae (q.v.). When fully grown each pupates in a rolled leaf, the adults occurring from July to September.

DAMAGE
Larvae cause slight distortion of the new growth but attacks are not important.

Anacampsis populella (Clerck) (**601–602**)
This species is essentially similar to *Anacampsis blattariella* but is associated with poplar (*Populus*) and willow (*Salix*). The larvae (up to 14 mm long) are greyish to yellowish grey, with prominent black pinacula and a black head and prothoracic plate. They feed within rolled leaves during the late spring and early summer. The dark grey adults (16–18 mm wingspan) occur in July and August.

599 Adult of *Anacampsis blattariella.*

600 Larva of *Anacampsis blattariella.*

601 Larva of *Anacampsis populella.*

602 Adult of *Anacampsis populella.*

603 Juniper webber moth (*Dichomeris marginella*).

604 Juniper webworms (*Dichomeris marginella*).

Dichomeris marginella (Fabricius) (**603–604**)

Juniper webber moth

larva = juniper webworm

A locally common pest of juniper (*Juniperus*), often causing damage to ornamental plants in gardens and nurseries. Widespread in Europe; also present in parts of North America.

DESCRIPTION

Adult: 15–16 mm wingspan; fore wings dark brown, with a white stripe along the anterior and hind margins; head and thorax white. **Larva:** up to 10 mm long; light brown, with darker dorsal and subdorsal stripes, the area between often slightly whitish; head and prothoracic plate brown to brownish black.

LIFE HISTORY

Adults occur in July and August, depositing eggs in groups on host plants. The larvae feed gregariously in May and June, webbing the shoots together with silken threads. Fully grown larvae pupate in cocoons spun amongst the webbing, and moths appear a few weeks later.

DAMAGE

Infested shoots are disfigured by webbing, much of the foliage dying and turning brown.

Hypatima rhomboidella (Linnaeus) (**605–606**)

syn. *H. conscriptella* (Hübner)

A minor pest of birch (*Betula*) and hazel (*Corylus*); occasionally found on nursery trees. Widely distributed in central Europe; also found in the more southerly parts of northern Europe.

DESCRIPTION

Adult: 13–16 mm wingspan; fore wings elongate, light grey, each with a conspicuous blackish, triangular blotch and a terminal streak; hind wings light grey. **Larva:** up to 12 mm long; green or pinkish brown, with small, black pinacula; head and prothoracic plate shiny black.

LIFE HISTORY

Adults occur in July, August and September. They often hide amongst dead leaves during the day, scurrying away if disturbed. The larvae feed on the leaves if birch and hazel in the following May and June, each sheltering within a rolled leaf or folded leaf edge. Fully grown individuals usually pupate amongst debris on the ground, and the adults emerge shortly afterwards. There is one generation annually.

DAMAGE

Infested leaves attract attention but damage caused is of no importance.

605 Adult of *Hypatima rhomboidella*.

606 Larva of *Hypatima rhomboidella*.

Family **BLASTOBASIDAE**

A small family of small moths with elongate wings. The larvae have relatively small abdominal spiracles.

Blastobasis lignea (Wollaston) (**607–608**)

A minor pest of yew (*Taxus baccata*); also reported attacking Norway spruce (*Picea abies*). Of subtropical origin but now widely distributed in Europe.

DESCRIPTION

Adult: 16–18 mm wingspan; fore wings elongate, yellowish brown, marked with black. **Larva:** up to 10 mm long; purplish brown and rather plump, with a blackish head and prothoracic plate, and small, inconspicuous pinacula.

LIFE HISTORY

Adults occur during the summer, depositing eggs in August on yew and, less commonly, on other evergreen hosts such as Norway spruce. Larvae feed from September onwards, each sheltering on a shoot within a cluster of dead or decaying needles held together with silk. The larvae survive the winter in these habitations, and complete their development in the spring or early summer. Although browsing directly on living tissue, the larvae also feed on decaying leaves and other dead vegetative matter.

DAMAGE

Larvae cause only slight damage to the needles, and do not affect the growth of plants.

607 Adult of *Blastobasis lignea*.

608 Larva of *Blastobasis lignea*.

Blastobasis decolorella (Wollaston) (**609–610**)
Straw-coloured apple moth

A subtropical species of Madeiran origin, now established in parts of Europe where it sometimes causes damage in orchards, attacking tree bark and the maturing fruits; the larvae also feed on various other trees and shrubs, including ornamentals such as *Rhododendron*. Adults and larvae are similar in appearance to those of *Blastobasis lignea*, but the former more uniformly coloured, the wings being mainly pale ochreous yellow, and the latter purplish brown.

609 Straw-coloured apple moth (*Blastobasis decolorella*).

610 Larva of straw-coloured apple moth (*Blastobasis decolorella*).

Family **TORTRICIDAE** (tortrix moths)

A large and important group of small, relatively broad-winged moths, including several well-known pests. The larvae have well-developed thoracic legs and five pairs of abdominal prolegs, each proleg bearing a complete circle of similarly sized crotchets; many species have an anal comb with which particles of frass are ejected from the anus. Larvae often feed in spun or rolled leaves, wriggling backwards rapidly if disturbed. Pupation takes place in the larval habitation or elsewhere, the pupa protruding from the cocoon following emergence of the adult.

Acleris rhombana (Denis & Schiffermüller) (**611–612**)
Rhomboid tortrix moth

An often common pest of trees and shrubs, including crab-apple (*Malus*), hawthorn (*Crataegus*), flowering cherry (*Prunus*), ornamental pear (*Pyrus calleryana* 'Chanticleer') and rose (*Rosa*); often present in gardens and nurseries. Widely distributed in central and northern Europe; also present in North America.

DESCRIPTION
Adult: 13–19 mm wingspan; fore wings with a sub-falcate tip, dark reddish brown to ochreous with darker markings, a reticulate pattern and, sometimes, a distinct central scale tuft; hind wings light grey, darker in female. **Egg:** 0.8 × 0.5 mm; flat, oval and yellowish green. **Larva:** up to 14 mm long; greyish green or yellowish green, with pale inconspicuous pinacula; prothoracic plate reddish brown or green; anal plate light green; anal comb yellowish; thoracic legs black. **Pupa:** 7–9 mm long; reddish brown or dark brown; cremaster short and broad.

LIFE HISTORY
Adults occur from August to October. Eggs are laid singly or in small batches on the bark of trunks and branches. They hatch in the spring. Larvae then invade the opening buds. Later they feed in webbed leaves, usually at the tips of the young shoots. Larvae also feed on blossom trusses. Pupation takes place in June or early July, usually in a cocoon spun in the soil. The pupal stage is protracted, and lasts for six to eight weeks or more.

DAMAGE
Restricted mainly to the loss or distortion of younger leaf tissue and blossoms but with little or no effect on tree growth.

Acleris schalleriana (Linnaeus) (613–614)

Locally common in association with guelder-rose (*Viburnum opulus*) and wayfaring tree (*V. lantana*); infestations sometimes occur on ornamentals. Holarctic. Widely distributed in mainland Europe; in the British Isles most frequent in southern England.

DESCRIPTION

Adult: 15–19 mm wingspan; fore wings greyish white, greyish brown or dark ochreous, with a more or less conspicuous dark costal blotch. **Larva:** up to 14 mm long; green to yellowish green, with a yellowish or brownish green head; prothoracic plate yellowish green or green. **Pupa:** 8–9 mm long; light brown.

LIFE HISTORY

Adults appear from late August or September onwards. They hibernate during the winter and reappear in the spring. Eggs are then laid on host plants and hatch in late May or June. Larvae feed within spun leaves until August. Pupation occurs in the larval habitation.

DAMAGE

Infested shoots are distorted but infestations are of little importance.

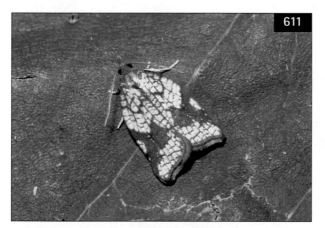

611 Rhomboid tortrix moth (*Acleris rhombana*).

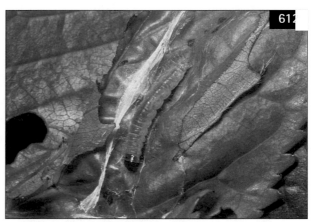

612 Larva of rhomboid tortrix moth (*Acleris rhombana*).

613 Adult of *Acleris schalleriana.*

614 Larva of *Acleris schalleriana.*

Acleris sparsana (Denis & Schiffermüller)

syn. *A. fagana* (Curtis); *A. reticulana* (Haworth)

A generally common but minor pest of beech (*Fagus sylvatica*) and sycamore (*Acer pseudoplatanus*), and often found on such hosts in parks and gardens; in mainland Europe, additional hosts include birch (*Betula*), oak (*Quercus*), poplar (*Populus*) and rowan (*Sorbus aucuparia*). Widespread in Europe.

DESCRIPTION

Adult: 18–22 mm wingspan; fore wings grey to white, marked with reddish brown and yellowish, often with an expanded costal blotch; hind wings grey. **Larva:** up to 16 mm long; light green, with inconspicuous, shiny pinacula; head brownish green or light green; prothoracic plate green, with a black mark on each side; anal plate green; thoracic legs green. **Pupa:** 8–10 mm long; light brown.

LIFE HISTORY

Adults appear in the late summer and autumn, hibernating in the winter and reappearing in the spring. Larvae occur from June to late July or early August. Young larvae spin flimsy webs on the underside of expanded leaves. Older individuals occur between two overlapping leaves which they spin together with silk, sheltering in a folded portion of the uppermost leaf. Fully fed individuals pupate in July or August, either within the larval habitation or amongst debris on the ground.

DAMAGE

Larval feeding is restricted to the leaves and, although sometimes noticeable, is of no economic importance.

Acleris variegana (Denis & Schiffermüller) (615–616)

Garden rose tortrix moth

A generally common pest of rosaceous plants, including flowering cherry (*Prunus*), but most common on rose (*Rosa*); also occurs on barberry (*Berberis*). Palaearctic. Widely distributed in Europe.

DESCRIPTION

Adult: 14–18 mm wingspan; fore wings whitish ochreous to purplish, variably suffused with grey, frequently with much of the basal half white and with distinct, often black, scale tufts; hind wings grey. **Egg:** 0.6 × 0.5 mm; pale yellowish to reddish. **Larva:** up to 14 mm long; light green or yellowish green; head and prothoracic plate yellowish brown or greenish brown; anal plate green; thoracic legs yellowish brown. **Pupa:** 7–8 mm long; light brown.

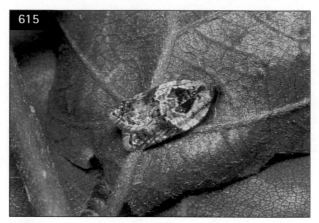

615 Female garden rose tortrix moth (*Acleris variegana*).

616 Larva of garden rose tortrix moth (*Acleris variegana*).

LIFE HISTORY

Adults occur from July to September. Eggs are then laid singly or in small batches on either side of leaves, usually along the midrib. They hatch in the following spring. The larvae feed on the young shoots from May to late June or early July, sheltering within loosely spun leaves or in folded leaf edges. Pupation occurs in the larval habitation or amongst fallen leaves.

DAMAGE

Larval habitations are unsightly and cause distortion.

617 Larva of *Acleris emargana.*

618 Adult of *Acleris emargana.*

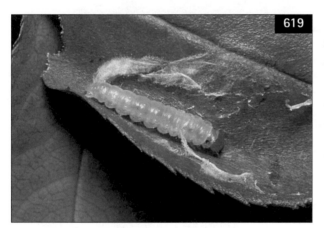

619 Broad-barred button moth (*Acleris laterana*).

620 Larva of broad-barred button moth (*Acleris laterana*).

Acleris emargana (Fabricius) (**617–618**)

A generally common species, associated with birch (*Betula*), poplar (*Populus*), pussy willow (*Salix caprea*) and other broad-leaved willows. Minor infestations are sometimes noted on nursery trees. The mainly green larvae (up to 15 mm long) feed within folded leaves or spun shoots during May and June. Damage also occurs on alder (*Alnus*) and hazel (*Corylus*). Adults (16–22 mm wingspan) are mainly brown, with a characteristic emargination on the costa of each fore wing. They appear from July onwards.

Acleris laterana (Fabricius) (**619–620**)

syn. *A. latifasciana* (Haworth)

Broad-barred button moth

This widely distributed and generally common species is a minor pest of rose (*Rosa*) and certain other trees and shrubs, but is more common on raspberry. The larvae (up to 15 mm long) are whitish green, with a pale yellowish-brown head; they feed between spun leaves in May and June, causing conspicuous but usually minor damage. The adults (15–20 mm wingspan) are mainly whitish to ochreous, marked with black, brown or red; they fly in August and September, and are often common in gardens.

Adoxophyes orana (Fischer von Röslerstamm) (**621–622**)

Summer fruit tortrix moth

This species is widely distributed in Europe and, since the 1950s, following a noticeable westerly extension of its range, has also become established as an important fruit pest in south-eastern England. Although associated mainly with fruit trees, the larvae also occur on various other hosts, including alder (*Alnus*), birch (*Betula*), hazel (*Corylus*), honeysuckle (*Lonicera*), poplar (*Populus*), rose (*Rosa*) and willow (*Salix*). They feed on buds and also web the foliage of the young shoots, but damage caused on ornamentals is unimportant. The larvae (up to 20 mm long) are yellowish green or greyish green to dark green, with small pale pinacula and a yellowish-brown head and prothoracic plate. They occur in two main broods, from June to August, and from September to May. The greyish-brown adults (15–22 mm wingspan) occur in June, with a larger flight extending from mid-August to September or October.

Aleimma loeflingiana (Linnaeus) (**623–624**)

This widely distributed and generally common woodland species is associated mainly with oak (*Quercus*) but also attacks hornbeam (*Carpinus betulus*) and maple (*Acer*). The larvae feed within folded or rolled leaves throughout May and usually pupate in June. Attacks are sometimes noted on young cultivated trees but are of no economic importance. The larvae (up to 15 mm long) are green to blackish green, with blackish-brown to black pinacula, head, prothoracic and anal plates. Adults, which occur in June and July, are mainly whitish brown to dark brown, with a dark margin at the base of the cilia.

Ancylis mitterbacheriana (Denis & Schiffermüller) (**625–626**)

A locally common pest of beech (*Fagus sylvatica*) and oak (*Quercus*). Eurasiatic. Widely distributed in Europe.

621 Larva of summer fruit tortrix moth (*Adoxophyes orana*).

622 Summer fruit tortrix moth (*Adoxophyes orana*).

623 Larva of *Aleimma loeflingiana*.

624 Male of *Aleimma loeflingiana*.

DESCRIPTION

Adult: 12–16 mm wingspan; fore wings mainly brownish orange to whitish, with a prominent reddish-brown or chocolate-brown dorsal patch, and a curved blackish streak towards the apex. **Larva:** up to 12 mm long; greyish green to yellowish green, with pale and relatively large pinacula; head yellowish brown to greenish brown, marked with black; prothoracic plate light green to yellowish brown, with a pair of blackish lateral patches towards the hind margin and blackish markings dorsally; anal plate light green, marked with blackish brown. **Pupa:** 6–8 mm long; light reddish brown.

LIFE HISTORY

Adults occur in May and June, and are most common in or close to beech and oak woodlands. Larvae inhabit individual pod-like shelters, formed out of a folded leaf, and devour the innermost surface. They feed from July to September, overwinter in the larval habitation and pupate in the following spring.

DAMAGE

Larval habitations attract attention when present on nursery or specimen trees, but feeding damage is superficial and of little or no importance.

Ancylis upupana (Treitschke) (**627–628**)

A locally distributed species, associated mainly with birch (*Betula*) but also attacking elm (*Ulmus*). Although generally uncommon, the larvae are found, occasionally, on nursery and specimen trees, sheltering between spun leaves and causing minor damage to the expanded foliage. They feed from July to September, overwintering when fully fed and pupating in the spring. Larvae (up to 12 mm long) are greenish grey to dark olive-brown, with dark pinacula (those on the thoracic segments being most prominent) and a black or brownish head and prothoracic plate. Adults (12–18 mm wingspan) are mainly brown, marked with silvery grey and brownish orange; they occur in May and June.

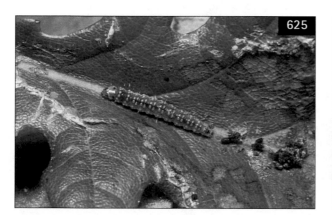

625 Larva of *Ancylis mitterbacheriana*.

626 Larval habitation of *Ancylis mitterbacheriana* on *Quercus*.

627 Larva of *Ancylis upupana*.

628 Adult of *Ancylis upupana*.

629 Female of *Archips crataegana*.

630 Larva of *Archips crataegana*.

631 Male fruit tree tortrix moth (*Archips podana*).

632 Larva of fruit tree tortrix moth (*Archips podana*).

Archips crataegana (Hübner) (629–630)

A minor pest of trees and shrubs, including birch (*Betula*), ash (*Fraxinus excelsior*), crab-apple (*Malus*), elm (*Ulmus*), flowering cherry (*Prunus*), lime (*Tilia*), oak (*Quercus*) and broad-leaved willows such as grey willow (*Salix cinerea*). Eurasiatic. Widely distributed in central and much of northern Europe.

DESCRIPTION

Adult female: 23–27 mm wingspan; fore wings brown, with darker, chocolate-coloured markings; hind wings brownish grey. **Adult male:** 19–22 mm wingspan; fore wings paler than in female. **Egg:** 0.6 × 0.4 mm; laid in a white mass. **Larva:** up to 23 mm long; dull greenish black, with black pinacula; head, prothoracic plate and anal plate shiny black; anal comb black, with 6–8 teeth. **Pupa:** 9–12 mm long; dull black; cremaster elongate.

LIFE HISTORY

Adults of this single-brooded species occur from late June to August. Eggs are laid in conspicuous batches of about 30 on the trunks and main branches, and then coated with a white substance which quickly hardens and disguises them as bird-droppings. Eggs hatch in April or early May. The tiny larvae are very active and rapidly climb the tree to begin feeding on the underside of the leaves. Later, each feeds inside a tightly rolled leaf edge, usually on fully expanded foliage at the shoot tips. Pupation occurs in a rolled leaf or between two spun leaves. Adults emerge a few weeks later.

DAMAGE

Feeding is restricted mainly to the expanded leaves and, unless larvae are very numerous, damage is insignificant.

Archips podana (Scopoli) (631–632)

Fruit tree tortrix moth

A generally common, polyphagous pest of trees and shrubs, including birch (*Betula*), crab-apple (*Malus*), flowering cherry (*Prunus*), hawthorn (*Crataegus*), Japanese spindle (*Euonymus japonica*), ornamental pear (*Pyrus calleryana* 'Chanticleer') and *Sorbus*; often present on garden trees and nursery stock. Eurasiatic; also present in North America, probably as an introduced species. Widely distributed in Europe.

DESCRIPTION

Adult female: 20–28 mm wingspan; fore wings purplish ochreous, each with a brown, reticulated pattern and darker markings, and a dark spot at the tip; hind wings brownish grey, suffused with orange apically. **Adult male:** 19–23 mm wingspan; fore wings purplish to purplish ochreous, each with dark reddish-brown, velvety markings and a dark spot at the tip; hind wings greyish, tinged with orange apically. **Egg:** 0.6–0.7 mm across; flat and almost circular; green and laid in a large raft-like batch. **Larva:** up to 22 mm long; light green to greyish green, with pale pinacula; head chestnut-brown or black; prothoracic plate chestnut-brown, with darker lateral and hind margins, a pale anterior margin and a pale, narrow mid-line; anal plate green or grey; thoracic legs brownish black or black; prothoracic spiracle elliptical and hindmost spiracle distinctly larger than the rest. **Pupa:** 9–14 mm long; dark yellowish brown to blackish brown.

LIFE HISTORY

Adults occur from June to September but are usually most abundant in July. Eggs are deposited on the leaves in flat, oval batches of about 50, with the shells overlapping like roof tiles. The eggs, which are extremely difficult to find as they closely match the colour of the leaves, hatch in about three weeks. The larvae feed on the foliage for a few weeks and then enter hibernation. Overwintered larvae become active in late March or April, immediately burrowing into the opening buds. Larvae later attack the foliage, each webbing two or more leaves together and sheltering between them, or forming a retreat by spinning a dead leaf to a healthy one or to a twig. Larval development is completed in May or June. Individuals then pupate within the larval habitation or within freshly spun leaves nearby. Adults emerge three or more weeks later. In favourable situations, when adult emergence and egg laying is particularly advanced, some larvae feed up and pupate, to produce a partial second generation of moths in the late summer or early autumn.

DAMAGE

Attacks on buds are potentially serious. Also, larval feeding or the presence of their webbing disfigures, and may also affect the development of, young shoots. Damage to fully expanded foliage is usually unimportant.

Archips rosana (Linnaeus) (633–634)

Rose tortrix moth

An often abundant but minor garden pest of trees and shrubs, including ornamentals such as crab-apple (*Malus*), flowering cherry (*Prunus*), rose (*Rosa*) and certain conifers. Eurasiatic; also occurs in North America. Widely distributed in Europe.

DESCRIPTION

Adult female: 17–24 mm wingspan; fore wings darker brown and markings more diffuse than in male; hind wings grey, suffused apically with orange-yellow. **Adult male:** 15–18 mm wingspan; fore wings light brown to purplish brown, with dark brown, often pinkish-tinged, markings; hind wings grey. **Egg:** 0.9 × 0.7 mm; flat, oval and greyish green; laid in a large raft. **Larva:** up to 22 mm long; light green to dark green, with pale pinacula; head and prothoracic plate light brown to black; anal plate green or light brown. **Pupa:** 9–11 mm long; dark brown.

633 Male rose tortrix moth (*Archips rosana*).

634 Larva of rose tortrix moth (*Archips rosana*).

LIFE HISTORY

Eggs are laid during August and September on the bark of host plants, and hatch in the following April. The larvae then feed in the buds; later, each larva inhabits a rolled leaf or spun shoot. Pupation occurs in the larval habitation from June onwards. Adults occur from July to early September.

DAMAGE

Infested leaves or shoots are disfiguring but of little or no importance.

Archips xylosteana (Linnaeus) (635–637)

Brown oak tortrix moth

Generally common in woodland habitats on trees, shrubs and certain herbaceous plants; minor infestations occur occasionally in gardens and nurseries on ash (*Fraxinus excelsior*), birch (*Betula*), crab-apple (*Malus*), elm (*Ulmus*), fir (*Abies*), hazel (*Corylus*), honeysuckle (*Lonicera*), *Hypericum*, lime (*Tilia*), oak (*Quercus*) and some other trees and shrubs. Eurasiatic. Present throughout much of Europe.

DESCRIPTION

Adult: 15–23 mm wingspan; fore wings whitish ochreous, with reddish-brown, variegated markings; hind wings greyish. **Larva:** up to 22 mm long; whitish grey, sometimes grey or dark bluish grey, with black pinacula; head shiny black; prothoracic plate black or dark brown, with a whitish mid-line and collar; anal comb present. **Pupa:** 9–12 mm long; dark brown or black; cremaster elongate.

LIFE HISTORY

Adults occur in July, eggs being laid in batches on the trunks or branches of various trees and shrubs. The eggs are then coated with a brownish secretion which camouflages them against the bark. They hatch in the following April or early May. Larva then attack the foliage, each inhabiting a rolled leaf. Individuals are fully grown in June; they then pupate in the larval habitation, and adults emerge a few weeks later.

DAMAGE

Larval habitations disfigure host plants and cause concern. However, feeding is confined mainly to fully expanded leaves and is, therefore, of little or no significance.

635 Female brown oak tortrix moth (*Archips xylosteana*).

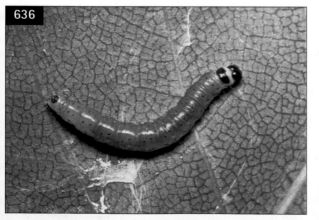

636 Larva of brown oak tortrix moth (*Archips xylosteana*).

637 Larval habitation of brown oak tortrix moth (*Archips xylosteana*).

Cacoecimorpha pronubana (Hübner) (638–640)
Carnation tortrix moth

Originally from southern Europe and nowadays a common greenhouse pest in many parts of Europe, including the British Isles where it was first reported in 1905. Ornamental hosts include bay laurel (*Laurus nobilis*), *Cupressus*, *Cytisus*, *Daphne*, *Dianthus*, false acacia (*Robinia pseudoacacia*), *Fuchsia*, *Grevillea*, honeysuckle (*Lonicera*), *Hypericum*, ivy (*Hedera*), Japanese spindle (*Euonymus japonica*), privet (*Ligustrum vulgare*) and many others. In favourable districts infestations occur outdoors. Also present in Asia Minor and North America.

DESCRIPTION

Adult female: 18–22 mm wingspan; fore wings pale orange-brown, reticulated with darker brown; hind wings mainly orange. **Adult male:** 12–17 mm wingspan; fore wings orange-brown, with variable reddish-brown and blackish markings; hind wings bright orange, each with a blackish border. **Egg:** flat and oval; light green, laid in a large, scale-like batch. **Larva:** up to 20 mm long; olive-green to bright green, paler below, and with slightly paler pinacula; head greenish yellow or yellowish brown, marked with dark brown; prothoracic and anal plates green, marked with dark brown; anal comb green, usually with six teeth. **Pupa:** 9–12 mm long; brownish black to black; cremaster elongate and tapered, with eight strong, hooked spines.

LIFE HISTORY

Adults appear mainly from April to October, but are most numerous from May to June, and in late August and September. The males have a characteristic, erratic flight and are very active in sunny weather. Eggs are laid on leaves in large groups of up to 200, the shells overlapping like roof tiles, and hatch in 2–3 weeks. Larvae at first browse on the surface of leaves, sheltering beneath a slight web. Later they feed in spun leaves, shoots or blossom trusses. Pupation occurs within the larval habitation, in a freshly folded leaf or amongst webbed foliage, and adults emerge shortly afterwards. The winter is usually passed as young larvae sheltering on the host in silken webs.

DAMAGE

The larvae are voracious feeders, and cause considerable harm to foliage, buds and flowers.

638 Female carnation tortrix moth (*Cacoecimorpha pronubana*).

639 Male carnation tortrix moth (*Cacoecimorpha pronubana*).

640 Larva of carnation tortrix moth (*Cacoecimorpha pronubana*).

Celypha lacunana (Denis & Schiffermüller) (**641–642**)

syn. *Olethreutes lacunana* (Denis & Schiffermüller)

Dark strawberry tortrix moth

Larvae of this widely distributed species are associated mainly with low-growing plants in damp habitats, and are well-known pests of strawberry crops. They are also sometimes noted on garden ornamentals, including marsh marigold (*Caltha palustris*). Occasionally, larva also occur on trees and shrubs, including birch (*Betula*), crab-apple (*Malus*), larch (*Larix*), privet (*Ligustrum vulgare*) and spruce (*Picea*). The larvae (up to 14 mm long) are mainly dark purplish brown to blackish brown, and extremely active if disturbed. They occur in early summer, with those of a second brood feeding from late August or September onwards; the latter overwinter whilst still small and complete their development in the spring. The greyish-ochreous to greenish-black adults (15–18 mm wingspan) appear from May onwards, with a second generation flying in August and September.

Clepsis spectrana (Treitschke) (**643–644**)

syn. *C. costana* sensu Fabricius

Straw-coloured tortrix moth

Most common in fenland and coastal habitats on herbaceous plants, including various horticultural crops; often a pest in greenhouses and nurseries on ornamentals such as *Cyclamen*, *Iris*, rose (*Rosa*) and conifers. Widely distributed in central and northern Europe.

DESCRIPTION

Adult: 15–24 mm wingspan (male usually noticeably smaller than female); fore wings pale ochreous to yellowish, with variable dark brown to blackish markings; hind wings light grey. **Egg:** orange, flat and oval. **Larva:** up to 25 mm long; brown to greyish olive-green, paler dorsally, with whitish pinacula; head and prothoracic plate shiny black or blackish brown; anal plate whitish brown, marked with black or brown; anal comb with 6–8 teeth. **Pupa:** 10–14 mm long; dull black; cremaster stout and elongate.

LIFE HISTORY

In the open, first-generation adults of this double-brooded species occur from early June to July, but sometimes earlier. Eggs are laid in small groups on the foodplant, and hatch 2–3 weeks later. The larvae then feed beneath a web or in the shelter of young, webbed leaves or 'capped' flowers. When fully grown, each pupates in a white, silken cocoon in the larval habitation, in webbed leaves or amongst dead leaves.

A second generation of adults appears in August and September. Larvae from these moths feed for a short time before hibernating. The larvae re-appear in early spring to continue feeding. Fully grown individuals eventually pupate in May or June. Some populations in greenhouses breed continuously, having evolved the ability to develop without a diapause phase.

DAMAGE

The larvae destroy flowers and flower buds. They also cause considerable damage to the foliage, particularly on the young shoots.

Cnephasia asseclana (Denis & Schiffermüller) (**645–646**)

syn. *C. interjectana* (Haworth); *C. virgaureana* (Treitschke)

Flax tortrix moth

Polyphagous on herbaceous plants, and sometimes a pest of garden and greenhouse-grown ornamentals such as *Chrysanthemum*, golden-rod (*Solidago virgaurea*), *Helenium*, *Pelargonium*, *Phlox*, primrose (*Primula vulgaris*), *Rudbeckia* and sweet pea (*Lathyrus odoratus*); also damaging to young spruce (*Picea*) trees. Widely distributed in Europe; also recorded in the Canary Islands and in Newfoundland.

DESCRIPTION

Adult: 15–18 mm wingspan; fore wings whitish grey, suffused with black and dark yellowish-brown, blackish-edged markings; hind wings greyish brown. **Egg:** greenish yellow; flat and oval. **Larva:** up to 14 mm long; grey or bluish white to dark cream or greyish green, with black pinacula; head light or yellowish brown, marked with black; prothoracic plate light brownish to dark brown, marked with black; anal comb with about six long teeth; thoracic legs light to dark brown; prolegs marked with grey or black.

LIFE HISTORY

Moths occur from June to August. Eggs are deposited, either singly or in small batches, on herbaceous plants, tree trunks, posts and other supports. They hatch in about three weeks. Larvae then spin small cocoons in suitable shelter nearby, having fed only on their egg shells. They then hibernate. Activity is resumed early in the following spring, the first-instar larvae mining leaves to feed in irregular, usually blotch-like mines. Later, each larva feeds amongst spun leaves or on a flower, spinning the petals down with silk to form a 'capped' blossom. If disturbed, the larva rolls into a tight 'C' and drops to the ground. Pupation takes place in June, in the folded edge of a leaf or amongst debris on the ground.

641 Larva of dark strawberry tortrix moth (*Celypha lacunana*).

642 Dark strawberry tortrix moth (*Celypha lacunana*).

643 Female straw-coloured tortrix moth (*Clepsis spectrana*).

644 Larva of straw-coloured tortrix moth (*Clepsis spectrana*).

645 Flax tortrix moth (*Cnephasia asseclana*).

646 Larva of flax tortrix moth (*Cnephasia asseclana*).

DAMAGE
Attacks are usually of only minor importance but infestations, particularly on flowers, may affect the marketability of commercial pot plants.

647 Allied shade moth (*Cnephasia incertana*).

648 Larval habitation of *Cnephasia stephensiana* on leaf of *Geranium*.

Cnephasia incertana (Treitschke) (647)
Allied shade moth

Larvae of this generally distributed and common species are polyphagous on herbaceous plants, and are sometimes found on ornamentals, including *Chrysanthemum*, *Geranium* and *Saxifraga*, as well as young conifers. They occur from early spring onwards, pupating in late May or early June. Adults appear in June and July. The larvae are similar to those of *Cnephasia asseclana* but usually darker and have no anal comb; the adults of both species are very similar, but the fore wings of *Cnephasia incertana* are slightly narrower.

Cnephasia longana (Haworth)
larva = omnivorous leaf tier

This species occurs mainly in coastal and chalkland sites, and is reported occasionally as a minor pest of greenhouse and outdoor flowers. The larvae are polyphagous but are most often associated with Asteraceae, including *Aster*, *Chrysanthemum* and *Hypochoeris*; they attack the young shoots and flower head, often feeding under a canopy of webbed-down petals. Larvae are rather plump (up to 18 mm long) and greenish grey or yellowish grey, with pale longitudinal lines along the back and sides, and a light brown head, prothoracic plate and anal plate. They occur from early spring onwards, pupating in June. The whitish-ochreous to brownish-ochreous adults (15–22 mm wingspan) emerge in July.

649 Adult of *Cnephasia stephensiana*.

Cnephasia stephensiana (Doubleday) (648–649)

Larvae of this widely distributed species are polyphagous on herbaceous plants, and are often pests of cultivated plants in gardens and greenhouses, especially Asteraceae. The larvae feed from April to June, at first mining the leaves but later living in spun or folded leaves beneath slight webs; flowers are also attacked. Fully grown individuals are 15–18 mm long and shiny grey or bluish grey to greenish grey, with large, black pinacula, a brown or black head, a mainly black prothoracic plate, black thoracic legs and a blackish-brown anal plate; there is no anal comb. Adults (18–22 mm wingspan) are mainly light grey, with blackish-edged, brownish-grey markings; they occur in July and August.

Croesia bergmanniana (Linnaeus) (**650–651**)

Generally common on wild and cultivated rose (*Rosa*); in mainland Europe also reported on alder buckthorn (*Frangula alnus*) and common buckthorn (*Rhamnus cathartica*). Present throughout central and northern Europe; also occurs in North America.

DESCRIPTION

Adult: 12–15 mm wingspan; head, thorax and fore wings bright yellow, the latter suffused with brownish orange and bearing often dark-edged silvery-grey markings; hind wings brownish grey. **Larva:** up to 10 mm long; greenish grey to yellowish white or yellow, with indistinct pinacula; head, thoracic legs, prothoracic and anal plates brownish black or black; anal comb present. **Pupa:** 7–8 mm long; dark brown, with a yellowish-brown or light brown abdomen.

LIFE HISTORY

Moths occur in June and July, often flying during the afternoon as well as at night. Eggs are laid on the stems of rose bushes, where they remain until the following spring. Larvae feed from April onwards, inhabiting tightly folded leaves or spun shoots and sometimes spinning a shoot tip to an adjacent flower bud. Pupation occurs in the larval habitation or in a folded leaf. Adults appear a few weeks later.

DAMAGE

If numerous, the larvae cause noticeable distortion and loss of young shoots, affecting growth and reducing flowering potential of the bushes.

Croesia holmiana (Linnaeus) (**652–653**)

An often common but minor pest of ornamental trees and shrubs, including blackthorn (*Prunus spinosa*), crab-apple (*Malus*), *Chaenomeles*, hawthorn (*Crataegus*) and rose (*Rosa*). Eurasiatic. Widely distributed in central and northern Europe.

DESCRIPTION

Adult: 12–15 mm wingspan; fore wings orange-yellow, suffused with dark reddish brown, each with a distinctive white triangular mark; hind wings dark grey.

650 Adult of *Croesia bergmanniana*.

651 Larva of *Croesia bergmanniana*.

652 Adult of *Croesia holmiana*.

653 Larva of *Croesia holmiana*.

Larva: up to 10 mm long; yellowish green; head light brown; prothoracic plate brown to black; thoracic legs brown. **Pupa:** 7–8 mm long; reddish yellow or dark yellow.

LIFE HISTORY

Adults occur in July and August, eggs being deposited on the bark of the shoots and small branches of host plants. Eggs hatch in the following spring. Larvae feed from May to June, each in a shelter formed from two leaves spun together. When fully fed, individuals pupate in the larval habitation or in a folded leaf. There is just one generation annually.

DAMAGE

Larvae cause minor damage to the leaves but infestations on cultivated plants are unimportant.

Croesia forsskaleana (Linnaeus) (654)
syn. *C. forskaliana* (Haworth)

Larvae of this widely distributed species feed from September to May or June on field maple (*Acer campestre*) and sycamore (*A. pseudoplatanus*); after hibernation individuals attack the unfurling leaves and flowers and, later, inhabit a longitudinally rolled leaf. Although infestations are often common in parks and gardens, damage caused to cultivated plants is unimportant. Larvae (up to 10 mm long) are yellow, with small, pale pinacula, a yellowish-green head and anal plate, and a greenish prothoracic plate. Adults (12–15 mm wingspan) are mainly yellow, the fore wings reticulated with brownish yellow, partly bordered with black, and more or less suffused centrally with dark grey. They occur in July and August.

Ditula angustiorana (Haworth) (655–656)

A locally common and polyphagous pest of trees, shrubs and herbaceous plants, including ornamentals such as beech (*Fagus sylvatica*), common box (*Buxus sempervirens*), juniper (*Juniperus*), larch (*Larix*), bay laurel (*Laurus nobilis*), pine (*Pinus*), flowering cherry (*Prunus*), *Rhododendron* and yew (*Taxus baccata*). Widely distributed in Europe; also present in parts of North Africa and North America.

DESCRIPTION

Adult female: 14–18 mm wingspan; fore wings pale ochreous brown, marked distinctly with dark purplish brown and black; hind wings dark brown. **Egg:** pale yellow, flat and almost circular; laid in a scale-like batch. **Larva:** up to 18 mm long; slender, pale yellowish green to brownish green or greyish green, but darker above, with light green pinacula; head greenish yellow or yellowish brown, marked with blackish brown; prothoracic plate yellowish green, light brown or dark brown; anal comb greenish or brownish, with four teeth; thoracic legs green, tipped with blackish brown; spiracles small, the hindmost twice the diameter of the others. **Pupa:** 8 mm long; light brown; cremaster elongate, with eight tightly hooked setae.

LIFE HISTORY

Adults occur in June and July, the males often flying in sunshine. Eggs are laid on the leaves in moderately large batches, and hatch in August. Young larvae feed on the foliage but in the autumn, whilst still small, they spin silken cocoons on the buds or spurs. They then hibernate. The larvae reappear early in the following spring. They then attack the buds and young leaves, each spinning tissue together with silk to form a shelter. The larvae are very active if disturbed, wriggling backwards and dropping to the ground. Pupation occurs in May or June in a cocoon spun in a folded leaf, in webbed foliage or amongst dead leaves on the ground.

DAMAGE

Attacked leaves are either grazed on one surface or bitten right through; attacked young growth is also distorted but attacks on ornamentals are rarely sufficiently numerous to cause economic injury.

654

654 Adult of *Croesia forsskaleana.*

655 Male of *Ditula angustiorana.*

656 Larva of *Ditula angustiorana.*

657 Cherry bark tortrix moth (*Enarmonia formosana*).

Enarmonia formosana (Scopoli) (**657**)
syn. *Laspeyresia woeberiana* (Denis & Schiffermüller)
Cherry bark tortrix moth

A locally common pest of flowering cherry (*Prunus*); sometimes also damaging to crab-apple (*Malus*), ornamental pear (*Pyrus calleryana* 'Chanticleer') and *Sorbus.* Widely distributed in Europe.

DESCRIPTION
Adult: 15–18 mm wingspan; fore wings more or less brown, with a purplish sheen, and with irregular, yellowish-orange markings and silvery-white costal striae; hind wings dark brown. **Egg:** 0.7 × 0.6 mm; flat and oval, whitish to reddish. **Larva:** up to 11 mm long; brownish to salmon-pink, with brownish pinacula; head light brown; prothoracic and anal plates light greyish brown. **Pupa:** 7–9 mm long; light brown; tip blunt.

LIFE HISTORY
Adults appear from May or early June to September, the extended emergence period giving the impression of two generations. The moths are active in sunshine, and often make repeated short flights to and from the branches or trunks of infested trees. Eggs are laid singly or in groups of two or three, usually on previously infested or otherwise injured parts of trees, such as frost-damaged or mechanically damaged bark, or adjacent to pruning wounds. They hatch in two or three weeks and the larvae then feed beneath the surface, excavating irregular, often deep galleries in the bark; the underlying cambium may also be damaged but tunnels do not extend into the wood. Larvae are usually fully grown by the following spring or early summer, passing through five instars. Each then pupates in a silken cocoon formed in the larval feeding gallery. Pupae remain protruding from the bark after the adult moths have emerged.

DAMAGE
A considerable quantity of gum exudes from infested parts of host trees and this, along with accumulations of light-brown frass and silken webbing forced out of cracks in the bark, may be one of the first indications of an attack. Cherry trees are often invaded near the base of the trunks, infestations producing cracks, swellings and cankers; severely damaged trees may be killed.

Epiblema cynosbatella (Linnaeus) (658)

A common species on wild and cultivated rose (*Rosa*). In mainland Europe various other plants, including ornamentals such as *Chaenomeles*, are also attacked. Widespread in Europe.

DESCRIPTION

Adult female: 16–22 mm wingspan; fore wings creamy white, marked with brownish black, bluish grey and light brown, and with a large blackish basal patch; hind wings dark grey; labial palps yellow. **Adult male:** similar to female but hind wings light grey. **Larva:** up to 18 mm long; reddish brown, with inconspicuous pinacula. **Pupa:** 9-11 mm long; blackish green.

LIFE HISTORY

Larvae attack the young shoots and flower buds of rose bushes, feeding mainly in April and May. They then pupate in the larval habitation, usually between two spun leaves or in a spun shoot. Adults occur in May and June.

DAMAGE

Damage to foliage is of little consequence, but infested shoots are distorted and flower buds destroyed.

Epiblema roborana (Denis & Schiffermüller) (659–660)

Generally common on wild and cultivated rose (*Rosa*). Eurasiatic. Widely distributed in Europe.

DESCRIPTION

Adult: 16–22 mm wingspan; fore wings ochreous white, suffused or marked with buff, dark brown, bluish grey and black, and each with a prominent brownish-black basal patch which extends along the front margin; hind wings whitish-grey. **Larva:** up to 18 mm long; reddish brown to dark brown, with inconspicuous brown pinacula; head yellowish brown; prothoracic plate brownish black or black. **Pupa:** 9–11 mm long; light brown.

LIFE HISTORY

Adults occur from late June to August, and are often abundant in the vicinity of rose bushes. Eggs are deposited on the bushes, but do not hatch until the following spring. Larvae feed throughout May and early June, each sheltering in a tightly spun shoot and devouring the innermost tissue. Pupation usually occurs in the larval habitation.

DAMAGE

Infested bushes are disfigured and new shoots distorted or destroyed. The larvae also damage flower buds.

Epiblema rosaecolana (Doubleday) (661–662)

Frequently a pest of cultivated species of rose (*Rosa*) but associated mainly with sweet briar (*R. rubiginosa*). Eurasiatic; also established as an introduced pest in North America. Widely distributed in Europe.

658 Adult of *Epiblema cynosbatella*.

659 Adult of *Epiblema roborana*.

660 Larva of *Epiblema roborana*.

DESCRIPTION
Adult: 16–20 mm wingspan; fore wings white, partly suffused and striated with brown and bluish grey, and marked with several prominent black spots; hind wings grey. **Larva:** up to 18 mm long; reddish brown or purplish brown above, but creamy along the sides and below, with small blackish pinacula; head yellowish brown; prothoracic plate blackish brown or black; anal plate brown; thoracic legs blackish brown. **Pupa:** 8–10 mm long; light brown.

LIFE HISTORY
Moths fly in June and July, and are often common in parks and gardens. Eggs laid on the foodplant hatch in the following spring. The larvae then feed within spun shoots during May and June; they also bore into the young shoots. Pupation occurs in the larval habitation or in a slight cocoon spun amongst debris on the ground.

DAMAGE
Infestations cause loss of young shoots and lead to a reduction in the number of flowers; they also disfigure bushes.

Epiblema trimaculana (Haworth) (**663–664**)
syn. *E. suffusana* (Duponchel)
A minor pest of hawthorn (*Crataegus*), and sometimes present on nursery stock. Palaearctic. Widely distributed in Europe.

DESCRIPTION
Adult: 15–17 mm wingspan; fore wings white, suffused and striated with brown and bluish grey, and marked with black; hind wings grey. **Larva:** up to 16 mm long; brown to reddish brown, with small inconspicuous pinacula; head brown; prothoracic plate brownish black to black; anal plate brown; thoracic legs blackish brown. **Pupa:** 8–9 mm long; brown.

661 Adult of *Epiblema rosaecolana*.

662 Larva of *Epiblema rosaecolana*.

663 Adult of *Epiblema trimaculana*.

664 Larva of *Epiblema trimaculana*.

LIFE HISTORY
Larvae shelter and feed within spun shoots of hawthorn from April to May. Fully fed individuals pupate in loose cocoons spun within the larval habitation, and adults emerge in June.

DAMAGE
Larvae cause slight distortion of young terminal shoots but infestations are unimportant.

Epichoristodes acerbella (Walker)
syn. *E. ionephela* (Meyrick)
African carnation tortrix moth
A native of Africa, sometimes introduced into Europe on imported carnation (*Dianthus caryophyllus*) cuttings. Although infestations are usually intercepted by Plant Health authorities, the pest has sometimes established itself in greenhouses in parts of northern Europe, including Denmark and Norway; also, introduced into parts of southern Europe, including France, Italy and Spain. Although mainly a pest of carnation, the larvae also feed on *Chrysanthemum*, rose (*Rosa*) and various other plants.

DESCRIPTION
Adult: 17–20 mm wingspan; fore wings reddish brown to yellowish, merging into dark reddish brown or blackish on the hind margin; lighter areas noticeably speckled with brownish black; hind wings light grey. **Larva:** up to 15 mm long; green to yellowish green, with dark green dorsal and subdorsal lines and whitish pinacula; head greenish brown, marked with brownish black; prothoracic plate green, marked with black along the lateral margin and above the prothoracic spiracle.

LIFE HISTORY
Eggs are laid on the leaves of carnation in elongate clusters of about 25, and hatch in about ten days. Larvae then feed for 4–8 weeks, sheltering within rolled leaves which they spin together with silk. Pupation occurs within a silken cocoon spun between the leaves, and adults emerge 2–4 weeks later. Breeding is continuous so long as conditions remain favourable.

DAMAGE
Larvae devour leaves and flowers, and also burrow into buds and stems. They often cause considerable destruction, and also disfigure and distort plants by spinning the leaves together.

Epinotia brunnichana (Linnaeus) (665–667)
A minor pest of birch (*Betula*) and hazel (*Corylus*), especially young trees; also occurs on broad-leaved willows such as grey willow (*Salix cinerea*). Widespread in Europe.

DESCRIPTION
Adult: 18–22 mm wingspan; fore wings white, each more or less suffused with brownish orange and usually marked with a pale, blackish-edged, dorsal patch; hind wings pale and brownish. **Larva:** up to 18 mm long; greenish grey, with black pinacula; head brownish black; prothoracic plate greenish brown, paler anteriorly; anal plate greenish; thoracic legs black.

LIFE HISTORY
Larvae feed during May and June, each inhabiting a transversely rolled leaf (cf. *Epinotia solandriana*, p. 274), and occur most commonly on birch trees. When fully fed they enter the soil to pupate in silken cocoons. Moths occur in July and August.

DAMAGE
The characteristic larval habitations attract attention, particularly if present on young trees, but damage caused is of no importance.

Epinotia immundana (Fischer von Röslerstamm) (668–670)
Locally common on alder (*Alnus*) and birch (*Betula*), and occasionally present on cultivated trees. The larvae feed from autumn to the following spring, attacking the buds and catkins. They eventually pupate in silken cocoons formed amongst leaf debris or in the ground, and adults occur from April to June. In favourable districts a small number of larvae feed during the summer, inhabiting characteristically rolled leaves; such larvae eventually pupate and produce a partial second generation of adults in August and September. Adults (12–14 mm wingspan) are whitish to greyish brown, heavily marked with black, with a pale angular patch dorsally on each fore wing. The larvae (up to 10 mm long) are translucent, greenish grey to yellowish grey, suffused above with red.

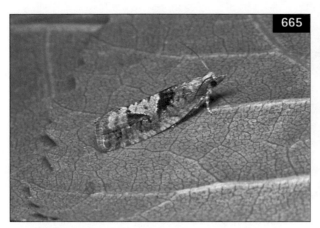

665 Adult of *Epinotia brunnichana*.

666 Larva of *Epinotia brunnichana*.

667 Larval habitation of *Epinotia brunnichana* on *Betula*.

668 Larval habitation of *Epinotia immundana* on *Alnus*.

669 Adult of *Epinotia immundana*.

670 Larva of *Epinotia immundana*.

671 Larval habitation of *Epinotia solandriana* on *Betula*.

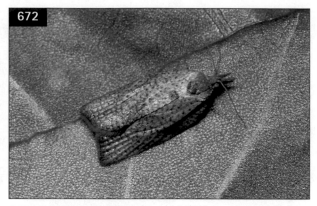

672 Female light brown apple moth (*Epiphyas postvittana*).

673 Male light brown apple moth (*Epiphyas postvittana*).

674 Larva of light brown apple moth (*Epiphyas postvittana*).

Epinotia solandriana (Linnaeus) (671)

Larvae of this widely distributed species feed in May and June on birch (*Betula*), hazel (*Corylus*) and pussy willow (*Salix caprea*), each inhabiting a longitudinally folded leaf (cf. *Epinotia brunnichana*, p. 272). Fully fed individuals are 15–18 mm long and dull grey to greenish grey, with grey or blackish-grey pinacula and a yellowish-brown, darker-mottled head and a yellowish-brown to dark brown prothoracic plate. The adults (16–21 mm wingspan) occur in July and August; there are various colour forms, the fore wings commonly being creamy white, suffused with brown, and with a dark dorsal blotch (in some forms the wings are mainly reddish brown and the blotch light brown or white); the hind wings are grey.

Epiphyas postvittana (Walker) (672–674)

Light brown apple moth

An Australian species, unknown in Europe until 1936 when breeding colonies were found on Japanese spindle (*Euonymus japonica*) in south-western England. The pest is now well established in Britain on various outdoor and greenhouse ornamentals, including azalea (*Rhododendron*), *Azara serrata*, bay laurel (*Laurus nobilis*), *Camellia*, cape figwort (*Phygelius capensis*), *Ceanothus*, Chilean fire-bush (*Embothrium coccineum* var. *longifolium*), daisy bush (*Olearia*), *Drimys*, firethorn (*Pyracantha*), *Fothergilla major*, honeysuckle (*Lonicera*), New Zealand tea tree (*Leptospermum scoparium*), *Pittosporum*, *Skimmia japonica* and many others.

DESCRIPTION

Adult female: 17–25 mm wingspan; fore wings pale yellowish brown, speckled with black, but sometimes noticeably darkened apically. **Adult male:** 16–21 mm wingspan; fore wings subdivided into a pale yellow to orange-yellow basal area and a dark brown to reddish-brown apical section; hind wings grey. **Larva:** up to 18 mm long; yellowish green, with pale pinacula; head light brown; prothoracic plate light greenish brown; anal plate brownish green; anal comb light green; thoracic legs brown. **Pupa:** 8–10 mm long; dark reddish brown.

LIFE HISTORY

In England there are two overlapping generations annually, the adults occurring from May to October, with peaks of activity in June and from August to September. Eggs are laid in small batches on the leaves of various host plants, and hatch in about ten days. The larvae inhabit spun leaves at the tips of the new shoots; they occur in June and July, and from September to April. Pupation takes place in the larval habitation in May and in August. Under protected conditions the lifecycle varies and adults may also emerge in late autumn or during the winter.

DAMAGE

Infestations affect the growth of the shoots and also spoil the appearance of ornamentals. Attacks are particularly damaging on young container-grown shrubs and pot plants.

Gypsonoma aceriana (Duponchel) (675–677)

Poplar shoot-borer moth

larva = poplar shoot-borer

A pest of poplar (*Populus*), especially balsam poplar (*P. balsamifera*), black poplar (*P. nigra*), Lombardy poplar (*P. nigra* 'Italica') and white poplar (*P. alba*), including those in nurseries, parks and gardens. Widely distributed in Europe but most common in more southerly areas; also present elsewhere, including eastern Turkey and North Africa.

DESCRIPTION

Adult: 12–15 mm wingspan; fore wings creamy white, marked with grey, each with a well-defined blackish basal patch; hind wings grey. **Larva:** up to 12 mm long; light brown, with a black or brownish-black head and prothoracic plate; anal plate brown. **Pupa:** 5–6 mm long; orange-brown.

LIFE HISTORY

Larvae tunnel within the young shoots and leaf stalks of poplar during May and June, the presence of a larva being disclosed by the appearance of a silken, frass-filled tube that protrudes through a small hole in the wall of the feeding gallery. Fully grown larvae pupate in the larval habitation or may vacate the shoots to pupate amongst litter on the ground or in crevices on the bark of host trees. The moths occur in July.

DAMAGE

Infested shoots are killed. This may subsequently result in the development of numerous lateral shoots, adversely affecting the quality and marketability of nursery stock.

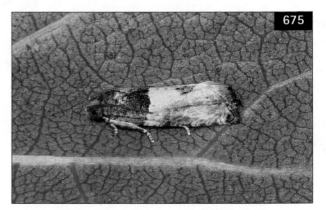

675 Poplar shoot-borer moth (*Gypsonoma aceriana*).

676 Poplar shoot-borer (*Gypsonoma aceriana*).

677 Frass tube produced by poplar shoot-borer (*Gypsonoma aceriana*).

Hedya dimidioalba (Retzius) (**678–679**)

syn. *H. nubiferana* (Haworth); *Argyroploce variegana* (Hübner)

An often common pest of trees and shrubs, including alder (*Alnus*), ash (*Fraxinus excelsior*), flowering cherry (*Prunus*), rose (*Rosa*) and *Sorbus*; particularly common on hawthorn (*Crataegus*) and often abundant on garden hedges. Eurasiatic; also occurs in North America.

DESCRIPTION

Adult: 15–21 mm wingspan; fore wings ochreous white apically, suffused with silver and ochreous grey, the remainder marbled with dark brown, bluish grey and black; hind wings brownish grey. **Egg:** 0.85 × 0.65 mm; oval, flat and iridescent. **Larva:** up to 20 mm long; olive-green to dark green, with black pinacula; head, prothoracic plate, anal pate and anal comb dark brown to black; thoracic legs black. **Pupa:** 8–10 mm long; dull black; cremaster tapered, with an apical tuft of eight hooked bristles.

LIFE HISTORY

Adults occur in June and July, and when at rest (in common with other species of *Hedya*) they closely resemble bird-droppings. Eggs are laid singly or in small groups, mainly on the underside of leaves, and hatch in about two weeks. Larvae then feed for several weeks before hibernating, each in a dense cocoon spun within a bark crevice or beneath old bud scales. Activity is resumed early in the following spring. The larvae then attack the opening buds, blossom trusses, foliage and young shoots. During their development, they often shelter between two leaves webbed together with silk. Fully grown individuals pupate in spun leaves in late May or June, and adults emerge 3–4 weeks later.

DAMAGE

Larvae contribute to damage done to leaves and blossoms by other species; they also tunnel into the young shoots and cause wilting or death of the tips.

Hedya ochroleucana (Frölich) (**680–681**)

Locally common on wild rose (*Rosa*) but also associated with cultivated bushes; adults commonly rest exposed on the foliage. The larvae feed mainly in the spring, from April to June, each inhabiting a bunch of spun leaves. Fully fed specimens (16–18 mm long) are dull greyish green to olive-green, with inconspicuous pinacula and a brownish-black to black head and prothoracic plate. They pupate within the larval habitation in May or June. Adults (16–23 mm wingspan) are mainly blackish to bluish grey and brownish grey, with the fore wings partly creamy white and often tinged with pink. They occur in June and July.

678 Adult of *Hedya dimidioalba*.

679 Larva of *Hedya dimidioalba*.

Lobesia littoralis (Humphreys & Westwood) (682–683)

Thrift tortrix moth

Locally common and often abundant in coastal habitats where thrift (*Armeria maritima*) is established; also occurs on thrift plants cultivated in gardens and nurseries. Widespread in Europe.

DESCRIPTION

Adult: 11–16 mm wingspan; fore wings white, suffused with grey and variably marked with brown and reddish brown; hind wings whitish grey. **Larva:** up to 10 mm long; greenish grey to yellowish grey, with inconspicuous pinacula; head and prothoracic plate brownish black or black; anal comb present; thoracic legs black. **Pupa:** 6–8 mm long; greenish brown.

LIFE HISTORY

This species is double brooded, adults occurring in June and July, and in the autumn. The moths are most active in sunny weather, and fly rapidly over the foodplant. Larvae feed amongst the vegetative shoots, each inhabiting a silken tube; they also attack the flower heads, to feed on the unripe seeds. Larvae of the first generation occur from April to May or early June, and those of the second in August. Pupation takes place in a dense silken cocoon spun amongst the leaves or hidden within a flower head.

DAMAGE

Larvae cause death of shoots and slight distortion of flower heads, but damage is of importance only on nursery plants.

680 Larva of *Hedya ochroleucana*.

681 Adult of *Hedya ochroleucana*.

682 Thrift tortrix moth (*Lobesia littoralis*).

683 Larva of thrift tortrix moth (*Lobesia littoralis*).

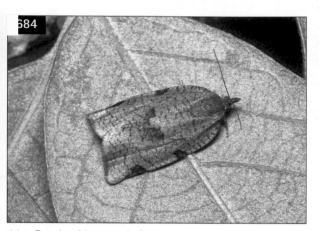

684 Female of *Lozotaenia forsterana*.

685 Larva of *Lozotaenia forsterana*.

686 Adult of *Orthotaenia undulana*.

687 Larva of *Orthotaenia undulana*.

Lozotaenia forsterana (Fabricius) (684–685)

A minor pest of cultivated plants, including bell flower (*Campanula*), cherry laurel (*Prunus laurocerasus*), ivy (*Hedera*), honeysuckle (*Lonicera*) and various conifers. Widely distributed in Europe.

DESCRIPTION

Adult: 20–29 mm wingspan, female larger than male; fore wings light greyish brown, with dark brown markings; hind wings grey. **Larva:** up to 25 mm long; dull greyish green, darker above, with a brown or black head; prothoracic plate green or brown; anal plate greenish, marked with black on each side; anal comb present; thoracic legs brown. **Pupa:** 12–14 mm long; dark brown.

LIFE HISTORY

Adults occur in late June and July. Eggs are laid on the foliage and hatch in September. Larvae then feed during the autumn before hibernating. Activity is resumed in April, each larva then living between two or more leaves spun together with silk. Pupation occurs in June, again between spun leaves.

DAMAGE

If numerous, larvae cause considerable defoliation but serious attacks rarely develop.

Orthotaenia undulana (Denis & Schiffermüller) (686–687)

A locally common but minor pest of trees and shrubs, including alder (*Alnus*), birch (*Betula*), elm (*Ulmus*), honeysuckle (*Lonicera*), juniper (*Juniperus*), maple (*Acer*), pine (*Pinus*), sea-buckthorn (*Hippophae rhamnoides*) and broad-leaved willows such as grey willow (*Salix cinerea*) and pussy willow (*S. caprea*); certain herbaceous plants are also attacked. Widely distributed in Europe; also present in North America.

DESCRIPTION
Adult: 15–20 mm wingspan; fore wings creamy white, marked with silvery grey, greyish yellow, olive-yellow and brownish black; hind wings greyish. **Larva:** up to 18 mm long; reddish brown to blackish brown, with inconspicuous, brown or reddish-brown pinacula; head, prothoracic plate and anal plate black.

LIFE HISTORY
Adults occur mainly in June and July. The larvae feed within spun leaves from April onwards, most becoming fully grown by late May or early June. They then pupate, either in a spun leaf or in the larval habitation, and adults appear shortly afterwards. There is just one generation annually. Adults and larvae of this species are often mistaken for those of *Celypha lacunana* (p. 264).

DAMAGE
The larvae cause slight foliage damage and disfigurement but infestations are unimportant.

Pandemis cerasana (Hübner) (**688–689**)
Barred fruit tree tortrix moth
An often common but minor pest of trees and shrubs in gardens and nurseries close to woodlands. Hosts include alder (*Alnus*), birch (*Betula*), crab-apple (*Malus*), elm (*Ulmus*), flowering cherry (*Prunus*), hazel (*Corylus*), lime (*Tilia*), oak (*Quercus*), rowan (*Sorbus aucuparia*), sycamore (*Acer pseudoplatanus*) and willow (*Salix*). Eurasiatic. Widely distributed in Europe.

DESCRIPTION
Adult: 16–24 mm wingspan; fore wings pale yellowish to yellowish brown, with light brown, chestnut-edged markings; hind wings greyish brown; antennae of male with a basal notch. **Egg:** flat and oval; laid in an oval, raft-like batch. **Larva:** up to 20 mm long; whitish green, but darker above, with light green pinacula; head light green to brownish green; prothoracic plate light green or yellowish green, with dark sides and hind edge; anal plate green, dotted with black; anal comb yellowish, with 6–8 teeth; first and hindmost spiracles elliptical and distinctly larger than the others (unlike *Pandemis heparana* and *P. corylana*). **Pupa:** 8–13 mm long; brown or blackish brown; cremaster longer than broad.

LIFE HISTORY
Adults occur from June to August, eggs being deposited on the leaves or branches of various trees and shrubs. Some eggs hatch after a few weeks but others not until the following spring. Young summer larvae feed on the foliage for a short time and then, whilst still small, spin silken retreats on the twigs in which they overwinter. Activity recommences at bud-burst, when overwintered eggs also hatch. Larvae feed in a rolled or folded leaf until May or early June. They then pupate in whitish cocoons, each spun in the larval habitation or in the shelter of a folded leaf.

DAMAGE
Leaf damage is usually unimportant as larvae rarely feed together in large numbers.

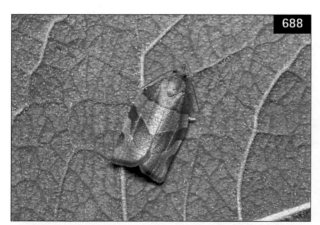

688 Female barred fruit tree tortrix moth (*Pandemis cerasana*).

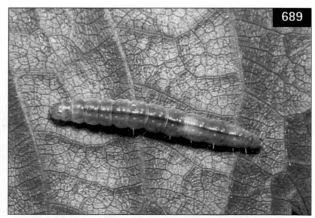

689 Larva of barred fruit tree tortrix moth (*Pandemis cerasana*).

Pandemis heparana (Denis & Schiffermüller) (690–691)

Dark fruit tree tortrix moth

Often common in woodland areas, nurseries and gardens, attacking trees and shrubs such as birch (*Betula*), crab-apple (*Malus*), flowering cherry (*Prunus*), *Forsythia*, honeysuckle (*Lonicera*), lime (*Tilia*) and willow (*Salix*). Eurasiatic. Widely distributed in Europe.

DESCRIPTION

Adult: 16–24 mm wingspan; fore wings reddish ochreous to reddish brown, with darker markings; hind wings dark brownish grey; antennae of male with a basal notch. **Egg:** flat and oval; laid in a small, raft-like batch. **Larva:** up to 25 mm long; bright green, with pale sides; head and prothoracic plate green to brown; anal plate green; anal comb whitish, usually with 6–8 teeth. **Pupa:** 10–12 mm long; brownish black; cremaster about as long as wide.

690 Female of dark fruit tree tortrix moth (*Pandemis heparana*).

LIFE HISTORY

Adults occur from late June to August or September. Eggs are laid on the upper surface of leaves of various trees and shrubs, usually in batches of 30–50, and hatch 2–3 weeks later. The larvae feed for a short time before hibernating in silken retreats spun on the twigs. They resume feeding in the following May and June, each sheltering within rolled leaves on the young shoots or under webs spun on the underside of expanded leaves. Pupation occurs in the larval habitation or in spun leaves near the tips of infested shoots.

DAMAGE

In addition to feeding on leaves, larvae sometimes destroy blossoms, but infestations are usually unimportant.

Pandemis corylana (Fabricius) (692)

Hazel tortrix moth

A locally common but minor pest of trees and shrubs, including ash (*Fraxinus excelsior*), dogwood (*Cornus*), hazel (*Corylus*) and oak (*Quercus*). Adults occur from July to September, eggs being laid on the bark of host plants where they overwinter until the following spring. Larvae feed on the foliage from May to July, inhabiting either a longitudinally folded leaf or a bunch of spun leaves. Pupation occurs in June within the larval habitation. Adults (18–24 mm wingspan) are similar in appearance to those of *Pandemis cerasana* but the fore wings are paler and more strongly reticulated. The larvae (up to 25 mm long) are green and slender-bodied, with a well-developed anal comb.

691 Larva of dark fruit tree tortrix moth (*Pandemis heparana*).

692 Hazel tortrix moth (*Pandemis corylana*).

693 Larval habitation of pine resin-gall moth (*Petrova resinella*) on *Pinus*.

694 Female Leche's twist moth (*Ptycholoma lecheana*).

Petrova resinella (Linnaeus) (693)

Pine resin-gall moth

A minor and locally common pest of pine (*Pinus*), the characteristic larval habitations occasionally attracting attention on young amenity trees. Adults occur in June, eventually depositing eggs on the young shoots. After egg hatch each larva bores into the base of a needle and also damages the surface of the stem, forming a groove from which resin exudes; the larva then feeds close to a bud whorl, protected by silk and a coating of resin which, at this stage, forms a pea-like gall. Feeding continues throughout the following year, the protective gall becoming greatly enlarged (*c.* 20–30 mm across) and obvious. Pupation occurs in the following spring, almost two years after eggs were laid. Adults (16–22 mm wingspan) have whitish fore wings, mottled with grey and blackish brown; the hind wings are brown, with white cilia.

Ptycholoma lecheana (Linnaeus) (694–695)

Leche's twist moth

A generally common, polyphagous species but associated mainly with oak woodlands; a minor pest of trees and shrubs including crab-apple (*Malus*), hawthorn (*Crataegus*), hazel (*Corylus*), oak (*Quercus*), poplar (*Populus*), sycamore (*Acer pseudoplatanus*) and willow (*Salix*); conifers such as fir (*Abies*), larch (*Larix*) and spruce (*Picea*) are also attacked. Eurasiatic. Widely distributed in Europe.

DESCRIPTION

Adult: 16–22 mm wingspan; fore wings blackish brown, suffused with greenish yellow, particularly basally, and with silver-metallic markings; hind wings blackish brown, with pale cilia. **Egg:** 1.0 × 0.7 mm, oval, flat and light green. **Larva:** up to 20 mm long; bluish green, with a darker line along the back and

695 Larva of Leche's twist moth (*Ptycholoma lecheana*).

another along each side; pale yellowish green below; pinacula yellowish; head yellowish brown to black; prothoracic plate pale yellow to yellowish brown, marked on sides with black; anal comb small; thoracic legs black. **Pupa:** 9–11 mm long; black; apical hook-like setae on cremaster forming a strong projection.

LIFE HISTORY

Adults occur in June and July. Eggs are deposited on various host plants and hatch in the summer. Young larvae then feed on the foliage for a few weeks and then hibernate, each in a silken cocoon spun on a twig or spur. The larvae reappear early in the following spring, to feed on the opening buds and young foliage. Later they feed in rolled leaves, and become fully grown in May or early June. Pupation occurs in a dense, white cocoon formed in the larval habitation or in a freshly rolled leaf.

DAMAGE

Larvae cause defoliation and damage to young shoots but are rarely sufficiently numerous to be a major problem.

696 Adult of *Ptycholomoides aeriferanus*.

697 Holly tortrix moth (*Rhopobota naevana*).

698 Holly leaf tier (*Rhopobota naevana*).

Ptycholomoides aeriferanus (Herrich-Schäffer) (**696**)

A minor pest of larch (*Larix*) in central and south-eastern Europe. Also present in Russia and Japan. Recorded in England in 1951, where it has since become widely distributed. Larvae feed in May and June, spinning a few needles together as a shelter. Fully grown individuals are green, with a light brown head and a yellowish-brown prothoracic plate. Adults (17–21 mm wingspan) occur from late June to August; the fore wings are pale golden yellow, suffused with dark brown and marked with brownish black; the hind wings are dark brown.

Rhopobota naevana (Hübner) (**697–698**)
Holly tortrix moth
larva = holly leaf tier

A generally common pest of trees and shrubs, including crab-apple (*Malus*) and hawthorn (*Crataegus*), and often present in parks and gardens; at least in the wild, most abundant on European holly (*Ilex aquifolium*). Eurasiatic; also present in North America. Widely distributed in Europe.

DESCRIPTION

Adult: 12–15 mm wingspan; fore wings notched below the apex, dark grey, marked with blackish and dark reddish brown, and with several whitish striae on the front margin; hind wings grey. **Egg:** 0.7×0.5 mm; flat and oval; whitish and translucent, becoming yellowish to reddish. **Larva:** up to 12 mm long; oily, yellowish green to greenish brown; head, prothoracic plate and anal plate brown; anal comb usually with two dark teeth; thoracic legs brown. **Pupa:** 5–7 mm long; yellowish brown; tip with four thorn-like spines and a small bump behind the anal slit.

LIFE HISTORY

This species overwinters as eggs laid singly on the smooth bark of trunks and branches. The eggs hatch in the spring. Larvae then feed within a tightly webbed shelter of young, tender leaves; on certain hosts they also feed within spun blossom trusses. Feeding is completed in June, each larva pupating in a white cocoon formed in a folded leaf or amongst dead leaves or debris on the ground. Adults emerge about three weeks later. Eggs are deposited in July and August.

DAMAGE

Larvae destroy young leaves and are particularly harmful when present on the new growth of clipped holly hedges.

Rhyacionia buoliana (Denis & Schiffermüller)
Pine shoot moth

An important forestry pest, especially on Austrian pine (*Pinus nigra* var. *nigra*), beach pine (*P. contorta*) and Scots pine (*P. sylvestris*), and also damaging to ornamentals. The fast-growing species bishop pine (*P. muricata*) and Monterey pine (*P. radiata*) are very susceptible. Eurasiatic. Widely distributed in Europe, and an introduced pest in North and South America.

DESCRIPTION

Adult: 16–24 mm wingspan; fore wings mainly orange or reddish orange, variably patterned with silvery-white stripes and flecks; hind wings greyish brown to dark brown. **Larva:** up to 18 mm long; body reddish brown, with small brownish-black or black pinacula; head and prothoracic plate brown or black; thoracic legs dark brown; anal plate yellowish brown, mottled with dark brown; no anal comb.

LIFE HISTORY

Adults occur from late June to mid-August. Eggs are laid singly or in small groups close to a whorl of young buds, and hatch 2–3 weeks later. Each larva then tunnels into the base of a needle, sheltered beneath an outer covering of silk. After 1–2 weeks the larva bores into a lateral bud to continue feeding. Resin soon exudes from the damaged bud, and the larva incorporates this in the walls of a protective silken tent spun amongst the invaded whorl of buds. Larvae overwinter within hollowed-out buds, and resume activity in the spring. They then invade buds on the developing shoots, again forming resinous, silken shelters. Larvae are fully grown in June, and pupation takes place in the larval habitation.

DAMAGE

Larvae cause death of buds, and damage is most important if the main shoot is affected. Death of all buds in the whorl results in the development of untidy secondary growth from adventitious buds; if just the leading shoot is killed, this is eventually replaced by one of the laterals, with minimal distortion. However, if the leading shoot is merely damaged superficially, it keels over but then recovers, to produce a characteristic 'posthorn'-shaped deformation of the main stem. The pest is a particular problem on young trees.

Spilonota ocellana (Denis & Schiffermüller) (699–700)
Bud moth

An often common pest of rosaceous trees and shrubs, including ornamentals; various other trees, including alder (*Alnus*), hazel (*Corylus*) and oak (*Quercus*), are also suitable hosts. Larch (*Larix*) is attacked by a distinct form – *Spilonota ocellana laricana*. Eurasiatic; also present in North America. Widely distributed in Europe.

DESCRIPTION

Adult: 12–16 mm wingspan; fore wings whitish, more or less suffused with grey, marked towards the apex with metallic bluish grey and black, and each with a dark, triangular dorsal spot and a blackish, angular basal patch; hind wings dark grey. **Egg:** flat and more or less circular; pale yellowish white. **Larva:** up to 12 mm long; dark purplish brown, with lighter pinacula (larch-feeding form light greyish brown, with indistinct pinacula); head, prothoracic plate and anal plate shiny black or blackish brown. **Pupa:** 6–7 mm long; brown, with outline of wing cases distinctly darker than abdomen; tip blunt.

699 Bud moth (*Spilonota ocellana*).

700 Larva of bud moth (*Spilonota ocellana*).

LIFE HISTORY

Adults occur from June to August. Eggs are deposited singly or in small batches on the leaves of host plants. Larvae feed from August onwards, each inhabiting a silken tube and eventually hibernating in a silken retreat spun on the bark, often close to a bud. Activity is resumed early in the following spring. The larvae then invade the opening buds and, later, each feeding within a tightly woven tent of young unfurling leaves. Pupation occurs in the larval habitation, usually in June.

DAMAGE

Direct damage to leaves and shoots is relatively unimportant, but attacked buds wither and die.

Syndemis musculana (Hübner) (701–702)
Autumn apple tortrix moth

A generally common but minor pest of birch (*Betula*), oak (*Quercus*), *Rubus* and other plants; infestations sometimes occur on nursery plants, including conifer seedlings. Eurasiatic. Widely distributed in Europe.

DESCRIPTION

Adult: 15–22 mm wingspan; fore wings white to greyish, suffused with black and marked conspicuously with dark brown. **Larva:** up to 22 mm long; olive-green to blackish brown but paler below, with distinct, pale pinacula; head yellowish brown to brownish orange; prothoracic plate greyish brown to yellowish brown, characteristically marked posteriorly with black; anal plate yellowish brown or greenish.

LIFE HISTORY

Adults occur mainly in May and June, depositing eggs on the leaves of host plants. The larvae feed from July onwards, typically inhabiting a folded leaf or tightly rolled tube of spun leaves. Feeding is completed in the autumn. Larvae then hibernate in the larval habitation or in a silken cocoon formed amongst fallen leaves, and pupate in the following spring.

DAMAGE

Larval habitations cause noticeable distortion of terminal shoots and affect the quality of young trees and shrubs.

Tortricodes alternella (Denis & Schiffermüller) (703–704)
syn. *T. tortricella* (Hübner)

Minor infestations of this widely distributed and locally common species sometimes occur on ornamental trees growing in the vicinity of mixed deciduous woodlands. The larvae (up to 15 mm long) are reddish brown, with mottled, yellowish-brown prothoracic and anal plates, and prominent white or yellowish-white pinacula. They feed in May and June on the leaves of hornbeam (*Carpinus betulus*) and oak (*Quercus*), each sheltered by a folded leaf edge or spun leaves. Larvae also attack various other trees, including birch (*Betula*), hawthorn (*Crataegus*), hazel (*Corylus*) and lime (*Tilia*). Fully grown larvae pupate in tough, silken cocoons formed in the soil or amongst debris on the ground, and remain *in situ* throughout the winter. The rather drab, brownish-coloured adults (19–23 mm wingspan), emerge in the following February or March, but later in more northerly areas. The males fly rapidly in sunny weather, whereas females usually remain at rest on tree trunks. Both sexes are active at night.

701 Adult of autumn apple tortrix moth (*Syndemis musculana*).

702 Larva of autumn apple tortrix moth (*Syndemis musculana*).

Tortrix viridana (Linnaeus) (**705–706**)

Green oak tortrix moth

An important, and sometimes abundant, pest of oak (*Quercus*) trees, especially English oak (*Q. robur*). Although primarily a forest pest, damage is also caused to trees in parks, gardens and nurseries. Widely distributed in Europe; also present in North Africa and parts of Asia.

DESCRIPTION

Adult: 17–24 mm wingspan; head, thorax and fore wings light green; hind wings light grey. **Egg:** flat and oval, pale yellow to brownish-orange. **Larva:** up to 18 mm long; greenish grey to light olive-green, with prominent blackish-brown or black pinacula; head blackish brown to black; prothoracic plate greenish brown to greyish, marked with blackish brown, often edged anteriorly with white and with a pale median line; anal plate green or black; anal comb with eight prongs; thoracic legs black. **Pupa:** 10–12 mm long; brownish black to black.

LIFE HISTORY

Adults are most numerous in June and July, and are most abundant in oak woodlands. They rest on host trees during the daytime but are readily disturbed. Eggs are laid in pairs on the bark, usually close to the leaf bases and where the shoots and small branches divide. They hatch in the following spring at about bud-burst. The larvae feed from late April to June, at first entering the opening buds but later inhabiting rolled or folded leaves. Pupation usually occurs in a folded leaf on the host tree, but larvae also descend on a silken thread and pupate on underlying plants. When larvae are present in large numbers they defoliate trees before becoming fully fed; in such circumstances they larvae might attack the foliage of other trees, shrubs and underlying herbage.

DAMAGE

Larval defoliation is redressed by the appearance of new leaves on Lammas shoots which develop in the summer.

703 Larva of *Tortricodes alternella*.

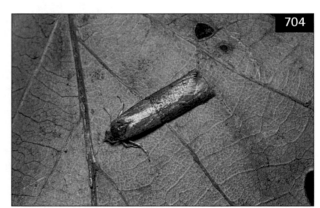

704 Adult of *Tortricodes alternella*.

705 Female green oak tortrix moth (*Tortrix viridana*).

706 Larva of green oak tortrix moth (*Tortrix viridana*).

Family **PYRALIDAE**

A variable family, the adults having moderately long, narrow bodies, prominent palps, relatively narrow fore wings and broad hind wings. The larvae usually have few body hairs and are often extremely active when disturbed.

Acrobasis consociella (Hübner) (**707–710**)

A locally common pest of oak (*Quercus*), and sometimes troublesome on young amenity trees and nursery stock. Widespread in Europe

DESCRIPTION
Adult: 19–22 mm wingspan; fore wings mainly greyish pink or crimson, suffused with white, purplish red, black and brownish black; hind wings light brownish grey. **Larva:** up to 15 mm long; dull greyish green to yellowish grey, with darker longitudinal lines; pinacula small and blackish; head yellowish brown; prothoracic plate yellowish brown, spotted with black; first thoracic segment with a pair of dark, whitish-bordered spots.

LIFE HISTORY
Adults occur mainly in July and August, but sometimes earlier. Eggs are deposited on oak trees, and hatch shortly afterwards. Larvae then feed briefly on the foliage before entering hibernation. Overwintered larvae reappear in the following year, groups of individuals inhabiting distinctive clusters of tightly webbed leaves. Such habitations are typically formed at the tips of young shoots, and are very noticeable on young trees from late May onwards. Fully grown larvae pupate in cocoons formed within the larval habitation,

707 Adult of *Acrobasis consociella*.

708 Larva of *Acrobasis consociella*.

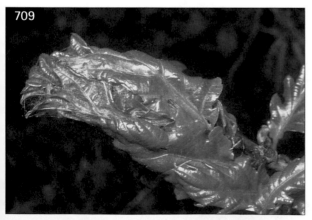

709 Larval habitation of *Acrobasis consociella* on *Quercus*.

710 *Acrobasis consociella* damage to shoots of *Quercus*.

and adults emerge shortly afterwards. Two generations annually are also reported.

DAMAGE

Infested shoots become distorted, and some may be completely defoliated. Although infestations on mature trees are of no consequence, damage may be of considerable significance on young ones.

Cydalima perspectalis (Walker) (711–713)
syn. *Diaphania perspectalis* (Walker)
Box-tree pyralid moth

An eastern Asian pest of box (*Buxus*), including common box (*B. sempervirens*). Discovered in Germany in the mid-2000s and subsequently found in several other parts of Europe (including Austria, Belgium, England, France, the Netherlands and Switzerland), where it is continuing to spread.

DESCRIPTION

Adult: 38–40 mm wingspan; wings usually white with a purplish sheen, and with a broad bronzy-brown border; body mainly white, with the head, prothorax and tip of abdomen bronzy brown; rarely, wings and body almost entirely brown. **Larva:** up to 38 mm long; body mainly green, patterned with white and yellow stripes and a series of black, white-rimmed spots; head black and shiny. **Pupa:** 15–20 mm long; initially green with darker markings but later turning brown.

LIFE HISTORY

Eggs are laid on the underside of leaves in flat raft-like batches of 10–20 and hatch about a week later. Larvae then feed on the leaves, partly sheltered by loose strands of silk webbing amongst which large quantities of frass accumulate. Larvae are fully grown in about three weeks and then pupate on the foodplant, each in a silken cocoon. Adults emerge a few weeks later. In Europe, the pest completes two or three generations annually and overwinters in the larval stage.

DAMAGE

Larvae cause extensive and rapid defoliation of box plants, both in nurseries and on established plants in parks and gardens. Severely damaged plants may be killed.

711 Box-tree pyralid moth (*Cydalima perspectalis*).

712 Larva of box-tree pyralid moth (*Cydalima perspectalis*).

713 Pupa of box-tree pyralid moth (*Cydalima perspectalis*).

714 Brown china-mark moth (*Elophila nymphaeata*).

715 Larva of jasmine moth (*Palpita vitrealis*).

Duponchelia fovealis (Zeller)
Marshland pyralid moth

An invasive Mediterranean species, associated mainly with aquatic and waterside plants; also present in the Canary Islands. Currently posing a threat to greenhouse-grown crops in other parts of the world, including northern Europe and North America. Cultivated ornamental hosts include the aquatics *Aponogeton*, *Ophiopogon* and water trumpet (*Cryptocoryne*), as well as *Begonia*, busy lizzie (*Impatiens*), *Cyclamen*, *Gerbera*, *Heuchera*, *Kalanchoe*, poinsettia (*Euphorbia pulcherrima*) and sea lavender (*Limonium*); cuttings of pithy shrubs such as elder (*Sambucus*) are also attacked.

DESCRIPTION
Adult: 19–21 mm wingspan; olive-brown, each fore wing with two pale, sinuous crosslines, the outer of which includes a characteristic U-shaped outwardly directed projection. **Larva:** up to 20 mm long; body semi-translucent, creamy white to pinkish white, with numerous greyish to blackish dorsal plates; head, prothoracic and anal plates shiny black.

LIFE HISTORY
Adults occur mainly in May and June, but also later in the year. Eggs are laid singly or in small groups, and hatch about a week later. Larvae then feed on hosts, their habitations typically protected by silken webbing. Individuals are fully grown in about a month. They then pupate in strong oval cocoons formed on host plants or in sheltered situations nearby. New adults appear a few weeks later. The pest appears unable to survive cold winters, and is unlikely to persist outdoors in northern Europe.

DAMAGE
Larvae attack the leaves and flowers of hosts. They also bore into the base and crowns of plants, which then wilt and may subsequently die.

Elophila nymphaeata (Linnaeus) (**714**)
Brown china-mark moth

A generally common pest of aquatic plants, including bur-reed (*Sparganium*), frog-bit (*Hydrocharis morus-ranae*), pondweed (*Potamogeton*) and water-lilies. Widespread in Europe.

DESCRIPTION
Adult: 20–22 mm wingspan; wings white, marked irregularly with brown. **Larva:** up to 25 mm long; creamy white, with a darker line down the back; head and prothoracic plate mainly brown.

LIFE HISTORY
Adults occur from June to August, depositing eggs in batches on the underside or at the edges of leaves of host plants. Eggs hatch in about two weeks, and the young larvae then burrow into the leaves to begin feeding. They reappear about three days later, each constructing a small, flat, oval case out of leaf fragments. Larvae continue to feed on the surface of leaves throughout July and August, replacing their case with a larger one as necessary. In early September the larvae move to the edge of the pond or stream, where they hibernate until the spring. Larval cases are often common on water-lily leaves. They also occur during larval development on submerged parts of host plants, the larvae surviving on air trapped within the confines of their habitations. When larvae are not feeding, cases float freely on the surface of the water. Fully fed larvae pupate in the early summer within silken cocoons, into which fragments of leaf are incorporated, spinning up to a convenient leaf or stem a few centimetres above the water level.

DAMAGE
Infested leaves are extensively holed and often become ragged. Damaged tissue also rots and becomes unsightly.

716 Jasmine moth (*Palpita vitrealis*).

717 Larva of *Udea prunalis*.

Palpita vitrealis (Rossi) (**715–716**)
syn. *P. unionalis* (Hübner)
Jasmine moth

A migratory, Asiatic, African and southern European species, the very active, mainly greenish to yellowish-green larvae feeding on the shoots of various members of the Oleaceae. Damage to the new shoots is often extensive and is sometimes noted in northern Europe on oleaceous plants such as jasmine (*Jasminum*); infestations are sometimes also found on decorative, container-grown olive (*Olea europaea*) trees imported into northern Europe from Italy and other countries where the pest is endemic. The adults (28–30 mm wingspan) are mainly white, with a brown leading edge to each fore wing. In countries such as the British Isles this species occurs as a rare, non-resident migrant and is not regarded as a pest.

718 Adult of *Udea prunalis*.

Trachycera suavella (Zincken)
syn. *Numonia suavella* (Zincken)
Porphyry knothorn moth

A local but widely distributed insect, associated mainly in the wild with blackthorn (*Prunus spinosa*) but occasionally troublesome on ornamental *Cotoneaster*. The purplish-red, greyish- to whitish-marked adults (23–25 mm wingspan) occur in July but, although being attracted to light, are rarely seen. The larvae feed during the spring within distinctive whitish, silken galleries spun on the shoots beneath the leaves; dead leaf fragments and particles of frass accumulate on these larval shelters, making them even more conspicuous. The larvae are dark chestnut-brown, with a brown head, black prothoracic and anal plates, and a pair of distinctive dark spots on the second thoracic and the eighth abdominal segments. Pupation takes place in June, in a greyish-white cocoon spun either within or alongside the larval habitation.

Udea prunalis (Denis & Schiffermüller) (**717–718**)
syn. *U. nivealis* (Fabricius)

A generally common species, associated mainly with herbaceous plants but also attacking the foliage of young ornamental trees and shrubs. The larvae feed within spun leaves; they occur briefly during the autumn and then hibernate, completing their development in the following spring. Individuals (up to 25 mm long) are greenish and shiny, with white subdorsal lines and a pale head. Adults (22–26 mm wingspan) are mainly brownish grey; they appear in June and July.

Family **PIERIDAE**

A family of mainly white- or yellow-winged butterflies. The larvae have setae on the head arising from raised tubercles; crotchets on the abdominal prolegs are of two sizes (biordinal).

Pieris brassicae (Linnaeus) (**719–720**)
Large white butterfly

Although of importance mainly as a pest of vegetable brassica crops, eggs of this often abundant species are sometimes deposited on ornamentals such as mignonette (*Reseda odorata*) and nasturtium (*Tropaeolum*). The larvae are gregarious, feeding in large groups and causing extensive damage to the leaves, particularly in the late summer. Individuals (up to 40 mm long) are light green to yellow, variably marked with black. The mainly white, black-marked adults (55–65 mm wingspan) occur in the spring but are more numerous in the summer.

Pieris rapae (Linnaeus) (**721–722**)
syn. *Artogeia rapae* (Linnaeus)
Small white butterfly

This well-known pest of vegetable brassicas also, occasionally, breeds on mignonette (*Reseda odorata*) and nasturtium (*Tropaeolum*). However, the larvae are usually present in only small numbers and, unlike those of the previous species, cause insignificant damage. Fully grown specimens are *c.* 30 mm long, green, finely speckled with black, with a yellow line along the back and yellow markings along the sides. The mainly white to yellowish-white adults (45 mm wingspan) appear in late April and May, with a second generation emerging in the summer.

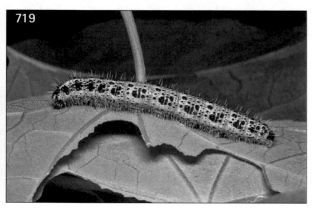

719 Larva of large white butterfly (*Pieris brassicae*).

720 Large white butterfly (*Pieris brassicae*) damage to *Tropaeolum*.

721 Larva of small white butterfly (*Pieris rapae*).

722 Female small white butterfly (*Pieris rapae*).

Family **LYCAENIDAE**

A large family of mainly small, brightly coloured butterflies, including 'blues', 'coppers' and 'hairstreaks'. The larvae are stumpy and somewhat slug-like in appearance, with a retractile head and the body coated in short hairs.

Cacyreus marshalli Butler
Geranium bronze butterfly

A southern African species, associated with various kinds of *Geranium* and *Pelargonium*; scented cultivars tend to be avoided. Recently established in the South of France, Italy, Portugal and Spain, having spread to mainland Europe via the Balearic Islands, and also found elsewhere, e.g. in Belgium, England, Germany, the Netherlands and Switzerland.

DESCRIPTION
Adult female: 18–27 mm wingspan; upper surface mainly bronzy brown, with hair fringes alternately chequered bronzy brown and white; underside greyish brown, with black, grey and white markings forming an irregular wavy pattern; hind wings each with a narrow, white-tipped tail-like projection. **Adult male:** 15–23 mm wingspan; coloration and form as in female. **Larva:** up to 13 mm long; mainly yellowish to light green, often with a partial pinkish tinge.

LIFE HISTORY
In southern Europe, eggs are laid from May to October. Most are deposited on flower buds and flower stalks, but some may be found elsewhere. Following egg hatch, larvae feed inside the flower buds, and both in or on the terminal shoots. They also attack the open flowers. Fully grown larvae eventually pupate on or near their foodplant, and adults emerge shortly afterwards. Depending on temperature, there are up to five generations annually.

DAMAGE
Infested flower buds and terminal shoots are destroyed, and damage caused is often extensive. Young plants may even be killed. Damaged tissue is also often invaded by secondary bacterial and fungal pathogens.

723 Female small eggar moth (*Eriogaster lanestris*).

724 Final-instar larva of small eggar moth (*Eriogaster lanestris*).

725 Larval tent of small eggar moth (*Eriogaster lanestris*).

Family **LASIOCAMPIDAE**

Medium-sized to very large moths with bipectinate antennae (most noticeable in males); males usually much smaller than females of the same species. The larvae are very hairy on both head and body; crotchets on the abdominal prolegs are of two sizes (biordinal).

Eriogaster lanestris (Linnaeus) (**723–725**)
Small eggar moth

In parts of mainland Europe a minor pest of amenity trees such as birch (*Betula*) and lime (*Tilia*); also occurs on certain other hosts, including blackthorn (*Prunus*

726 Male lackey moth (*Malacosoma neustria*).

727 Larva of lackey moth (*Malacosoma neustria*).

spinosa), hawthorn (*Crataegus*), oak (*Quercus*) and broad-leaved willows such as grey willow (*Salix cinerea*). In some countries (e.g. England and Ireland) the insect is restricted mainly to wild blackthorn and hawthorn hedgerows, and is not regarded as a pest. Widely distributed in central and northern Europe.

DESCRIPTION
Adult female: 42 mm wingspan; wings thinly scaled, greyish brown to pale reddish brown, with a pale wavy crossline and, on each fore wing, a pale sub-central spot; abdomen with a greyish anal hair tuft. **Adult male:** 32 mm wingspan; similar to female but darker and without the anal hair tuft; antennae strongly bipectinate. **Egg:** olive-green, oval. **Larva:** up to 50 mm long; black or greyish black, with reddish to whitish hairs and a series of brown, yellowish-edged patches along the back; head black; early instars are darker and recently moulted final-instar larvae brightly coloured, with a gingery-brown head.

LIFE HISTORY
Adults appear in February and March, females depositing batches of eggs on the twigs of host plants. The eggs are then covered with hairs from the anal tuft. Larvae fed from May to June or July, living gregariously and constructing dense, silken webs. Pupation occurs in large, yellowish-white to reddish-brown cocoons, usually formed in the ground. Most adults emerge in the following year but some individuals remain in the pupal stage through two or more winters.

DAMAGE
Larvae cause considerable defoliation and the webs are disfiguring, but infestations rarely affect plant growth.

Malacosoma neustria (Linnaeus) (726–729)
Lackey moth
An often common pest of trees and shrubs, including alder (*Alnus*), birch (*Betula*), *Cotoneaster*, crab-apple (*Malus*), elm (*Ulmus*), firethorn (*Pyracantha*), hawthorn (*Crataegus*), flowering cherry (*Prunus*), lilac (*Syringa*), rose (*Rosa*) and willow (*Salix*); sometimes of importance in nurseries, parks and gardens. Eurasiatic. Present throughout much of Europe, except for the extreme north.

DESCRIPTION
Adult: 30–40 mm wingspan; wings and body pale ochreous to dark brown; each fore wing with two crosslines, often (particularly in female) enclosing a darker band. **Egg:** 0.5 mm across; cylindrical, laid in a large batch. **Larva:** up to 50 mm long; greyish blue, with a white dorsal stripe, and with orange-red, black-edged stripes running along the back and sides; body clothed in reddish-brown hairs; head blue, with two black spots. **Pupa:** 18 mm long; brownish black and hairy.

LIFE HISTORY
Moths occur from late July to September, each female depositing about 100–200 eggs in a characteristic band (6–14 mm wide) around a twig or spur of a host plant. Each egg mass, protected by a clear varnish-like coating secreted by the egg-laying female, remains *in situ* throughout the winter. The eggs hatch in the spring, usually in late April and May. The young larvae are blackish but soon become more brightly coloured. They feed gregariously in a communal web or 'tent'. These larval tents, which are very conspicuous and may exceed 30 cm in length, are gradually extended as the larvae and their feeding areas grow. Larvae are fully fed

728 Egg band of lackey moth (*Malacosoma neustria*).

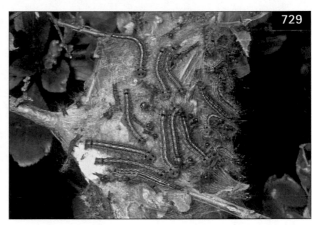

729 Larval tent of lackey moth (*Malacosoma neustria*) on *Prunus.*

by late June or early July. They then pupate in white or yellowish, double-walled cocoons spun between leaves, in bark fissures or amongst herbage on the ground. Adults emerge about three weeks later.

DAMAGE
Defoliation is often severe, with infested branches or even whole trees stripped of leaves and covered in webbing. Growth, particularly of young trees, may be seriously affected.

Poecilocampa populi (Linnaeus) (730–731)
December moth
Widely distributed and locally common, the larvae feeding from late April to June on trees such as birch (*Betula*), crab-apple (*Malus*), flowering cherry (*Prunus*), hawthorn (*Crataegus*), lime (*Tilia*), oak (*Quercus*) and poplar (*Populus*). The larvae are sometimes present on ornamental plants but damage caused is not of significance. Fully grown larvae (*c.* 45 mm long) are distinctly downy, mainly greyish and black above, with a red crossline just behind the head, but ochreous and black below. The moths (35–45 mm wingspan) are mainly brown and rather hairy, the wings marked with grey and ochreous; males have strongly bipectinate antennae. Adults fly during the winter, and are most numerous in November and December.

730 Larva of December moth (*Poecilocampa populi*).

731 Male December moth (*Poecilocampa populi*).

Family **THYATIRIDAE**

Achlya flavicornis (Linnaaeus) (**732–734**)
syn. *Polyploca flavicornis* (Linnaeus)
Yellow-horned moth

A generally common species, associated with birch (*Betula*); found occasionally on ornamental and amenity trees in the vicinity of birch woodlands. The larvae feed during June and July, sheltering during the daytime in a folded leaf. Young specimens are blackish olive, marked with small, pale pinacula, and are superficially tortrix-like. Older individuals (up to 33 mm long) are whitish to pale greenish white, more or less suffused with dark olive-green, and marked with black and white spots; the head is mainly yellowish brown to reddish brown. When fully grown, they enter the soil to pupate in flimsy cocoons. Adults are mainly greenish grey, with dark grey or black markings. They occur in March and April.

732 Young larva of yellow-horned moth (*Achlya flavicornis*).

Family **GEOMETRIDAE**

A very large family of mainly slender-bodied moths with relatively large, usually broad, wings; most species are rather weak fliers. The larvae, commonly called 'looper caterpillars' or 'loopers', usually have just two functional pairs of abdominal prolegs (those on the third, fourth and fifth abdominal segment being rudimentary or absent) and progress with a characteristic looping gait; some are brightly coloured but many are cryptic and, when at rest, bear a close resemblance to twigs.

Abraxas grossulariata (Linnaeus) (**735–736**)
Magpie moth

The characteristic creamy-white and black-spotted larvae of this generally distributed species are sometimes damaging to the foliage of ornamental trees and shrubs, including crab-apple (*Malus*), flowering cherry (*Prunus*) and flowering currant (*Ribes sanguineum*), hawthorn (*Crataegus*), hazel (*Corylus*) and spindle (*Euonymus*). The mainly black, white and yellow adults (35–40 mm wingspan) occur in July and August. Larvae feed from August onwards, individuals hibernating during the winter months and completing their development in the following May or June. Although capable of causing noticeable defoliation, numbers of larvae on ornamental plants are usually small and damage is rarely, if ever, of significance.

Agriopis aurantiaria (Hübner) (**737**)
syn. *Erannis aurantiaria* (Hübner)
Scarce umber moth

This widespread and locally common species attacks various trees and shrubs, including beech (*Fagus sylvatica*), birch (*Betula*), crab-apple (*Malus*), hornbeam (*Carpinus betulus*) and oak (*Quercus*), the larvae causing minor damage to the foliage. Larvae (up

733 Larva of yellow-horned moth (*Achlya flavicornis*).

734 Male yellow-horned moth (*Achlya flavicornis*).

to 35 mm long) are greyish to yellowish or brownish, marked with purplish lines and blackish patches; they occur from April to late May or early June. Adults occur mainly in October and November. Males (35–40 mm wingspan) have pale golden-yellow fore wings, marked with purplish speckles and crosslines. In females the wings are reduced to greyish stubs (2–4 mm long); the body (*c*. 8 mm long) is yellowish brown, mottled with

black, and marked dorsally with numerous, yellow scales.

Agriopis marginaria (Fabricius) (**738–739**)
syn. *Erannis marginaria* (Fabricius)
Dotted border moth

Larvae of this generally common species frequently occur on ornamental trees and shrubs, including alder (*Alnus*), birch (*Betula*), crab-apple (*Malus*), hornbeam (*Carpinus betulus*), oak (*Quercus*) and willow (*Salix*). They feed on the foliage from April to late May or early June and then pupate in the soil. Adults appear in the following March or April. Larvae (up to 30 mm long) are rather slender, olive-green to brownish, with pale patches along the sides and, often, a series of blackish x-shaped markings along the back. Adult males (30–35 mm wingspan) are mainly yellowish brown, the fore wings marked with dark crosslines and a characteristic row of black dots along the outer margin. Females (7–10 mm long) are greyish brown, mottled with black; the wings are reduced to short stubs, the hind wings being the longer pair.

735 Magpie moth (*Abraxas grossulariata*).

736 Larva of magpie moth (*Abraxas grossulariata*).

737 Larva of scarce umber moth (*Agriopis aurantiaria*).

738 Larva of dotted border moth (*Agriopis marginaria*).

739 Male dotted border moth (*Agriopis marginaria*).

740 Larva of mottled beauty moth (*Alcis repandata*).

741 Male mottled beauty moth (*Alcis repandata*).

742 Male March moth (*Alsophila aescularia*).

743 Larva of March moth (*Alsophila aescularia*).

Alcis repandata (Linnaeus) (**740–741**)
syn. *Boarmia repandata* (Linnaeus)
Mottled beauty moth

An often common but minor pest of trees and shrubs, including birch (*Betula*), elm (*Ulmus*), hawthorn (*Crataegus*), hazel (*Corylus*) and sycamore (*Acer pseudoplatanus*). The larvae (up to 40 mm long) are elongate and mainly brown. They feed throughout the summer, and complete their development in the following spring. Specimens are sometimes found on ornamental trees in parks, gardens and nurseries but they cause only slight damage. The greyish-white to dark yellowish-grey adults (40–44 mm wingspan) occur from June to July. In some seasons, there may be a partial second generation.

Alsophila aescularia (Denis & Schiffermüller) (**742–743**)
March moth

An often common pest of ornamental trees and shrubs, including beech (*Fagus sylvatica*), birch (*Betula*), crabapple (*Malus*), flowering cherry (*Prunus*), hawthorn (*Crataegus*), hornbeam (*Carpinus betulus*), lilac (*Syringa*), oak (*Quercus*), privet (*Ligustrum vulgare*) and rose (*Rosa*). Eurasiatic. Present throughout most of Europe.

DESCRIPTION

Adult female: wingless; body 8 mm long, shiny greyish brown, with a large anal tuft of hair. **Adult male:** 25–30 mm wingspan; fore wings light grey to brownish grey, with lighter crosslines; hind wings light grey. **Egg:** dark brown, laid in a large batch that encircles a twig. **Larva:** up to 25 mm long; light green, with a darker green dorsal line and yellowish lines along the sides, including one below the spiracles; body relatively

744 Female lilac beauty moth (*Apeira syringaria*).

745 Male peppered moth (*Biston betularia*).

narrow, with a vestigial pair of prolegs on the fifth abdominal segment (cf. *Operophtera brumata*, p. 306). **Pupa:** 10 mm long; brown and stumpy; cremaster with two curved, divergent spines.

LIFE HISTORY
Adults occur from mid-February to mid-April. The males, which are readily attracted to light, are commonly found at rest during the daytime on walls, fences and other surfaces; the flightless females, however, are rarely seen. Eggs, which are deposited in compact bands around the twigs of host plants, hatch in April. The larvae then feed on the foliage of various trees and shrubs, and on the blossoms of early-flowering hosts. When fully grown, in late May or June, larvae enter the soil and each pupates in a silken cocoon.

DAMAGE
Larvae damage unopened buds but they cause most harm to the young foliage. They also attack the blossoms of flowering trees and shrubs, biting holes into the petals and destroying the stamens as well as other floral parts.

Apeira syringaria (Linnaeus) (**744**)
syn. *Phalaena syringaria* (Linnaeus)
Lilac beauty moth
Although not a significant pest, the unusual larvae of this species are sometimes noticed on cultivated elder (*Sambucus*), honeysuckle (*Lonicera*), lilac (*Syringa*) and privet (*Ligustrum vulgare*). Larvae, which are most often found in May, are yellowish brown, marked with red and purplish red, with a slightly hairy dorsal surface, two small projections on the second and third abdominal segments and a distinctive, forked projection on the fourth. They feed briefly in the autumn before

746 Larva of peppered moth (*Biston betularia*).

hibernating and resuming their activity in the spring. Fully grown larvae pupate on the host plant, usually in early June, each in a pendulous, silken cocoon to which the cast skin of the larva is also attached. The adults (35–40 mm wingspan) are tawny yellow, marked with purplish white and red; they appear in June and July, often with a partial second generation in the autumn.

Biston betularia (Linnaeus) (**745–746**)
Peppered moth
Generally common on various trees and shrubs, including beech (*Fagus sylvatica*), birch (*Betula*), crab-apple (*Malus*), elm (*Ulmus*), flowering cherry (*Prunus*), larch (*Larix*) and rose (*Rosa*); larvae also occur on herbaceous plants such as *Chrysanthemum* and pot marigold (*Calendula officinalis*). Eurasiatic. Widely distributed in Europe.

DESCRIPTION
Adult: 42–55 mm wingspan; body and wings white, peppered with black; entirely black (ab. *carbonaria*)

and intermediate (ab. *insularia*) forms also occur; male antennae strongly bipectinate. **Egg:** 0.7 × 0.5 mm; whitish green. **Larva:** up to 50 mm long; brown or green, with pinkish markings and reddish spiracles; body stick-like, with a pair of dark purplish prominences on the fifth abdominal segment; head purplish brown, with a distinct central cleft. **Pupa:** 20–22 mm long; blackish brown, terminating in a spike.

LIFE HISTORY
Moths fly in May, June and July, and deposit eggs on a wide range of host plants. Larvae occur from July to September or October; they feed on the foliage but usually remain at rest during the daytime, mimicking a shoot or broken twig, with the body held straight out at an angle of about 45 degrees. Fully grown larvae enter the soil to pupate. Adults emerge in the following year.

DAMAGE
Larvae cause noticeable defoliation, particularly if present on herbaceous plants. Damage to trees and shrubs is usually unimportant since the bulk of feeding occurs relatively late in the season.

Bupalus piniaria (Linnaeus) (747–748)
Bordered white moth
Generally common in the vicinity of coniferous woodlands, mainly on Scots pine (*Pinus sylvestris*); also occurs on common silver fir (*Abies alba*), Douglas fir (*Pseudotsuga menziesii*), larch (*Larix*) and Norway spruce (*Picea abies*). A notorious forestry pest; attacks on coniferous ornamentals and nursery stock are less frequent and usually of little or no importance. Widespread in Europe, except for the extreme north.

DESCRIPTION
Adult female: 32–35 mm wingspan; wings mainly orange or orange-yellow, suffused with brown, and the underside conspicuously flecked with white. **Adult male:** 32–35 mm wingspan; wings whitish yellow to yellowish, bordered and variably suffused with brownish black; antennae strongly bipectinate. **Egg:** 0.5 mm long; light green. **Larva:** up to 30 mm long; bright green, with white or yellowish-white, longitudinal stripes. **Pupa:** 10–15 mm long; shiny brown.

LIFE HISTORY
Larvae feed from June or July to October or November. They occur mainly on mature conifers but are sometimes found on nursery trees. Fully grown larvae pupate in the soil at the base of host trees, and adults appear in the following May or June.

DAMAGE
Larvae cause extensive defoliation, such depredations being most severe on sandy-soil sites in low-rainfall areas. Attacks on ornamentals are seldom, if ever, of significance.

Campaea margaritata (Linnaeus) (749–750)
Light emerald moth
Larvae of this widely distributed woodland species feed mainly on birch (*Betula*), crab-apple (*Malus*), elm (*Ulmus*), hornbeam (*Carpinus betulus*) and oak (*Quercus*); they also occur in small numbers on ornamental trees in gardens and nurseries. The delicate, light green moths (38–45 mm wingspan) appear in June and July, depositing eggs in batches on the underside of

747 Female bordered white moth (*Bupalus piniaria*).

748 Larva of bordered white moth (*Bupalus piniaria*).

leaves. The larvae feed from September onwards, lying stretched out flat against the twigs when at rest. Unlike most other species, they continue to feed during the winter, removing bark from the young shoots to expose the pale wood; they also feed on the dormant buds. In the spring, when about 20 mm long, the larvae attack the expanding buds and unfurling leaves. Most individuals pupate by the end of May. Fully grown larvae (30–35 mm long) are greyish to greenish brown or purplish brown, with orange, brown-rimmed spiracles; they possess a characteristic skirt-like fringe of pale bristles along either side of the body and, unlike most geometrids, three pairs of functional prolegs; younger larvae are generally greyish, often with T-shaped markings along the back.

Chloroclysta truncata (Hufnagel) (**751–752**)
syn. *Cidaria truncata* (Hufnagel)
Common marbled carpet moth

Minor infestations of this generally common and extremely varied species occur occasionally on herbaceous ornamentals such as *Geranium* and *Pelargonium*. The larvae also feed on seedling trees, including maple (*Acer*). Although most often reported on outdoor plants, minor attacks sometimes occur in greenhouses. The characteristic larvae (up to 35 mm long) are slender and greenish, with a darker dorsal line and a pair of yellowish subdorsal lines along the back (there is sometimes also a complete or interrupted reddish stripe along the somewhat warty sides); the last body segment bears a pair of pointed projections. The larvae feed from September to April, hibernating during the winter months, but continuing to feed if in heated greenhouses. Larvae of the summer generation occur from late June to late July or early August. Fully grown individuals pupate amongst withered leaves on the host plant or amongst debris on the ground. The mainly black to brownish adults (32–35 mm wingspan) occur in May and June, and in the autumn.

749 Light emerald moth (*Campaea margaritata*).

750 Larva of light emerald moth (*Campaea margaritata*).

751 Common marbled carpet moth (*Chloroclysta truncata*).

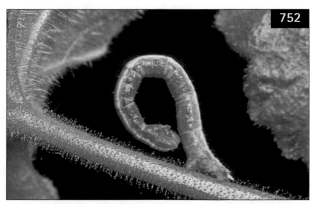

752 Larva of common marbled carpet moth (*Chloroclysta truncata*).

Colotois pennaria (Linnaeus) (**753–754**)
Feathered thorn moth

Adults of this common woodland insect occur in October and November. The females deposit oblong batches of 100–200 olive-green, smooth-shelled eggs along the shoots of various trees and shrubs, including established ornamentals and nursery stock. Hosts include birch (*Betula*), crab-apple (*Malus*), flowering cherry (*Prunus*), hawthorn (*Crataegus*), hornbeam (*Carpinus betulus*), larch (*Larix*), oak (*Quercus*), poplar (*Populus*) and willow (*Salix*). The eggs hatch in the following spring. Larvae feed from April to June and eventually pupate in the soil. Fully grown larvae (40–45 mm long) are stout-bodied and purplish grey to slate-grey, with faint, yellowish diamond-shaped marks down the back and similarly coloured spots along the sides; the ninth abdominal segment bears two red-tipped projections. Adults (40 mm wingspan) are mainly light orange to pale yellowish; males have strongly bipectinate antennae.

Crocallis elinguaria (Linnaeus) (**755–756**)
Scalloped oak moth

A polyphagous and generally distributed species, the larvae feeding from April to June on the leaves of various deciduous trees and shrubs, including blackthorn (*Prunus spinosa*), crab-apple (*Malus*), flowering cherry (*Prunus*) and honeysuckle (*Lonicera*); attacks sometimes occur on such plants in parks and gardens and also, occasionally, on nursery stock. The larvae are capable of causing noticeable defoliation, but are usually present in only small numbers so that damage caused is of little or no significance. The twig-like larvae (up to 45 mm long) are greyish yellow to greyish black, tinged with purple, with a dark diamond-like pattern along the back and a slight, blackish-edged elevation on the eighth abdominal segment. Fully fed individuals pupate in the soil. The mainly whitish-yellow to ochreous-brown adults (32–40 mm wingspan) occur in July and August.

Deileptenia ribeata (Clerck) (**757–758**)
syn. *D. abietaria* (Denis & Schiffermüller);
Boarmia ribeata (Clerck)
Satin carpet moth

A widely distributed species, associated with various trees but most often encountered on fir (*Abies*), larch (*Larix*), spruce (*Picea*) and yew (*Taxus baccata*). The greyish-brown larvae sometimes cause slight damage to ornamentals but they are usually present in only small numbers. They feed from August onwards, hibernating during the winter and completing their development in the following June. Adults (40–48 mm wingspan) are pale

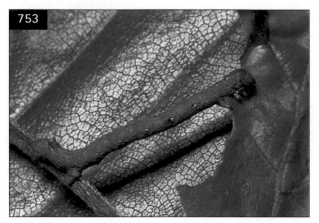

753 Larva of feathered thorn moth (*Colotois pennaria*).

754 Male feathered thorn moth (*Colotois pennaria*).

yellowish grey to black; they fly from late June to early August.

Ectropis bistortata (Goeze) (**759–760**)
syn. *Boarmia bistortata* (Goeze)
Engrailed moth

A generally common, double-brooded species. The pale greenish-grey, shoot-like larvae (up to 30 mm long) feed on various trees and shrubs, including beech (*Fagus sylvatica*), birch (*Betula*) and oak (*Quercus*). They occur from May onwards, and sometimes cause minor damage to the foliage of ornamentals. The pale yellowish-grey adults (38–40 mm wingspan) occur in the early spring and again in mid-summer.

755 Larva of scalloped oak moth (*Crocallis elinquaria*).

756 Scalloped oak moth (*Crocallis elinquaria*).

757 Larva of satin carpet moth (*Deileptenia ribeata*).

758 Male satin carpet moth (*Deileptenia ribeata*).

759 Larva of engrailed moth (*Ectropis bistortata*).

760 Engrailed moth (*Ectropis bistortata*).

761 Larva of November moth (*Epirrita dilutata*).

762 November moth (*Epirrita dilutata*).

763 Female mottled umber moth (*Erannis defoliaria*).

764 Male mottled umber moth (*Erannis defoliaria*).

765 Larva of mottled umber moth (*Erannis defoliaria*).

Epirrita dilutata (Denis & Schiffermüller) (761–762)
November moth

A generally common woodland species and sometimes a minor pest of trees and shrubs, including birch (*Betula*), crab-apple (*Malus*), elm (*Ulmus*), hawthorn (*Crataegus*), maple (*Acer*) and oak (*Quercus*). The larvae occur in small numbers on cultivated plants, browsing on the foliage from April to June, and contributing to damage caused by various other spring-feeding species. Fully grown larvae (*c.* 30 mm long) are rather plump and bright velvet-green (but greenish white below), with small, pale pinacula, reddish-brown spiracles, a creamy-white subspiracular line and faint whitish lines down the back; the back sometimes also bears a series of purplish-red, diamond-shaped markings. Adults (32–38 mm wingspan) are dull greyish to black, with wavy crosslines. They occur mainly in October and November.

766 Larva of common emerald moth (*Hemithea aestivaria*).

767 Common emerald moth (*Hemithea aestivaria*).

Erannis defoliaria (Clerck) (**763–765**)
Mottled umber moth

A generally common pest of ornamental trees and shrubs, including birch (*Betula*), elm (*Ulmus*), flowering cherry (*Prunus*), hazel (*Corylus*), honeysuckle (*Lonicera*), hornbeam (*Carpinus betulus*), lime (*Tilia*), oak (*Quercus*), poplar (*Populus*) and rose (*Rosa*); also a pest of deciduous forest trees and fruit trees. Widely distributed in Europe; also present in Canada.

DESCRIPTION
Adult female: wingless; body 10–15 mm long, mottled with black, yellow and, sometimes, white scales. **Adult male:** 35–38 mm wingspan; fore wings pale yellow to reddish brown, more or less finely peppered with black, and often variably decorated with dark cross markings. **Egg:** 0.9 × 0.5 mm; oval, yellowish to greyish; shell almost smooth. **Larva:** up to 35 mm long; reddish brown, with yellow or creamy-white areas on the sides of the first to seventh abdominal segments. **Pupa:** 12–14 mm long; dark yellowish brown; cremaster with a short, bifid tip.

LIFE HISTORY
Adults occur from mid-October to mid-January. The colourful males are sometimes noticed at rest during the daytime on trees, fences and walls but the wingless females are rarely seen. Eggs, which are laid in bark crevices, hatch in early April. The larvae then feed ravenously, usually resting fully exposed on the leaves or shoots. They are easily dislodged from the foodplant. They then remain temporarily suspended by a silken thread, with the head and thorax bent back at an angle to the abdomen. In June, fully fed larvae pupate in the soil a few centimetres below the surface.

DAMAGE
Larvae cause considerable defoliation. Attacks on the buds and blossoms of trees and shrubs are also of importance, and complete trusses may be destroyed.

Hemithea aestivaria (Hübner) (**766–767**)
Common emerald moth

A widely distributed species, the larvae feeding during the autumn on low-growing plants and then hibernating. In the spring, the larvae attack the foliage of various trees and shrubs. Minor damage is then sometimes caused to nursery plants and garden ornamentals, including birch (*Betula*), hawthorn (*Crataegus*), lime (*Tilia*), oak (*Quercus*), rose (*Rosa*) and willow (*Salix*). Fully grown larvae (*c*. 22 mm long) are dull green, marked with black and reddish brown; the head and first thoracic segment are distinctly notched, and the body surface noticeably roughened. The green, angular-winged adults (24–28 mm wingspan) occur in late June and July.

Hydriomena furcata (Thunberg) (**768–769**)
syn. *Cidaria furcata* (Thunberg); *Larentia sordidata* (Fabricius)
July highflier moth
Widespread and generally common in and around open woodlands, and in association with hedgerows. Larvae (up to 25 mm long) are stout-bodied, brownish to black, with partly whitish intersegmental rings and whitish lines along the back and sides. They feed between spun leaves in May and June, and are sometimes noted on ornamental or nursery trees and shrubs, especially hazel (*Corylus*) and willow (*Salix*); attacks have also occurred on conifers, including Sitka spruce (*Picea sitchensis*). The rather drab, greyish-green to brownish-yellow, round-winged adults (28–30 mm wingspan) occur in July and August.

Lomaspilis marginata (Linnaeus) (**770–771**)
Clouded border moth
A generally distributed and often common species, associated with aspen (*Populus tremula*), black poplar (*P. nigra*), common hazel (*Corylus avellana*) and various kinds of willow (*Salix*). The larvae feed on the foliage from June onwards. They often occur on cultivated plants but damage caused is unimportant. Fully fed individuals (18–20 mm long) are yellowish green, with paired dark green lines along the back and a purplish-brown blotch on the last abdominal segment; the head is green, marked with purplish brown. Adults (22–24 mm wingspan) are mainly white to yellowish white, with an irregular greyish-black border and, sometimes, a partial or complete median band on each wing. The moths are most numerous in May and June; in parts of mainland Europe, there are two generations annually.

768 Larva of July highflier moth (*Hydriomena furcata*).

769 July highflier moth (*Hydriomena furcata*).

770 Larva of clouded border moth (*Lomaspilis marginata*).

771 Clouded border moth (*Lomaspilis marginata*).

Lycia hirtaria (Clerck) (772–773)

Brindled beauty moth

A widely distributed species. Larvae attack the foliage of various trees and shrubs, including alder (*Alnus*), beech (*Fagus sylvatica*), crab-apple (*Malus*), elm (*Ulmus*), hawthorn (*Crataegus*) and lime (*Tilia*), and often occur on garden ornamentals and nursery stock. Larvae are large (up to 55 mm long), stout-bodied, purplish grey to reddish brown (the latter colour also forming wavy lines down the back), marked with dark speckles; there is also a yellow crossline just behind the head, and yellow spots on several of the abdominal segments. They feed from May to July, and then pupate in the soil. The rather hairy, greyish to blackish, ochreous-marked adults (40–45 mm wingspan) appear in March and April.

Menophra abruptaria (Thunberg) (774–775)

syn. *Hemerophila abruptaria* (Thunberg)

Waved umber moth

The mainly greyish-brown, stick-like larvae (up to 40 mm long) of this widely distributed and locally common species feed from May to August on jasmine (*Jasminum*), lilac (*Syringa*), privet (*Ligustrum vulgare*), rose (*Rosa*) and certain other garden ornamentals. However, they are not of importance as pests. Pupation occurs in a shallow depression formed by the larva on a branch of the host plant, the pupa hidden within a cocoon camouflaged by masticated fragments of bark. Adults (35–40 mm wingspan) are pale to whitish ochreous, suffused with brown. They occur in April and May.

772 Larva of brindled beauty moth (*Lycia hirtaria*).

773 Male brindled beauty moth (*Lycia hirtaria*).

774 Larva of waved umber moth (*Menophra abruptaria*).

775 Male waved umber moth (*Menophra abruptaria*).

776 Scalloped hazel moth (*Odontopera bidentata*).

777 Female of winter moth (*Operophtera brumata*).

Odontopera bidentata (Clerck) (776)
syn. *Gonodontis bidentata* (Clerck)
Scalloped hazel moth

The larvae of this polyphagous, widely distributed species occur from June to October on various trees and shrubs. They often attack cultivated plants, including beech (*Fagus sylvatica*), birch (*Betula*), fir (*Abies*), hawthorn (*Crataegus*), ivy (*Hedera*), larch (*Larix*) and oak (*Quercus*), but are usually present only in small numbers. Individuals (up to 50 mm long) are dark green to greyish or purplish, with yellowish and light brown lozenge-shaped marks down the back. When fully grown, they pupate in the soil. The mainly light greyish-brown adults (38–42 mm wingspan) appear in the following April, May or June.

778 Male winter moth (*Operophtera brumata*).

Operophtera brumata (Linnaeus) (777–779)
Winter moth

A generally common and often destructive pest of ornamental trees and shrubs, including *Cotoneaster*, crab-apple (*Malus*), dogwood (*Cornus*), elm (*Ulmus*), flowering cherry (*Prunus*), hawthorn (*Crataegus*), hazel (*Corylus*), hornbeam (*Carpinus betulus*), lilac (*Syringa*), lime (*Tilia*), ornamental pear (*Pyrus calleryana* 'Chanticleer'), *Rhododendron*, rose (*Rosa*), spruce (*Picea*), sycamore (*Acer pseudoplatanus*) and willow (*Salix*); also an important pest in woodlands and orchards. Widely distributed in central and northern Europe.

DESCRIPTION

Adult female: wings reduced to stubs; body 5–6 mm long, dark brown, mottled with greyish yellow. **Adult male:** 22–28 mm wingspan; fore wings light greyish brown, with darker wavy crosslines; hind wings light grey. **Egg:** 0.5 × 0.4 mm; oval, with a pitted surface;

779 Larva of winter moth (*Operophtera brumata*).

780 Female northern winter moth (*Operophtera fagata*).

781 Larva of northern winter moth (*Operophtera fagata*).

pale yellowish green when newly laid but soon becoming orange-red. **Larva:** up to 25 mm long; rather plump, light green, with a dark green dorsal stripe and several whitish or creamy-yellow stripes along the back and sides, including a pale yellow line passing through the spiracles (cf. *Alsophila aescularia*, p. 297). **Pupa:** 7–8 mm long; brown and stumpy; cremaster with a pair of laterally directed spines.

LIFE HISTORY
Adults occur from October to January but are most abundant in November and December. The males are active at night and are strongly attracted to light. They often rest openly on walls and fences during the daytime but the spider-like females hide on tree trunks and are less often seen. Mating takes place at night, copulating pairs occurring on the trees with the males characteristically standing head downwards, with the wings held outwards like a settled butterfly. Eggs, about 100–200 per female, are deposited singly in crevices in the bark. They hatch in late March or April and the newly emerged larvae then attack the buds, blossoms and expanding leaves. At this stage, the minute larvae are also blown about in the wind, each on a silken thread; they are then often carried into previously uninfested nurseries and gardens from adjacent woodland trees and hedges. Feeding continues until late May or early June, the rather sluggish larvae spinning two leaves loosely together with silk, or feeding within the blossom trusses, sheltered by the overlying petals or calyxes. Fully grown larvae drop to the ground and enter the soil to pupate in flimsy cocoons about 8–10 cm below the surface.

DAMAGE
Larvae cause considerable defoliation, and often completely strip the leaves from the branches of heavily infested host plants. Attacks on the buds and blossoms are also of importance; complete trusses may be destroyed.

Operophtera fagata (Scharfenberg) (**780–781**)
syn. *Cheimatobia boreata* (Hübner)
Northern winter moth
Adults of this widely distributed species occur in October and November. They are similar in appearance and habits to those of *Operophtera brumata*, but the females have slightly longer wings and the males are lighter in colour. Larvae (up to 21 mm long) are mainly green, with greyish-white lines along the back and sides, black spiracles and a black head. They feed on the foliage of host plants in May and June and, if numerous, cause noticeable defoliation. Attacks are most frequently established on beech (*Fagus sylvatica*) and birch (*Betula*) but also develop on other trees and shrubs; minor infestations are sometimes noted on ornamentals and nursery stock.

782 Larva of brimstone moth (*Opisthograptis luteolata*).

783 Brimstone moth (*Opisthograptis luteolata*).

784 Larva of swallow-tailed moth (*Ourapteryx sambucaria*).

785 Swallow-tailed moth (*Ourapteryx sambucaria*).

Opisthograptis luteolata (Linnaeus) (**782–783**)
Brimstone moth

A generally common species, the larvae sometimes feeding on rosaceous ornamental trees and shrubs, including crab-apple (*Malus*), flowering cherry (*Prunus*) and hawthorn (*Crataegus*). They cause damage to the foliage during the spring, summer and autumn but are usually present only in small numbers. Individuals (up to 25 mm long) are stout-bodied, green to brownish and twig-like, with four pairs of abdominal prolegs. The mainly yellow, orange-marked adults (32–35 mm wingspan) occur from April to August but are most numerous in May and June.

Ourapteryx sambucaria (Linnaeus) (**784–785**)
Swallow-tailed moth

Generally common, but a minor pest. The elongate (up to 60 mm long), greyish-brown, purplish-marked, twig-like larvae feed on the foliage of various trees and shrubs, including elder (*Sambucus*), flowering cherry (*Prunus*), hawthorn (*Crataegus*) and ivy (*Hedera*). Although sometimes infesting ornamentals, the larvae are rarely numerous. Older individuals, feeding in the spring, usually attract attention because of their size. The mainly pale yellow, butterfly-like adults (55–60 mm wingspan) appear in July.

Peribatodes rhomboidaria (Denis & Schiffermüller) (**786–787**)
Willow beauty moth

A generally common species, associated with various trees and shrubs. The larvae sometimes occur on ornamentals and nursery stock, including birch (*Betula*), *Clematis*, hawthorn (*Crataegus*), ivy (*Hedera*), lilac (*Syringa*), privet (*Ligustrum vulgare*), rose (*Rosa*) and yew (*Taxus baccata*), but cause only slight damage. They feed in the late summer and again in the following spring, each larva eventually pupating in a strong, silken cocoon formed on a twig or small branch of the foodplant. Individuals (up to 35 mm long) are reddish brown, mottled with ochreous, with faint diamond-shaped markings along the back. The greyish to brownish-grey adults (40 mm wingspan) appear in July and August.

786 Larva of willow beauty moth (*Peribatodes rhomboidaria*).

787 Male willow beauty moth (*Peribatodes rhomboidaria*).

788 Larva of pale brindled beauty moth (*Phigalia pilosaria*).

789 Male pale brindled beauty moth (*Phigalia pilosaria*).

Phigalia pilosaria (Denis & Schiffermüller) (788–790)

syn. *Apocheima pilosaria* (Denis & Schiffermüller)

Pale brindled beauty moth

A generally distributed species, the greyish-brown to reddish-brown, twig-like (up to 40 mm long) larvae feeding in the spring on the foliage of various trees and shrubs, including birch (*Betula*), crab-apple (*Malus*), elm (*Ulmus*), flowering cherry (*Prunus*), hawthorn (*Crataegus*), hornbeam (*Carpinus betulus*), larch (*Larix*), lime (*Tilia*), oak (*Quercus*), poplar (*Populus*), rose (*Rosa*) and broad-leaved willows such as grey willow (*Salix cinerea*) and pussy willow (*S. caprea*). They sometimes attack such hosts in gardens, parks and nurseries but damage caused is slight. Adults occur from January to March. Males (40–42 mm wingspan) are grey, tinged with greenish or brownish and suffused with dark grey or brown; females (12 mm long) are stout-bodied and virtually wingless.

790 Female pale brindled beauty moth (*Phigalia pilosaria*).

791 Larva of early thorn moth (*Selenia dentaria*).

792 Early thorn moth (*Selenia dentaria*).

793 Larva of early moth (*Theria primaria*).

794 Female early moth (*Theria primaria*).

795 Garden carpet moth (*Xanthorhoe fluctuata*).

Selenia dentaria (Fabricius) (**791–792**)
syn. *S. bilunaria* (Esper)
Early thorn moth

A generally abundant species, associated with various trees and shrubs including alder (*Alnus*), birch (*Betula*), crab-apple (*Malus*), flowering cherry (*Prunus*) and willow (*Salix*). The orange-brown to reddish-brown, twig-like larvae (35–40 mm long when fully grown) often feed on nursery stock and ornamentals; they occur from May to June and, less commonly, in August and September. The butterfly-like, whitish-yellow to yellowish-orange adults (32–40 mm wingspan) appear in April and early May, with a small second generation active in July and August.

Theria primaria Haworth (**793–794**)
syn. *T. rupicraparia* (Denis & Schiffermüller)
Early moth

A generally common species, the greenish to greenish-brown, white-striped larvae (up to 20 mm long) feeding on the leaves of hawthorn (*Crataegus*) throughout April and May. Various kinds of *Prunus*, especially blackthorn (*P. spinosa*), are also attacked. Although associated mainly with hedgerows, minor infestations are sometimes noted on nursery stock. The moths appear in January and February. Males (30–32 mm wingspan) are mainly light greyish brown; the virtually wingless females are grey, with their rather angular wing stubs each marked with a dark crossband.

Xanthorhoe fluctuata (Linnaeus) (**795**)
syn. *Cidaria fluctuata* (Linnaeus)
Garden carpet moth

Larvae of this generally common species feed on the foliage of various Brassicaceae from June onwards. They are sometimes found in small numbers on wallflower (*Cheiranthus cheiri*) but damage caused to cultivated plants is of no significance. Individuals (up to 32 mm long) are mainly yellowish green to greyish brown, with the body tapering gradually towards the small, pointed, black-marked head. Adults (26–29 mm wingspan) are mainly whitish to greyish, with darker markings on the fore wings. They occur most abundantly in May and June, and in August and September, resting openly during the daytime on walls, fences and other surfaces.

Family **SPHINGIDAE** (hawk moths)

Large-bodied, strong-flying moths with elongate fore wings. The larvae are stout-bodied, with a characteristic horn arising from the eighth abdominal segment.

Daphnis nerii (Linnaeus) (**796–798**)
Oleander hawk moth

A mainly non-resident migratory hawk moth, arriving in southern Europe annually from Africa and southern Asia. Further north a rare summer migrant. Host plants include oleander (*Nerium oleander*) and periwinkle (*Vinca*).

DESCRIPTION
Adult: 90–110 mm wingspan; wings and body deep olive-green, with irregular whitish and pinkish-white markings. **Larva:** up to 120 mm long; body usually

796 Oleander hawk moth (*Daphnis nerii*).

797 Final-instar larva of oleander hawk moth (*Daphnis nerii*).

798 Early-instar larva of oleander hawk moth (*Daphnis nerii*).

green, with the thoracic segments and last abdominal segments suffused with yellow, and the abdomen with a white dorsal stripe extending to the caudal horn; third thoracic segment with a pair of white, black-rimmed eye-spots; spiracles black; caudal horn mainly yellowish orange, short and bulbous (but long and whip-like in early instars).

LIFE HISTORY
Immigrant moths appear in southern Europe in June. Eggs are deposited singly on either side of the leaves of host plants, and hatch within 1–2 weeks. Larvae then feed on flowers and young leaves and eventually pupate, each in a flimsy cocoon spun on the ground amongst dried leaves and other debris. A further generation of adults appears in August, and these often migrate further northwards into central and, occasionally, into northern Europe. The autumn brood of larvae eventually pupates; however, the pupae do not survive European winters, except in a few favourable areas where the insect is an established resident.

DAMAGE
Larvae, particularly in their later instars, cause considerable defoliation. However, they are rarely sufficiently numerous to be of economic importance.

NOTE
Larvae of other migrant hawk moths from tropical or subtropical areas also occur on occasions on ornamental plants in southern Europe (and rarely further north). These include death's head hawk moth (*Acherontia atropos*) on Oleaceae and Solanaceae, silver-striped hawk moth (*Hippotion celerio*) on *Fuchsia*, and striped hawk moth (*Hyles livornica*) on both *Fuchsia* and Virginia creeper (*Parthenocissus*).

Deilephila elpenor (Linnaeus) (**799–800**)
syn. *Pergesa elpenor* (Linnaeus)
Elephant hawk moth
Generally common in the wild on bedstraw (*Galium*), rose-bay (*Chamaenerion angustifolium*) and willow-herbs (*Epilobium*) but also a minor pest of cultivated plants such as busy lizzie (*Impatiens*) and *Fuchsia*, both outdoors and in greenhouses. Palaearctic. Present throughout Europe, except for the extreme north.

DESCRIPTION
Adult: 60–70 mm wingspan; body olive-brown, marked with pink; fore wings olive-brown, with a pinkish-grey subterminal line and costa; hind wings bright pink, basally black. **Larva:** up to 85 mm long; brown or green, the abdominal segments speckled with

black; second and third abdominal segments distinctly swollen, and each marked with a pair of lilac-centred, black, eye-like patches; head and thoracic segments retractable; caudal horn relatively short. **Pupa:** 40–45 mm long; brown, speckled with darker brown.

LIFE HISTORY
Moths occur mainly in June, but sometimes also later in the year, depositing eggs singly on the leaves of host plants. Larvae feed during July and August; they may also occur in the autumn. Individuals often bask in sunshine. If disturbed they immediately retract the head and thoracic segments and dilate the anterior abdominal segments to display their eye-like markings. Fully grown larvae pupate in fragile, silken cocoons formed on or just below the surface of the ground.

DAMAGE
Larvae cause considerable defoliation, particularly in their later instars, but are usually present in only small numbers.

Laothoe populi (Linnaeus) (**801–802**)
syn. *Amorpha populi* (Linnaeus)
Poplar hawk moth
Widespread and generally common on poplar (*Populus*) and various kinds of willow (*Salix*); in mainland Europe also associated with ash (*Fraxinus excelsior*), birch (*Betula*) and crab-apple (*Malus*). The larvae feed on the foliage from July to September. They are often noticed on cultivated plants, particularly at the final stages of larval development, but are usually present in small numbers. Individuals (*c*. 70 mm long when fully grown) are green or bluish green, densely speckled with yellow, with seven lateral pairs of oblique yellow stripes, the hindmost extending into a yellowish-green, often red-tipped, caudal horn. Adults (70–85 mm wingspan) are mainly greyish brown, suffused with pinkish brown, with a rusty-red basal patch on each hind wing. They fly in May and June. A partial second generation sometimes arises in the autumn.

Mimas tiliae (Linnaeus) (**803–804**)
Lime hawk moth
Larvae of this widely distributed species feed mainly on lime (*Tilia*). In mainland Europe, alder (*Alnus*), ash (*Fraxinus excelsior*), birch (*Betula*), elm (*Ulmus*), oak (*Quercus*) and walnut (*Juglans*) are also attacked. The larvae feed from June or July onwards. They often occur on garden and amenity trees but do not cause significant damage. Individuals (50–60 mm long when fully grown) are green, with seven oblique yellow stripes along each side and red spiracles; the caudal horn is

799 Elephant hawk moth (*Deilephila elpenor*).

800 Larva of elephant hawk moth (*Deilephila elpenor*).

801 Poplar hawk moth (*Laothoe populi*).

802 Larva of poplar hawk moth (*Laothoe populi*).

803 Lime hawk moth (*Mimas tiliae*).

804 Larva of lime hawk moth (*Mimas tiliae*).

blue above but red and yellow below. The larvae are most often noticed in August or September when they have completed feeding and are about to enter the soil to pupate, individuals at this stage becoming tinged with purple or purplish brown. Adults (60–75 mm wingspan) occur in May and June, and are greyish ochreous to pinkish ochreous, suffused with greenish grey; the fore wings usually possess an irregular, dark olive-green median crossband.

Smerinthus ocellata (Linnaeus) (**805–806**)
Eyed hawk moth
Generally distributed and locally common. The larvae feed on crab-apple (*Malus*) and willow (*Salix*), and are often associated with nursery stock and young garden trees. They feed during July and August, rapidly defoliating the branches. Larvae (60–70 mm long when fully grown) are mainly green, with seven oblique white stripes along the sides and a bluish caudal horn. The adults (75–85 mm wingspan), which occur from May to July, have mainly greyish-brown to brown fore wings; the hind wings are brown, suffused with red, each with a large, eye-like mark.

Sphinx ligustri Linnaeus (**807–808**)
Privet hawk moth
Associated with ash (*Fraxinus excelsior*), lilac (*Syringa*), privet (*Ligustrum vulgare*) and snowberry (*Symphoricarpos rivularis*); other hosts, at least in mainland Europe, include elder (*Sambucus*), *Forsythia* and *Viburnum*. Palaearctic. Widely distributed in Europe.

DESCRIPTION
Adult: 100–120 mm wingspan; fore wings mainly greyish brown to blackish brown, with black markings, and pinkish tinged basally; hind wings pale pinkish white, with broad, blackish bands; thorax black, with whitish tegulae; abdomen yellowish brown, with pink and black crossbands. **Larva:** up to 80 mm long; mainly green, with seven oblique white, purple-edged stripes on each side; caudal horn black above and yellow below.

LIFE HISTORY
Moths occur in June and July, depositing eggs singly on the leaves and stems of host plants. The larvae feed on the foliage during July and August. Fully grown individuals enter the soil to pupate, burrowing down well below the surface before excavating a suitable chamber. Although adults normally emerge in the followings summer, the pupal stage sometimes extends over two winters.

DAMAGE
Larvae rapidly defoliate shoots but, as the number of individuals present is usually small, damage caused is of no importance.

805 Larva of eyed hawk moth (*Smerinthus ocellata*).

806 Eyed hawk moth (*Smerinthus ocellata*).

807 Privet hawk moth (*Sphinx ligustri*).

808 Larva of privet hawk moth (*Sphinx ligustri*).

Family **NOTODONTIDAE**

Medium-sized to large, plump-bodied, often downy moths. The larvae are variable in appearance, often with the anal prolegs modified into a pair of raised filaments; crotchets on the abdominal prolegs are of two sizes (biordinal).

Cerura vinula (Linnaeus) (**809–812**)

Puss moth

Generally common on poplar (*Populus*) and willows such as grey willow (*Salix cinerea*) and pussy willow (*S. caprea*), and often associated with trees in gardens, parks and nurseries; in mainland Europe also found occasionally on birch (*Betula*). Eurasiatic. Widely distributed in Europe.

DESCRIPTION

Adult: 65–80 mm wingspan; body fluffy and greyish white to yellowish grey, the thorax marked with black spots and the abdomen marked with black crossbars; wings greyish white, marked with greyish black, and the veins yellowish, edged with black. **Egg:** 1.8 mm across; hemispherical and purplish red. **Larva:** up to 70 mm long; bright green, with a broad, purplish, white-edged (sometimes yellowish-edged) dorsal stripe, widest and often reaching to the prolegs on the fourth abdominal segment; third thoracic segment humped; anal prolegs modified into a pair of fang-like appendages, each enclosing a red, extendible filament; head brownish, marked with purplish or black; young larva reddish brown, with a pair of broad, spinose projections on the first thoracic segment. **Pupa:** 25–30 mm long; black and stumpy.

LIFE HISTORY

Adults occur from May to July. Eggs are laid singly, but more frequently in twos or trees, on the upper surface of leaves, and hatch in about ten days. Larvae occur from June or July onwards, and often feed in pairs. There are five larval instars, the distinctive thoracic projections of the first-instar larva persisting to the third instar but then diminishing (fourth instar) and finally disappearing. The older larvae rest fully exposed on the shoots but, in

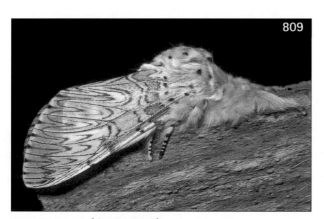

809 Puss moth (*Cerura vinula*).

810 Larva of puss moth (*Cerura vinula*).

811 First-instar larva of puss moth (*Cerura vinula*).

812 Third-instar larva of puss moth (*Cerura vinula*).

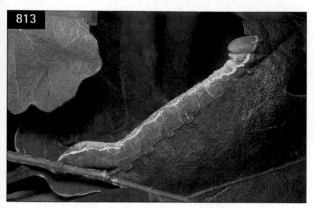

813 Larva of lunar marbled brown moth (*Drymonia ruficornis*).

814 Male lunar marbled brown moth (*Drymonia ruficornis*).

spite of their size, are easily overlooked. The presence of such larvae is often discovered only following a close examination of branches immediately above large pellets of frass which have accumulated on the ground beneath infested trees. Larvae complete their development in August or early September. They then wander away from the branches onto tree trunks or nearby fence posts where they construct tough, brown cocoons of silk and masticated wood. The larvae then pupate. Adults emerge in the following summer.

DAMAGE
Larvae cause considerable defoliation but, except on small trees, damage is unimportant.

Drymonia ruficornis (Hufnagel) (**813–814**)
Lunar marbled brown moth

Larvae of this widely distributed species feed on the leaves of oak (*Quercus*) from June to August; they are found occasionally on specimen trees in parks and gardens, mostly in southerly areas, but cause only slight damage. Fully fed individuals (*c.* 40 mm long) are bluish green to whitish green, marked with two thin, yellow lines down the back and a wider spiracular line along each side. Adults (37–40 mm wingspan) are whitish, marked with grey and greyish black. They are most numerous in May.

Furcula furcula (Clerck) (**815–816**)
Sallow kitten moth

Relatively common on willow (*Salix*), but of minor importance; also found on various kinds of *Populus*, including aspen (*P. tremula*), and birch (*Betula*). Eurasiatic; also present in North America. Widely distributed in Europe.

DESCRIPTION
Adult: 34–38 mm wingspan; fore wings pale greyish white, marked with light grey (the markings partly edged with black and orange-yellow), and with blackish dots along the outer margin between the veins; hind wings mainly whitish to greyish white. **Larva:** up to 35 mm long; light green, flecked with yellow, and with a broad, greyish-pink to purplish, yellow-edged dorsal stripe that widens to become saddle-like on the fourth abdominal segment; anal prolegs modified into eversible filamentous appendages; head purplish brown.

LIFE HISTORY
Adults emerge in late May and early June, depositing eggs singly or in twos or threes on the upper surface of leaves. Larvae feed from June onwards. When full grown, each forms a tough, relatively flat cocoon in a crevice or a hollow in the bark of the foodplant. The cocoons are constructed of silk and masticated wood, and are difficult to distinguish from the surrounding plant tissue. They remain intact until adults emerge in the following spring. In favourable districts a second generation of adults emerges in August; their larvae then feed in the late summer and autumn.

DAMAGE
Larvae cause slight defoliation but infestations are of no significance.

Phalera bucephala (Linnaeus) (**817–819**)
Buff-tip moth

A generally common pest of trees and shrubs, including alder (*Alnus*), beech (*Fagus sylvatica*), birch (*Betula*), elm (*Ulmus*), flowering cherry (*Prunus*), hazel (*Corylus*), hornbeam (*Carpinus betulus*), lime (*Tilia*), oak (*Quercus*), pussy willow (*Salix caprea*), rose (*Rosa*), sweet chestnut (*Castanea sativa*) and *Viburnum*;

815 Sallow kitten moth (*Furcula furcula*).

816 Larva of sallow kitten moth (*Furcula furcula*).

817 Buff-tip moth (*Phalera bucephala*).

818 Larva of buff-tip moth (*Phalera bucephala*).

often present on amenity trees, ornamentals and nursery stock. Widely distributed in Europe.

DESCRIPTION

Adult: 55–70 mm wingspan; fore wings ash-grey to silvery grey, with dark brown and reddish-brown markings and a large, pale yellow apical blotch; hind wings pale yellow to whitish. **Egg:** 1 mm across; hemispherical; pale bluish white above, with a dark central spot; green below. **Larva:** up to 60 mm long; downy and yellow, with several incomplete, black longitudinal lines and orange intersegmental crosslines; head black. **Pupa:** 25–28 mm long; dark purplish brown; cremaster with two pairs of short spines.

LIFE HISTORY

Adults occur from late May to July or early August, depositing eggs in batches of about 50 or more on the underside of leaves. The eggs hatch in about two weeks. At first, larvae feed on the lower epidermis; later, they devour entire leaves and are fully grown by the autumn. They then enter the soil to pupate in earthen chambers, where they overwinter. Although not forming webs,

819 Group of young buff-tip larvae (*Phalera bucephala*) on *Betula*.

buff tip larvae feed gregariously until the final stages of their development.

DAMAGE

Larvae rapidly defoliate shoots and branches, and attacks on young trees and shrubs are of particular importance.

Family **DILOBIDAE**

A small group of medium-sized moths. The larvae have dorsal setae arising from raised verrucae.

Diloba caeruleocephala (Linnaeus) (**820–821**)
Figure of eight moth

Widely distributed and common, the larvae occasionally infesting ornamental trees and shrubs, including *Cotoneaster*, crab-apple (*Malus*), hawthorn (*Crataegus*) and *Sorbus*. They feed from spring to June or early July, causing minor damage to the leaves. Although several individuals may occur on one and the same host, damage is rarely of significance. Larvae (*c*. 35 mm long when fully grown) are greyish blue, with an incomplete yellow dorsal stripe, a yellow stripe along each side, and prominent black verrucae (each bearing a short, black spine). Adults (30–35 mm wingspan) have brownish-grey fore wings, each with a large, pale greenish-yellow, irregular 'figure of eight' mark. Eggs are deposited in small groups on the trunks and branches of host plants in October and November.

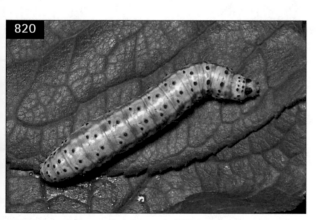

820 Larva of figure of eight moth (*Diloba caeruleocephala*).

821 Figure of eight moth (*Diloba caeruleocephala*).

Family **THAUMETOPOEIDAE**

Medium-sized, downy moths. Larvae are gregarious web-formers, and characteristically migrate in large masses, forming single-file, head-to-tail processions. Body hairs of third-instar and older larvae are poisonous, and considered a public health hazard.

Thaumetopoea pityocampa (Denis & Schiffermüller) (**822–823**)
Pine processionary moth

A pest of pine (*Pinus*); also recorded on cedar (*Cedrus*) and larch (*Larix*). Present in central Europe and throughout the Mediterranean basin.

DESCRIPTION
Adult: 30–45 mm wingspan; fore wings whitish grey with darker cross markings; hind wings white. **Larva:** up to 40 mm long; body bluish black above, whitish below, with gingery tufts of hairs on the back arising from reddish-brown verrucae, and greyish-white hairs at the sides; head black.

LIFE HISTORY
Moths occur from mid-June to the end of August, the flight period in hotter areas being delayed until late summer. Eggs are laid in long batches around a pair of needles, commencing at the base, and hatch 4–6 weeks later. The young larvae feed gregariously within a loosely spun web. They then overwinter, again in communal webs, and complete their development in the following year. Fully grown larvae migrate in large masses from host trees to suitable sites on the ground, where they eventually pupate. The duration and timing of larval development varies considerably, depending on temperature; in particularly favoured districts larvae are fully fed in January, but in cooler regions not until the end of June. The lifecycle is also recorded extending over two years.

DAMAGE
Larvae cause extensive defoliation and also disfigure host trees with their webbing.

Thaumetopoea processionea (Linnaeus)
Oak processionary moth

A sporadically important pest of oak (*Quercus*) in central and southern Europe; also now established in northern Europe, including Belgium, southern England, northern France, Germany and the Netherlands. Alternative hosts, utilized when oak is in short supply, include beech (*Fagus sylvatica*), birch (*Betula*), common hazel (*Corylus avellana*), hornbeam (*Carpinus betulus*), sweet chestnut (*Castanea sativa*) and walnut (*Juglans*).

822 Larvae of pine processionary moth (*Thaumetopoea pityocampa*). (Ingaret Howells)

823 Larvae of pine processionary moth (*Thaumetopoea pityocampa*) in procession. (Ingaret Howells)

DESCRIPTION

Adult: 30–40 mm wingspan; fore wings light grey with darker, but indistinct, blackish-edged crossbands; hind wings mainly white to greyish white. **Larva:** up to 40 mm long; body bluish grey below, coated in long whitish hairs, and with numerous tufts of reddish-brown hairs arising from reddish verrucae; the first to eighth abdominal segments each marked with a velvet-black patch dorsally; head brownish black.

LIFE HISTORY

Adults occur in July and August, eggs being laid in batches on the bark of host trees. The eggs are then coated with greyish scales from the female's body. They are thus well camouflaged and remain hidden throughout the winter. Larvae feed from bud burst onwards, at first attacking the swelling and opening buds, but later attacking the unfurling or unfurled foliage. They are very gregarious and inhabit communal retreats that, in their early stages consist of closely spun leaves. Later, distinctive webs up to a metre in length are formed on the trunk and larger branches of host trees. Fully grown larvae usually pupate on the host tree within the communal web, each in a yellowish or reddish-brown cocoon.

DAMAGE

Larvae cause extensive defoliation; infested trees are also disfigured by webbing.

Family **LYMANTRIIDAE**

Medium-sized moths with hairy bodies, the males with strongly bipectinate antennae. The larvae are hairy, with a pair of eversible dorsal glands on the abdomen or with brush-like tufts of hairs arising from the first to fourth abdominal segments.

Calliteara pudibunda (Linnaeus) (**824–825**)
syn. *Dasychira pudibunda* (Linnaeus)
Pale tussock moth
The colourful, mainly green or yellow, very hairy larvae (up to 50 mm long) of this widely distributed species are associated mainly with hop (*Humulus lupulus*) but also infest ornamental trees and shrubs, including beech (*Fagus sylvatica*), birch (*Betula*), elm (*Ulmus*), flowering cherry (*Prunus*), hazel (*Corylus*), hornbeam (*Carpinus betulus*), oak (*Quercus*), poplar (*Populus*), pussy willow (*Salix caprea*) and walnut (*Juglans*). They feed from July to October but are of no importance. The mainly grey adults (45–55 mm wingspan) occur in May and June.

Euproctis chrysorrhoea (Linnaeus) (**826–828**)
syn. *E. phaeorrhoeus* (Haworth)
Brown-tail moth
A locally distributed pest of shrubs such as blackberry (*Rubus fruticosus*), blackthorn (*Prunus spinosa*), hawthorn (*Crataegus*) and sea-buckthorn (*Hippophae rhamnoides*). At least in mainland Europe, attacks also occur on many other plants, including ash (*Fraxinus excelsior*), *Cotoneaster*, crab-apple (*Malus*), dogwood (*Cornus*), elder (*Sambucus*), false acacia (*Robinia pseudoacacia*), firethorn (*Pyracantha*), flowering cherry (*Prunus*), lilac (*Syringa*), ornamental pear (*Pyrus calleryana* 'Chanticleer') and privet (*Ligustrum vulgare*). Palaearctic; also introduced into North America. Widely distributed in central and southern Europe but more restricted in northern Europe.

DESCRIPTION
Adult: 35–42 mm wingspan; wings white, but fore wings of male sometimes with a few black dots; head and thorax white and fluffy; abdomen dark brown, with a large anal tuft of hair. **Larva:** up to 40 mm long; blackish grey, with tufts of gingery-brown hairs arising from brownish verrucae, two rows of bright red marks down the back, a series of downy white patches towards each side and bright, orange-red glands on the sixth and seventh abdominal segments; head black. **Pupa:** 15–18 mm long; brownish black and hairy.

824 Larva of pale tussock moth (*Calliteara pudibunda*).

825 Male pale tussock moth (*Calliteara pudibunda*).

LIFE HISTORY
Adults occur in July and August, eggs being laid in elongate batches on leaves or stems and then covered with brown hairs from the female's anal tuft. The eggs hatch from mid-August to early September. The young larvae then construct a stong, silken, web ('tent') in which they shelter during inclement weather. Larvae feed on the foliage in decreasing numbers until the end of October and then hibernate. Activity is resumed in the following April, the larvae appearing on the outside of their tent in increasing numbers, to bask in the spring sunshine. However, little or no feeding occurs until May. Young foliage is then devoured ravenously. As the larvae grow, they wander further and further from their communal tent, spinning additional, less substantial webs and establishing trails of silk along the branches. In the later stages of development, when about 25 mm long, they often become solitary and wander away to feed elsewhere. Individuals are fully grown by late June. They then spin silken cocoons between the leaves, either singly or in groups and pupate. Adults emerge about two weeks later.

826 Female brown-tail moth (*Euproctis chrysorrhoea*).

827 Larva of brown-tail moth (*Euproctis chrysorrhoea*).

828 Web of brown-tail moth (*Euproctis chrysorrhoea*) on *Cotoneaster*.

829 Female yellow-tail moth (*Euproctis similis*).

830 Male yellow-tail moth (*Euproctis similis*).

831 Larva of yellow-tail moth (*Euproctis similis*).

DAMAGE

The larvae are voracious feeders, particularly in their later instars, and often cause considerable defoliation of roadside hedges; their urticating hairs also constitute a public nuisance, sometimes requiring local authorities to conduct eradication campaigns. On ornamentals, feeding damage may be of considerable significance; the larval tents are also disfiguring.

Euproctis similis (Fuessly) (**829–831**)
syn. *Porthesia similis* (Fuessly)
Yellow-tail moth

Locally common on trees and shrubs, including beech (*Fagus sylvatica*), birch (*Betula*), crab-apple (*Malus*), flowering cherry (*Prunus*), hawthorn (*Crataegus*), oak (*Quercus*), pussy willow (*Salix caprea*), rose (*Rosa*) and *Viburnum*. Minor infestations often occur on garden trees and nursery stock. Eurasiatic. Widely distributed in Europe.

DESCRIPTION

Adult female: 40–45 mm wingspan; wings white; body with a bulbous, yellow tip. **Adult male:** 30–40 mm wingspan; wings white, with black markings on the trailing edge of each fore wing; antennae strongly bipectinate; body with a conspicuous yellow anal hair tuft. **Larva:** up to 35 mm long; velvet-black, prominently marked with white, and with an interrupted bright red, black-centred dorsal stripe; sixth and seventh abdominal segments each with an eversible orange dorsal gland; body hairs long and whitish; head black. **Pupa:** 15–16 mm long; dark brown; plump and slightly hairy.

LIFE HISTORY

Adults occur in late July and August. Eggs are laid in batches on the twigs of host plants and then coated with hairs from the female's anal tuft. They hatch 7–10 days later. Larvae then feed gregariously for a short while before spinning small (*c.* 6 mm long), roughly oval, greyish cocoons under flakes of bark or in other sheltered positions. Individuals hibernate within these cocoons, and become active again in the spring. The larvae continue to feed on the foliage until about June. They then pupate in oval, greyish-brown, silken cocoons which incorporate numerous body hairs. Adult moths emerge a few weeks later.

DAMAGE

Larvae cause slight defoliation but numbers present on ornamentals are usually small.

832 Male white satin moth (*Leucoma salicis*).

Leucoma salicis (Linnaeus) (**832**)

syn. *Stilpnotia salicis* (Linnaeus)

White satin moth

A local pest of poplar (*Populus*) and willows such as grey willow (*Salix cinerea*) and pussy willow (*S. caprea*); other hosts in mainland Europe include birch (*Betula*), crab-apple (*Malus*) and snowy mespilus (*Amelanchier laevis*). Eurasiatic; introduced into North America. Widely distributed in Europe.

DESCRIPTION

Adult: 45–55 mm wingspan; wings satin-white to creamy white and thinly scaled; male much smaller than female and with antennae strongly bipectinate. **Egg:** 0.65–0.85 mm diameter; light green. **Larva:** up to 45 mm long; hairy and mainly reddish brown, with a bright red and distinctive white patch on each segment, and a blue subspiracular line; pinacula orange-red to orange; head blackish grey to black.

LIFE HISTORY

Moths appear in July or August. Eggs are laid in batches on the twigs of host plants and then coated with a white secretion incorporating hairs from the female's abdomen. Larvae emerge from late August onwards. They feed for a short time and then spin small webs in which to hibernate. Activity is resumed in the spring, usually in April. The larvae feed ravenously, often in groups, and complete their development in late June or early July. They then pupate in silken cocoons spun in crevices in the bark of host trees. Adults emerge a few weeks later.

DAMAGE

Larvae destroy the foliage and, if numerous, cause severe defoliation. This affects the growth and appearance of host trees.

Lymantria dispar (Linnaeus) (**833–835**)

Gypsy moth

An important and often destructive pest of trees and shrubs, especially oak (*Quercus*) but also beech (*Fagus sylvatica*), common hazel (*Corylus avellana*), hornbeam (*Carpinus betulus*) and many others. Palaearctic. Widely distributed in mainland Europe, from mid-Sweden to the Mediterranean; extinct in Britain but has occurred in recent years. An introduced pest in North America.

DESCRIPTION

Adult female: 50–70 mm wingspan; wings mainly brownish white, with irregular greyish to blackish cross-markings. **Adult male:** 35–50 mm wingspan; wings greyish brown, with blackish markings; antennae strongly bipectinate. **Larva:** up to 75 mm long; grey to greyish yellow, with darker markings; body with long, brown hairs arising from prominent verrucae; verrcuae on the thoracic segments and first two abdominal segments blue, those on the third to eighth abdominal segments brownish red.

LIFE HISTORY

Adults occur from late July onwards, the flight period varying according to local conditions. The females do not fly but crawl about on the trunks and branches of host trees, sometimes also gliding to the ground from the tops of tall trees. Adult males are most active at mid-day, flying about rapidly in their search for newly emerged females. After mating, each female deposits a large batch of eggs on the trunk or the underside of a branch; the eggs are then camouflaged by a spongy coating that incorporates numerous body hairs. The eggs hatch in the following April or May. The young larvae eventually wander away to feed on the bursting buds and expanding leaves. Larvae develop through six instars, becoming fully grown in about 2–3 months. They then pupate on the host plant or elsewhere. Adults emerge 2–3 weeks later.

DAMAGE

Larvae cause considerable defoliation, and infestations are particularly severe on small trees and on those growing in light, sunny positions.

833 Female gypsy moth (*Lymantria dispar*).

834 Male gypsy moth (*Lymantria dispar*).

835 Larva of gypsy moth (*Lymantria dispar*).

836 Male black arches moth (*Lymantria monarcha*).

837 Female vapourer moth (*Orgyia antiqua*) depositing eggs on pupal cocoon.

Lymantria monacha (Linnaeus) (836)
Black arches moth

Polyphagous on trees and shrubs, but associated mainly with conifers such as fir (*Abies*), pine (*Pinus*) and spruce (*Picea*); often a serious forestry pest. Eurasiatic. Widespread in mainland Europe; in Britain restricted to southern England and parts of Wales, and associated mainly with oak (*Quercus*).

DESCRIPTION
Adult female: 45–50 mm wingspan; body and fore wings mainly white, irregularly marked with black, the abdomen tinged with pink; hind wings mainly light greyish brown. **Adult male:** 37–42 mm wingspan; similarly coloured to female but antennae strongly bipectinate. **Larva:** up to 35 mm long; dark grey, with an irregular brownish-black, black-edged dorsal stripe interrupted by a whitish, black-centred, mark on the third thoracic segment and by whitish, sometimes red-centred, patches on the fourth to sixth abdominal segments; pale hair tufts arise from pinacula along the dorsal and spiracular lines.

LIFE HISTORY
Eggs are laid during the summer on the bark of host trees, either singly or in pairs, and hatch in the following spring. Larvae feed on the foliage until June or early July. They then pupae in silken cocoons spun in bark crevices, and adults emerge in late July or August.

DAMAGE
Leaf damage is usually insignificant but severe infestations result in considerable defoliation which affects the vigour of host trees.

838 Male vapourer moth (*Orgyia antiqua*).

839 Larvae of vapourer moth (*Orgyia antiqua*).

Orgyia antiqua (Linnaeus) (**837–839**)
Vapourer moth

An often abundant pest of trees and shrubs in nurseries, parks and gardens, feeding indiscriminately on various ornamentals; most important damage is likely on younger plants, especially buddleia (*Buddleja*), *Camellia*, *Ceanothus*, crab-apple (*Malus*), firethorn (*Pyracantha*), flowering cherry (*Prunus*), heather (*Erica*), *Rhododendron* and rose (*Rosa*); infestations also occur on conifers, including Douglas fir (*Pseudotsuga menziesii*), fir (*Abies*), larch (*Larix*) and pine (*Pinus*). Holarctic. Widespread in Europe.

DESCRIPTION
Adult female: 10–15 mm long and virtually wingless; body dark yellowish grey, fat and sack-like. **Adult male:** 25–33 mm wingspan; wings ochreous brown or chestnut-brown; fore wings with darker markings and with a large white spot near the hind angle. **Egg:** 0.9 mm across; rounded, brownish grey to reddish grey, with a central spot and a dark rim-like band. **Larva:** up to 35 mm long; greyish or violet, with red, black and yellow markings; body very hairy, including four brush-like tufts of yellow or greyish hairs on the back, long blackish pencil-like tufts near the head and similar brownish tufts near the tail. **Pupa:** shiny brownish black and rather hairy.

LIFE HISTORY
Adults occur from July to September or October, the males flying in sunshine in search of newly emerged, unmated and flightless females. The latter are sluggish and, on emerging from the pupa, each remains alongside the pupal cocoon upon which, after mating, about 100–300 eggs are laid in a conspicuous batch. The eggs hatch intermittently in the following spring. Larvae then wander away to feed on the foliage of the same or nearby plants. They sometimes rest together on flowers or foliage whilst undergoing the change from one instar to the next but soon become solitary. Fully grown individuals eventually pupate in silken cocoons, incorporating body hairs, spun on a twig, branch or trunk of the host plant or on a suitable support such as a nearby fence or wall. Adults emerge shortly afterwards. In favourable conditions there is a second generation in the autumn.

DAMAGE
Larvae cause noticeable defoliation and also damage buds and flowers, but infestations are rarely significant.

Family **ARCTIIDAE** (ermine moths and tiger moths)

Medium-sized to large, often brightly coloured moths. The larvae are very hairy but the head is virtually hairless; the body hairs arise in tufts from large verrucae.

Arctia caja (Linnaeus) (**840–841**)
Garden tiger moth

A generally common but minor pest of herbaceous plants, mainly in weedy situations; also sometimes associated with seedling trees and shrubs. North Palaearctic. Widespread in Europe.

DESCRIPTION
Adult: 60–75 mm wingspan; fore wings chocolate-brown, irregularly marked with creamy white; hind wings orange-red, with several blue blotches. **Egg:** 0.8 mm across; hemispherical, glossy, yellowish to green. **Larva:** up to 60 mm long; mainly blackish, with a thick coat of long, gingery, often pale-tipped hairs;

840 Garden tiger moth (*Arctia caja*).

841 Larva of garden tiger moth (*Arctia caja*).

head black. **Pupa:** 22–28 mm long; black to brownish black; cremaster with several distorted bristles.

LIFE HISTORY

Adults fly in July and August. Eggs are deposited in large batches on the underside of leaves of herbaceous plants, especially weeds such as dandelion (*Taraxacum officinale*), dock (*Rumex*) and plantain (*Plantago*); they hatch from August onwards. The larvae (commonly known as 'woolly bears') feed during August and September and then hibernate, resuming activity in the following spring. They are then often found sunning themselves on the foliage of low plants or on nearby fences and walls. Larvae are fully grown by mid- or late June. They then pupate in yellowish cocoons spun amongst debris on the ground.

DAMAGE

Foliage damage is indiscriminate and usually of no significance; however, on rare occasions, larvae may be locally abundant and then cause extensive defoliation.

Callimorpha dominula (Linnaeus) (**842–843**)

Scarlet tiger moth

This widely distributed, distinctly local species is polyphagous on herbaceous plants, but most abundant on comfrey (*Symphytum officinale*), green alkanet (*Pentaglottis sempervirens*) and hounds-tongue (*Cynoglossum officinale*). In areas where colonies are established, older larvae sometimes feed in the spring on ornamental herbaceous plants and young trees, including ash (*Fraxinus excelsior*), blackthorn (*Prunus spinosa*), elm (*Ulmus*), flowering cherry (*Prunus*), oak (*Quercus*), rowan (*Sorbus aucuparia*) and willow (*Salix*). Although causing slight damage to the leaves, this attractive insect is not of pest status and specimens found on cultivated plant should not be destroyed. The young larvae feed briefly in the late summer or early autumn, before hibernating, and complete their development in the following spring. Fully grown larvae are *c.* 40 mm long and mainly black, marked prominently with white and bright yellow. The spectacular, bright red, yellow, white and black adults occur in June and July.

Hyphantria cunea (Drury) (**844–848**)

American white moth

larva = fall webworm

An important North American pest, first noted in Europe in 1940 in Hungary; subsequently found in certain other countries, including Austria, France, Italy and the former Yugoslavia. The larvae attack various trees and shrubs, including ash (*Fraxinus excelsior*), European hop-hornbeam (*Ostrya carpinifolia*), horse chestnut (*Aesculus hipposcastanum*), lime (*Tilia*), maple (*Acer*), rose (*Rosa*), *Sorbus*, southern nettle-tree (*Celtis australis*) and tree of heaven (*Ailanthus altissima*), but are most common on box elder (*Acer negundo*) and mulberry (*Morus*).

DESCRIPTION

Adult: 26–30 mm wingspan; mainly white, the fore wings sometimes flecked with black; male with noticeably bipectinate antennae. **Egg:** light green, laid in a large batch. **Larva:** up to 35 mm long; varying from yellow or yellowish green to brown, with tufts of whitish hairs arising from black verrucae; spiracles white, ringed with black; head shiny black. **Pupa:** 10–12 mm long; shiny blackish brown; cremaster with twelve hooks.

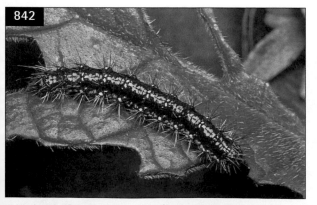

842 Larva of scarlet tiger moth (*Callimorpha dominula*).

843 Scarlet tiger moth (*Callimorpha dominula*).

LIFE HISTORY

Adults of the first generation occur in the spring from April onwards, the egg-laying females each depositing several hundred eggs on the spurs or on the underside of the expanded leaves; the egg batches are then partly covered with whitish hairs. The larvae are gregarious. They feed ravenously on the foliage from May to July, sheltering during the day within a large but flimsy communal web. When fully fed, the larvae wander away to pupate on the foodplant, each in a slight, greyish-brown cocoon. Moths of a second generation appear in July and August, eventually giving rise to larvae which complete their development in the autumn. These second-generation larvae usually pupate within bark crevices or amongst dead leaves, and then overwinter.

DAMAGE

Host trees are clothed in webbing and extensively defoliated. Attacks are often severe on hedgerows as well as on shade or ornamental trees in urban and suburban areas.

844 Female American white moth (*Hyphantria cunea*).

845 Male American white moth (*Hyphantria cunea*).

846 Egg batch of American white moth (*Hyphantria cunea*).

847 Fall webworm (*Hyphantria cunea*).

848 Web of fall webworms (*Hyphantria cunea*).

849 Ruby tiger moth (*Phragmatobia fuliginosa*).

850 Larva of ruby tiger moth (*Phragmatobia fuliginosa*).

851 Larva of white ermine moth (*Spilosoma lubricipeda*).

852 Young larva of white ermine moth (*Spilosoma lubricipeda*).

853 Male white ermine moth (*Spilosoma lubricipeda*).

854 Larva of buff ermine moth (*Spilosoma luteum*).

855 Male buff ermine moth (*Spilosoma luteum*).

Phragmatobia fuliginosa (Linnaeus) (**849–850**)
Ruby tiger moth

This locally common species is associated mainly with wild plants such as dandelion (*Taraxacum officinale*), dock (*Rumex*), golden-rod (*Solidago virgaurea*) and plantain (*Plantago*), but is sometimes a minor pest in flower borders; slight damage may also be caused to greenhouse-grown ornamentals. Adults (30–35 mm wingspan) are mainly reddish brown, with the fore wings thinly scaled centrally and the hind wings pink, suffused with dark grey; the abdomen is red, marked with black. The moths occur mainly during May and June, eggs being laid in batches on the foliage of various herbaceous plants. Larvae feed from late May or June onwards, most completing their development in the autumn and then overwintering. Individuals reappear in the following spring; they do not feed but, instead, spin brown, silken cocoons amongst foliage or debris on the ground and then pupate. Adults appear a few weeks later. Some larvae develop more rapidly than normal, become fully grown by August and give rise to a partial second generation of adults in September. The larvae (up to 35 mm long) are blackish, with a red dorsal stripe, red spots subdorsally and yellowish, reddish or brownish hair tufts arising from large, grey verrucae.

Spilosoma lubricipeda (Linnaeus) (**851–853**)
syn. *Spilosoma menthastri* (Esper)
White ermine moth

Larvae of this widespread and generally common species feed on low-growing plants from late July to September. They sometimes attack cultivated plants in herbaceous borders, grazing indiscriminately on the foliage, but damage caused is of no importance. Minor infestations sometimes occur on ornamental shrubs such as elder (*Sambucus*), honeysuckle (*Lonicera*), lilac (*Syringa*) and privet (*Ligustrum vulgare*). Larvae (35–40 mm long when fully grown) are dark brown, with a red or orange dorsal stripe and tufts of brown hairs arising from black verrucae; young specimens are dull greenish to yellowish green, with very prominent verrucae. Pupation occurs in silken cocoons formed amongst debris on the ground. The mainly white or creamish-white, black-spotted adults (35–45 mm wingspan) appear in the following May or June.

Spilosoma luteum (Hufnagel) (**854–855**)
syn. *Spilarctica lutea* (Hufnagel)
Buff ermine moth

Widespread and common, the greyish-brown, hairy larvae occurring mainly on weeds such as dandelion (*Taraxacum officinale*) and dock (*Rumex*) but also sometimes attacking the foliage of herbaceous garden plants. They feed during the summer and pupate in the autumn. The pale yellow to ochreous, black-marked adults (36–42 mm wingspan) appear in the following June.

Family NOCTUIDAE

A large family of mostly medium-sized, stout-bodied, dull-coloured moths, often with a kidney-shaped mark (reniform stigma) and a small circle on each fore wing (this pattern is sometimes highlighted). The larvae of most species have five pairs of abdominal prolegs but groups with a reduced number also occur; crotchets are of one size (uniordinal), arranged in a half-circle.

Acronicta psi (Linnaeus) (**856–857**)
Grey dagger moth
A generally common but minor pest of trees and shrubs, including alder (*Alnus*), birch (*Betula*), *Cotoneaster*, crab-apple (*Malus*), flowering cherry (*Prunus*), hawthorn (*Crataegus*), lime (*Tilia*), ornamental pear (*Pyrus calleryana* 'Chanticleer'), rose (*Rosa*), rowan (*Sorbus aucuparia*) and willow-leaved pear (*Pyrus salicifolia*); larvae also feed on herbaceous plants. Eurasiatic. Widely distributed in Europe.

DESCRIPTION
Adult: 38–42 mm wingspan; fore wings light grey, with black markings; hind wings greyish. **Larva:** up to 35 mm long; greyish black, with a broad, white stripe along either side and a broad yellow dorsal band, the latter bordered by a blue stripe (interrupted by red, black-edged spots); a prominent, black, pointed hump on the first abdominal segment and a small black hump on the eighth; head black. **Pupa:** 15 mm long; brown and rather slender, tapering towards the tip; cremaster with several strong spines.

LIFE HISTORY
Adults are present from late May onwards, depositing eggs on the foliage of host plants; they complete one generation in northerly areas but two elsewhere. Larvae occur from June to September or October. When fully fed, each pupates in a greyish-brown cocoon constructed amongst dead leaves on the foodplant or hidden in a crevice on the bark.

DAMAGE
Defoliation is usually unimportant but is sometimes of significance on nursery plants.

Acronicta aceris (Linnaeus) (**858–859**)
Sycamore moth
The colourful larvae of this locally common species are associated mainly with horse chestnut (*Aesculus hippocastanum*), and they often occur on such trees in urban areas. They also feed on other hosts, including sycamore (*Acer pseudoplatanus*), *Laburnum*, lime (*Tilia*) and rose (*Rosa*). Adults (40–45 mm wingspan) are mainly whitish to light grey, with blackish-marked fore wings and whitish hind wings. Larvae are *c.* 35 mm long when fully grown; the body, which bears long tufts of yellow or reddish hairs, is mainly yellowish brown, with a distinct series of white, black-bordered marks along the back. Adults occur from June to July or early August, and the larvae from July to September.

Acronicta alni (Linnaeus) (**860–861**)
Alder moth
A widely distributed but local species, the larvae feeding in July and August on the foliage of various trees and shrubs, including alder (*Alnus*), beech (*Fagus sylvatica*), birch (*Betula*), elm (*Ulmus*), flowering cherry (*Prunus*), lime (*Tilia*), maple (*Acer*), poplar (*Populus*), rose (*Rosa*), *Sorbus* and willow (*Salix*). The larvae usually occur singly, often resting fully exposed on the upper surface of a leaf. They are sometimes noticed on garden trees or nursery stock, their striking and unusual appearance immediately attracting attention, but they cause little damage and are of no consequence. Older individuals (up to 35 mm long) are black and yellow, with distinctive spatulate body hairs. The mainly grey, brownish-grey and blackish-marked adults (32–40 mm wingspan) appear in May and June.

856 Grey dagger moth (*Acronicta psi*).

857 Larva of grey dagger moth (*Acronicta psi*).

858 Sycamore moth (*Acronicta aceris*).

859 Larva of sycamore moth (*Acronicta aceris*).

860 Larva of alder moth (*Acronicta alni*).

861 Alder moth (*Acronicta alni*).

Acronicta megacephala (Denis & Schiffermüller) (862–863)
Poplar grey moth

This species occurs mainly on black poplar (*Populus nigra*), but also attacks other kinds of poplar as well as pussy willow (*Salix caprea*). The larvae are often found on such trees in urban areas. Adults occur from May to July, eggs being deposited singly on the leaves of host plants. The larvae feed from July to September, and typically rest on the leaves during the daytime with the head turned back alongside the abdomen. Fully grown larvae (*c*. 35 mm long) are black, finely speckled with yellowish grey, and marked with orange or reddish spots; there is a large pale patch on the seventh abdominal segment; the body is also partly clothed in long, fine hairs which arise in tufts from lateral pairs of brownish verrucae. The moths (40 mm wingspan) are mainly grey.

Acronicta rumicis (Linnaeus) (864–865)
Knotgrass moth

A generally distributed and common species, the larvae feeding on various plants from July onwards. The larvae occasionally damage the foliage of bedding plants and other cultivated hosts but attacks are of minor importance. Individuals overwinter as pupae in subterranean cocoons, and adults emerge in May or June; in favourable areas there is also an autumn generation. Fully grown larvae (*c*. 35 mm long) are brownish grey to black, with red markings and two lateral series of white patches along the back, and lateral tufts of light, brownish hairs; the subspiracular line is creamy white, interrupted by raised red verrucae; the first and eighth abdominal segments are mainly black and slightly humped; the head is black. Adults (38 mm wingspan) are mainly blackish to blackish grey; the fore wings are variably marked with grey and each has a more or less distinct paler subcostal spot; the hind wings are light brown.

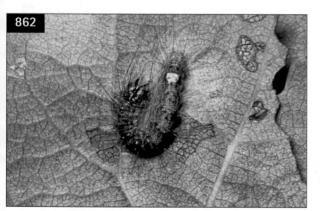

862 Young larva of poplar grey moth (*Acronicta megacephala*).

863 Poplar grey moth (*Acronicta megacephala*).

864 Larva of knotgrass moth (*Acronicta rumicis*).

865 Knotgrass moth (*Acronicta rumicis*).

Acronicta tridens (Denis & Schiffermüller) (866–867)

Dark dagger moth

Larvae of this generally distributed species occur on various trees and shrubs but are associated most frequently with rosaceous hosts. They feed from July onwards but damage is limited to the foliage and of little or no importance. Fully grown individuals (*c*. 40 mm long) are mainly black, marked with red and white along the back and sides; there is a black peg-like hump on the first abdominal segment and a smaller but broader swelling on the eighth; the body hairs are long and mainly blackish. Adults are virtually identical in appearance to those of *Acronicta psi*; they occur in June and July with, in favourable conditions, a partial second generation in September.

Agrotis segetum (Denis & Schiffermüller) (868–869)

Turnip moth

An important horticultural pest, sometimes affecting herbaceous ornamentals such as China aster (*Callistephus chinensis*), *Chrysanthemum*, *Dahlia*, Michaelmas daisy (*Aster*), *Petunia*, *Phlox*, pot marigold (*Calendula officinalis*), primrose (*Primula vulgaris*) and *Zinnia*; seedling trees and shrubs are also attacked. Eurasiatic. Widespread in Europe.

DESCRIPTION

Adult: 38–44 mm wingspan; fore wings whitish brown, with brownish black or blackish markings; hind wings pearly whitish; male antennae distinctly bipectinate.
Larva: up to 35 mm long; plump-bodied, glossy greyish brown, with a yellowish or pinkish tinge and indistinct darkish lines along the back; pinacula black.
Pupa: 18–20 mm long; light reddish brown; cremaster with two divergent spines.

866 Larva of dark dagger moth (*Acronicta tridens*).

867 Dark dagger moth (*Acronicta tridens*).

868 Male turnip moth (*Agrotis segetum*).

869 Larva of turnip moth (*Agrotis segetum*).

870 Larva of heart & dart moth (*Agrotis exclamationis*).

871 Heart & dart moth (*Agrotis exclamationis*).

LIFE HISTORY

Adults occur from late May or June or early July, depositing eggs on various weeds and cultivated plants. The eggs hatch in about 1–3 weeks, depending on temperature. Young larvae then browse on the foliage of host plants, this feeding habit persisting for only the first two instars. Older individuals inhabit the soil and become typical sluggish 'cutworms', attacking the roots, crowns and underground portions of plant stems. Cutworms are active at night, resting during the day in the soil close to their hosts. Most larvae are fully fed in the autumn but they do not normally pupate until the following spring; under favourable conditions, however, some individuals develop more rapidly and give rise to a small second generation of adults in the autumn.

DAMAGE

Larvae graze or burrow into corms, crowns, roots and tubers, causing plants to wilt; also, stems of plants may be girdled or completely severed, typically at about soil level. Infestations are most severe on light soils and in hot, dry conditions (cf. slug damage, p. 441), and tend to be most significant on younger, unirrigated, slower-growing hosts. Attacks in nurseries sometimes upset the establishment and growth of seedling trees.

Agrotis exclamationis (Linnaeus) (870–871)
Heart & dart moth

Generally abundant and sometimes a minor pest of young trees and shrubs in nurseries. The larvae occur from July onwards, at first feeding briefly on the foliage of host plants but later, in common with related species such as *Agrotis segetum*, attacking the roots and adopting a typical 'cutworm' habit. Larvae (up to 38 mm long) are dull brownish to greenish brown, with dark-edged longitudinal markings along the back and relatively large, black spiracles. They complete their development in the autumn but usually do not pupate until the following spring. Adults occur in June and July. Under favourable conditions, however, some larvae pupate, to produce a partial second generation of moths in the autumn. Adults (38–40 mm wingspan) are whitish brown to dark brown, each fore wing including a reniform stigma and a dart-like mark; the hind wings are light greyish brown in females but whitish in males.

Allophyes oxyacanthae (Linnaeus) (872–873)
syn. *Meganephria oxyacanthae* (Linnaeus)
Green-brindled crescent moth

Larvae of this widely distributed and often common species feed on rosaceous trees and shrubs, including blackthorn (*Prunus spinosa*), *Cotoneaster*, crab-apple (*Malus*) and hawthorn (*Crataegus*). Eggs laid in the previous autumn hatch in the early spring. Larvae attack the leaf buds and, later, the expanded foliage. Individuals are rather plump (*c.* 45 mm long when fully grown), greyish brown to reddish brown, with small, black-edged, pinkish-orange markings; the eighth abdominal segment is slightly humped and bears two pairs of pale projections. Feeding is completed by late May or early June. Larvae then enter the soil and

872 Larva of green-brindled crescent moth (*Allophyes oxyacanthae*).

873 Green-brindled crescent moth (*Allophyes oxyacanthae*).

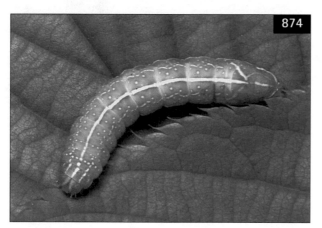

874 Larva of copper underwing moth (*Amphipyra pyramidea*).

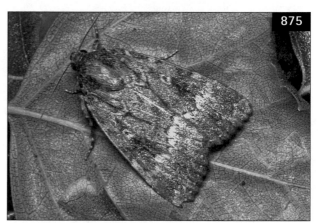

875 Copper underwing moth (*Amphipyra pyramidea*).

eventually pupate in strong, silken cocoons. The mainly brown, partly greenish-dusted adults (38–40 mm wingspan) closely resemble tree bark; they fly in October and November.

Amphipyra pyramidea (Linnaeus) (**874–875**)
Copper underwing moth
Locally common in parkland and woodland areas, the moths occurring in greater or lesser numbers from late July to September or October. The larvae feed from April to June, attacking the foliage of birch (*Betula*), flowering cherry (*Prunus*), hornbeam (*Carpinus betulus*), oak (*Quercus*), rose (*Rosa*) and willow (*Salix*) as well as various other trees and shrubs; they sometimes cause minor damage to such plants in gardens and nurseries, but are not important pests. Fully grown individuals (up to 45 mm long) are rather plump, and green to whitish green, dotted with white or pale yellow; there are three incomplete white lines along the back and one, partly bordered above with yellow, along each side; the eighth abdominal segment is humped, with a distinct horn-like apex. Adults (45–55 mm wingspan) have brownish fore wings, marked with black and pale yellowish grey; the hind wings are coppery red.

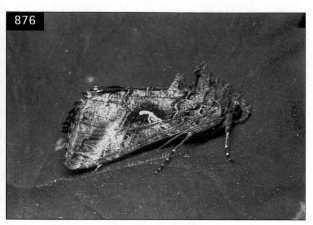

876 Silver Y moth (*Autographa gamma*).

877 Larva of silver Y moth (*Autographa gamma*).

Autographa gamma (Linnaeus) (876–877)
syn. *Plusia gamma* (Linnaeus)
Silver Y moth

An important horticultural pest, attacking greenhouse-grown and outdoor plants, including azalea (*Rhododendron*), carnation (*Dianthus caryophyllus*), *Chrysanthemum*, *Meconopsis*, *Pelargonium* and snapdragon (*Antirrhinum*). Holarctic. A major migratory species, in Europe spreading annually from its permanent breeding grounds around the Mediterranean to many western and northern areas.

DESCRIPTION
Adult: 35–45 mm wingspan; fore wings greyish brown to velvet-black, suffused with whitish grey, often tinged with purplish and each bearing a silver, Y-shaped mark; hind wings light brown, with a broad brownish-black or blackish border. **Larva:** up to 45 mm long; green to blackish green, with pale irregular lines along the back and a whitish or yellowish spiracular line; three pairs of abdominal prolegs. **Pupa:** 12–19 mm long; black or blackish brown, and unpunctured; cremaster bulbous and with curved, hook-tipped spines.

LIFE HISTORY
This insect is unable to survive the winter in north-western Europe, except in heated greenhouses. Nevertheless, annual infestations are commonplace, the first immigrant moths usually arriving in May or June. Eggs are laid singly or in small groups on the leaves of host plants, and hatch in 1–2 weeks according to temperature. The larvae feed mainly at night, attacking the leaves, buds and flowers. Individuals are fully grown in about a month. They then spin flimsy silken cocoons on the foodplant and pupate. Adults emerge a week or two later. The moths sometimes fly during daytime, but are usually most active at dusk, feeding avidly on the nectar secreted by many wild and cultivated flowers. Breeding is continuous under favourable conditions; however, most adults reared in northern Europe in the autumn do not breed locally but migrate south to more favourable areas.

DAMAGE
Leaf damage is often extensive. However, this is less important than direct damage to buds and flowers, which often results in the production of malformed blooms or the development of blind shoots.

Conistra vaccinii (Linnaeus) (**878**)
Chestnut moth

A generally distributed species, the larvae feeding on the foliage and flowers of various trees, shrubs and herbaceous plants, including alder (*Alnus*), elm (*Ulmus*), lime (*Tilia*), maple (*Acer*), oak (*Quercus*), poplar (*Populus*), *Sorbus* and willow (*Salix*); the larvae sometimes cause minor damage in gardens and nurseries but are not important pests. Adults appear in September and October, but do not deposit eggs until after emerging from hibernation in the following spring. Eggs hatch within a couple of weeks. Larvae feed on the foliage and flowers of host plants from April onwards. Fully grown individuals (*c.* 32 mm long) are stout-bodied and mainly greyish brown or greyish green, with whitish dorsal pinacula and indistinct dorsal and subdorsal lines. When feeding is completed, usually in July, larvae enter the soil and construct silken cocoons in which, after a lengthy period of aestivation, they eventually pupate. Adults (30–35 mm wingspan) are broad-bodied, with a distinctly flattened abdomen; the fore wings are usually glossy chestnut-red, with a prominent black spot in the reniform stigma; brownish variegated forms also occur; the hind wings are pinkish brown to greyish brown.

Cosmia trapezina (Linnaeus) (**879–880**)
syn. *Calymnia trapezina* (Linnaeus)
Dun-bar moth

A generally common but minor pest of trees and shrubs, including ash (*Fraxinus excelsior*), beech (*Fagus sylvatica*), birch (*Betula*), common buckthorn (*Rhamnus cathartica*), crab-apple (*Malus*), elm (*Ulmus*), flowering cherry (*Prunus*), hawthorn (*Crataegus*), hazel (*Corylus*), hornbeam (*Carpinus betulus*), maple (*Acer*), oak (*Quercus*), poplar (*Populus*), pussy willow (*Salix caprea*) and rowan (*Sorbus aucuparia*). The larvae often occur on such plants in gardens, parks and nurseries. They feed from April to June, but cause only slight damage. In their later developmental stages they often devour the larvae of other moths. Individuals (up to 30 mm long) are bright green, with three white lines along the back, a yellowish one along each side and yellowish intersegmental bands; the pinacula are small and black, each at least partly edged with white. Adults (25–32 mm wingspan) are mainly whitish grey to yellowish grey, with a pinkish tinge and a more or less darkened central band on each fore wing. They occur in July and August.

878 Chestnut moth (*Conistra vaccinii*).

879 Larva of dun-bar moth (*Cosmia trapezina*).

880 Dun-bar moth (*Cosmia trapezina*).

Earias clorana (Linnaeus) (881–882)

syn. *Earias chlorana* (Linnaeus)

Cream-bordered green pea moth

A widely distributed but local, wetland species breeding mainly in willow beds, primarily on osier (*Salix viminalis*). Other willows, including almond willow (*S. triandra*), and pussy willow (*S. caprea*), are also suitable hosts. The larvae attack the terminal shoots, feeding during July and August within closely webbed clusters of leaves. Larvae (*c.* 17 mm long when fully fed) are greenish white to greyish white, marked with reddish brown and orange; there are large, dark-brown pinacula on the second and eighth abdominal segments. Pupation occurs in boat-shaped, parchment-like cocoons spun on the shoots. The tortrix-like adults (20–24 mm wingspan), which appear in the following May and June, have pea-green fore wings and white hind wings (cf. *Tortrix viridana*, p. 285).

881 Larva of cream-bordered green pea moth (*Earias clorana*).

882 Pupal cocoon of cream-bordered green pea moth (*Earias clorana*).

Euplexia lucipara (Linnaeus) (883)

Small angle-shades moth

A generally common species, the larvae feeding on herbaceous plants and shrubs, and sometimes also attacking cultivated ferns such as *Dryopteris*. When present on ferns, the larvae graze the fronds but damage caused is rarely significant. Fully grown larvae are 30–35 mm long, rather stout-bodied, purplish brown or green, with a V-shaped mark on each segment and a whitish subspiracular line, whitish pinacula, black spiracles and a pair of white spots on the slightly humped eighth abdominal segment; the head is small, glossy and pale greenish. The larvae feed from August to September, eventually pupating in silken and earthen cocoons formed in the soil. Most individuals overwinter as pupae but, in favourable conditions, there is a partial second generation. The purplish to light reddish-brown adults (30–35 mm wingspan) occur mainly in June and July.

Gortyna flavago (Denis & Schiffermüller) (884)

Frosted orange moth

An occasional pest of herbaceous ornamentals, including *Chrysanthemum*, *Dahlia*, foxglove (*Digitalis purpurea*), hollyhock (*Alcea rosea*), lupin (*Lupinus*), mullein (*Verbascum*) and pot marigold (*Calendula officinalis*); attacks also occur on woody hosts such as elder (*Sambucus*), lilac (*Syringa*) and willow (*Salix*). Holarctic. Widely distributed in Europe.

DESCRIPTION

Adult: 32–42 mm wingspan; fore wings yellow, marked with brown, purplish brown and orange-yellow; hind wings pale ochreous. **Larva:** up to 35 mm long; pale yellow, with very large, black pinacula and black spiracles; head yellowish brown; prothoracic and anal

883 Small angle-shades moth (*Euplexia lucipara*).

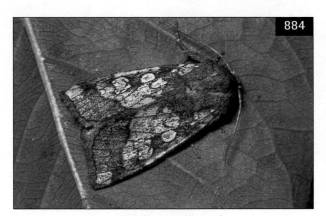

884 Frosted orange moth (*Gortyna flavago*).

885 Larva of scarce bordered straw moth (*Heliothis armigera*).

plates blackish brown. **Pupa:** 17–20 mm long; elongate, and yellowish brown; cremaster with a short pair of divergent spines.

LIFE HISTORY
Adults occur from late August to October, and eggs are deposited in groups at the base of suitable host plants. Larvae feed from late March or April to July or August, each tunnelling within a stem. If necessary, a larva will move to an adjacent plant in order to complete its development. Fully fed larvae eventually pupate within their feeding galleries, typically just above ground level. There is one generation each year.

DAMAGE
Shoots of infested plants wilt and may die.

Hadena bicruris (Hufnagel)
syn. *H. capsincola* (Denis & Schiffermüller);
Harmodia bicruris (Hufnagel)
Lychnis moth
A widely distributed but minor pest. The yellowish-brown, blackish-marked larvae (up to 35 mm long) feed mainly on the seeds of sweet william (*Dianthus barbatus*) and red campion (*Silene dioica*), but sometimes also damage the buds and seed heads of cultivated carnation (*D. caryophyllus*). Adults (30–40 mm wingspan) are greyish brown to blackish, the fore wings variegated with whitish and yellowish-white markings. They occur from early June onwards, and the larvae feed in June and July; in favourable situations larvae also occur from August to September.

Heliothis armigera (Hübner) (**885**)
Scarce bordered straw moth
A polyphagous, tropical or subtropical pest of many crops, including ornamentals such as garden geranium (*Pelargonium zonale*), pink (*Dianthus plumarius*) and rose (*Rosa*). Larvae also feed on various weeds. Present throughout Africa, Asia and Oceania; also established elsewhere, including the warmer parts of Europe (e.g. Bulgaria, southern France, Greece, Hungary, Italy, Portugal, Spain and the former Yugoslavia). A rare migrant to northern Europe; sometimes intercepted as larvae on imported plants or plant products.

DESCRIPTION
Adult: 30–40 mm wingspan; fore wings mainly ochreous to dark, purplish brown, with darker markings; hind wings creamy white, each with a broad, blackish peripheral band. **Larva:** up to 40 mm long; extremely variable in appearance, ranging from green or ochreous to purplish brown, with numerous, often whitish, sinuous lines running down the body; head olive-brown, and freckled.

LIFE HISTORY
In southern Europe, there are two or more generations annually, the larvae requiring both adequate warmth and moisture to successfully complete their development. Adults occur from May to late October, and the pest overwinters in the pupal stage.

DAMAGE
Larvae bore into buds and flowers, and also browse on foliage. Damage on cultivated plants may be extensive.

886 Rosy rustic moth (*Hydraecia micacea*).

887 Larva of rosy rustic moth (*Hydraecia micacea*).

Hydraecia micacea (Esper) (**886–887**)
Rosy rustic moth

A minor pest of robust herbaceous plants, including ornamentals such as *Chrysanthemum*, *Dahlia*, *Iris*, snapdragon (*Antirrhinum*) and sunflower (*Helianthus annuus*). Holarctic. Widely distributed in Europe but most common in coastal areas.

DESCRIPTION

Adult: 40–45 mm wingspan; fore wings reddish brown, somewhat darker centrally; hind wings pale, with a dark crossline. **Larva:** up to 45 mm long; slender-bodied, dull pinkish and translucent, with a slightly darker dorsal line and several brownish-black pinacula, each bearing a pinkish-brown hair; head and prothoracic plate yellowish brown.

LIFE HISTORY

Adults are most numerous in the autumn, depositing eggs close to the ground on the stems of grasses and certain other plants. The eggs hatch in the following spring, usually in late April or May. The young larvae then bore into the stems of suitable host plants, tunnelling downwards into the crowns, roots or rhizomes. Larval development is completed in July or August, fully fed individuals pupating in the soil a few centimetres below the surface. Adults emerge a few weeks later.

DAMAGE

Attacked plants are weakened and may wilt and die. Infestations are usually most severe on or near field headlands and on weedy sites.

Lacanobia oleracea (Linnaeus) (**888–890**)
syn. *Mamestra oleracea* (Linnaeus)
Tomato moth

An often common pest of herbaceous plants, including ornamentals growing outdoors and in greenhouses;

most significant damage occurs on *Chrysanthemum* and *Dianthus*. Eurasiatic. Widely distributed in Europe.

DESCRIPTION

Adult: 35–45 mm wingspan; fore wings reddish brown to purplish brown, with a small, yellowish stigma and a whitish subterminal line; hind wings pale brownish grey. **Egg:** greenish, hemispherical, slightly ribbed and reticulated. **Larva:** up to 40 mm long; green, yellowish brown or brown, finely speckled with white and, less densely, with black; dorsal and subdorsal lines pale, the spiracular line broad, and yellow or orange-yellow, darkly edged above; spiracles white, ringed with black; head light brown. **Pupa:** 16–19 mm long; dark brown to black, and coarsely punctured; cremaster with a pair of blunt-tipped spines.

LIFE HISTORY

Adults emerge outdoors from late May or early June onwards but the first individuals often appear from late January onwards in heated greenhouses. Eggs are deposited in large batches of 30–200 on the underside of leaves, and hatch in 1–2 weeks. Larvae then feed in groups on the lower surface of leaves. After their second instar, they disperse and tend to occur singly, each larva feeding voraciously and becoming fully grown a few weeks later. Larvae are most abundant from July to September. Pupation takes place in a flimsy cocoon, either in the soil or amongst debris or attached to a suitable surface. In favourable conditions there is a second generation of adults in the autumn; under glass there is commonly a second generation in the summer and a partial third in the autumn.

DAMAGE

Although young larvae merely graze away the lower surface of leaves, older larvae bite completely through the leaf and often reduce the foliage to a skeleton of

888 Tomato moth (*Lacanobia oleracea*).

889 Larva of tomato moth (*Lacanobia oleracea*) – green form.

890 Larva of tomato moth (*Lacanobia oleracea*) – brown form.

891 Larva of Blair's shoulder knot moth (*Lithophane leautieri hesperica*).

major veins. Leaves, stems and flowers are often damaged severely, particularly in greenhouses, attacked plants being rendered unmarketable if not completely destroyed.

Lithophane leautieri hesperica Boursin (891–892)

Blair's shoulder knot moth

A migratory, Mediterranean insect associated with common juniper (*Juniperus communis*) and Italian cypress (*Cupressus sempervirens*). In recent years this moth has greatly extended its range. It now occurs abundantly in many more northerly areas, including much of southern England, where it was first reported in 1951 and has since become adapted to Leyland cypress (*Cupressocyparis leylandii*) and Monterey cypress (*Cupressus macrocarpa*). The larvae (up to 35 mm long) are green, with a pair of broad white subdorsal stripes, a similar pair of lateral stripes, white pinacula and the abdominal spiracles surrounded by purplish-red patches. They feed mainly on the young growth, developing from February or March onwards. Fully

892 Blair's shoulder knot moth (*Lithophane leautieri hesperica*).

grown individuals enter the soil in July, each spinning a flimsy cocoon in which they pupate after an extended prepupal stage. The adults (40–45 mm wingspan) are mainly grey, irregularly marked with black. In northerly areas they occur in October and November.

893 Cabbage moth (*Mamestra brassicae*).

894 Larva of cabbage moth (*Mamestra brassicae*).

Mamestra brassicae (Linnaeus) (**893–895**)
syn. *Barathra brassicae* (Linnaeus)
Cabbage moth

A generally common and often important pest of herbaceous ornamentals; infestations frequently occur in greenhouses on *Alstroemeria*, *Chrysanthemum*, *Dianthus* and various other plants. Young trees and shrubs, including birch (*Betula*), flowering cherry (*Prunus*), hawthorn (*Crataegus*), larch (*Larix*), oak (*Quercus*) and pussy willow (*Salix caprea*) are also attacked. Eurasiatic. Widely distributed in Europe.

895 Young larva of cabbage moth (*Mamestra brassicae*).

DESCRIPTION

Adult: 38–45 mm wingspan; fore wings greyish brown to blackish brown, with pale, often black-edged markings; hind wings brownish grey. **Egg:** 0.8 mm across; hemispherical and distinctly ribbed; whitish, with a dark central spot and girdle. **Larva:** up to 45 mm long; green, greenish brown, brownish green or blackish brown; dorsal line black, the subdorsal lines comprising segmentally arranged blackish bars, the pair on the eighth abdominal segment meeting to form a saddle-like mark; spiracular line pale; spiracles white with black rims; a slight hump on the eighth abdominal segment; head light brown; young larva light green, with whitish longitudinal lines, and often marked with yellow intersegmentally. **Pupa:** 17–22 mm long; reddish brown and finely punctured; cremaster with two, often pale, hooked spines.

LIFE HISTORY

Although the moths occur at any time of year, they are most frequent in June and July, and from the end of August to late September. Eggs are deposited in large batches on the leaves of various plants, and hatch in 8–15 days. Larvae then feed from late June or July onwards. The young larvae browse on the surface of leaves; older individuals feed more extensively, and also often damage buds and flowers. Fully fed larvae pupate in flimsy subterranean cocoons. This species is basically single brooded but, mainly under glass, breeding may be sufficiently rapid for the completion of two generations annually.

DAMAGE

Infestations on greenhouse ornamentals are most important, and damage caused to buds, leaves and open blooms renders plants unmarketable.

Melanchra persicariae (Linnaeus) (**896–897**)
syn. *Mamestra persicariae* (Linnaeus)
Dot moth

Generally common in gardens and nurseries on ornamentals such as *Anemone*, *Dahlia* and lupin (*Lupinus*); minor infestations also occur on trees and shrubs, including alder (*Alnus*), birch (*Betula*), crab-apple (*Malus*), larch (*Larix*) and pussy willow (*Salix caprea*). Holarctic. Widely distributed in Europe.

DESCRIPTION

Adult: 38–48 mm wingspan; fore wings bluish black, each with a prominent, white, kidney-shaped stigma;

896 Dot moth (*Melanchra persicariae*).

897 Larva of dot moth (*Melanchra persicariae*).

898 Gothic moth (*Naenia typica*).

899 Larva of gothic moth (*Naenia typica*).

hind wings greyish brown. **Larva:** up to 45 mm long; light green or light brown, with darker chevron-like markings on the back and sides, and a thin, pale, dorsal stripe; prothoracic plate with distinct dorsal and subdorsal lines; eighth abdominal segment with a bluntly pointed hump. **Pupa:** 22–24 mm long; dark chestnut-brown; cremaster with two divergent, barb-tipped spines.

LIFE HISTORY
Adults occur form June to August. Eggs are laid on leaves of various plants, and hatch in about eight days. Larvae feed slowly from July onwards, and usually complete their development in September or October. They then enter the soil to pupate in flimsy cocoons. Moths emerge in the following summer.

DAMAGE
Larvae devour large amounts of foliage and rapidly strip the leaves from host plants. But injury is usually important only on young plants or where numbers of larvae are large.

Naenia typica (Linnaeus) (**898–899**)
Gothic moth
An occasionally troublesome pest of herbaceous plants, shrubs and young trees, including *Chrysanthemum*, crab-apple (*Malus*), flowering cherry (*Prunus*) and *Rhododendron*; infestations also occur under glass, as on *Chrysanthemum*, *Fuchsia*, *Geranium* and other pot plants. Eurasiatic. Widespread in Europe.

DESCRIPTION
Adult: 36–46 mm wingspan; fore wings whitish brown, suffused with blackish brown, each marked with pale stigmata and crosslines, the reniform stigma enclosing a pale line; hind wings brownish grey. **Larva:** up to 45 mm long; greyish brown, flecked with darker brown, the sides marked with pale oblique streaks and a pinkish, undulating spiracular line edged above by black; second and third thoracic segments with a pair of creamy-white spots; seventh and eighth abdominal segments each with distinctive, black, oblique markings; pinacula whitish; head pale brownish, marked with brown. **Pupa:** 15–18 mm long; dark chestnut-brown; cremaster with a pair of downwardly curved, convergent spines.

LIFE HISTORY

Adults occur in June and July. Eggs are laid on the foliage of various plants, and hatch in 10–14 days. Young larvae then feed gregariously from late July or August onwards. The larvae hibernate throughout the winter, reappearing in the following spring. They then feed singly, becoming fully grown in May. Pupation takes place in the soil in flimsy cocoons.

DAMAGE

Larvae cause defoliation which, if extensive, may be of importance, particularly on greenhouse-grown chrysanthemums; flowers are also destroyed.

Noctua pronuba (Linnaeus) (900–902)

syn. *Triphaena pronuba* (Linnaeus)

Large yellow underwing moth

A generally common pest of low-growing plants, including herbaceous ornamentals such as *Anemone*, *Chrysanthemum*, *Dahlia*, *Dianthus*, pot marigold (*Calendula officinalis*) and primrose (*Primula vulgaris*); seedling trees and shrubs in nurseries are also attacked. Although mainly an outdoor pest, infestations also occur on greenhouse-grown plants. Palaearctic. Present throughout Europe.

DESCRIPTION

Adult: 50–60 mm wingspan; fore wings yellowish brown, greyish brown to dark rusty brown; hind wings yellow, with a blackish-brown border. **Egg:** hemispherical, creamy white to purplish grey. **Larva:** up to 50 mm long; greyish brown, dull yellowish or greenish, with three pale lines along the back, the outer pair bordered inwardly with short blackish bars on each abdominal segment; spiracles white with black rims; head relatively small, light brown, marked with black. **Pupa:** 22 mm long; plump and reddish brown; cremaster with two strong, divergent spines.

LIFE HISTORY

Adults occur from mid-June to August. Although active mainly at night, the moths are readily disturbed during the daytime; they then career wildly through the air before resettling, the flash of colour displayed by the hind wings making them particularly obvious. Eggs are laid in large, neat batches, often on the leaves of monocotyledonous plants such as *Gladiolus* and *Iris*. They hatch in 2–3 weeks, earlier or later depending on temperature. Larvae occur from July onwards. They feed both above ground on leaves and flowers, and in the soil on roots and crowns. A few individuals feed up rapidly to produce a partial second generation of adults

900 Large yellow underwing moth (*Noctua pronuba*).

901 Egg batches of large yellow underwing moth (*Noctua pronuba*) on leaves of *Iris*.

902 Larva of large yellow underwing moth (*Noctua pronuba*).

in the autumn; however, most do not complete their development until the following May. Pupation takes place in a subterranean, earthen cell but without forming a cocoon. Unlike *Agrotis segetum* (pp. 333–4), larvae of this species are little affected by weather conditions and frequently cause damage to crops in cool, wet seasons.

DAMAGE
In addition to weakening or killing plants by attacking the roots and crowns, larvae are also directly harmful to the buds, foliage, flowers and stems, and sometimes destroy entire shoots or inflorescences.

Nycteola revayana (Scopoli) (**903**)
syn. *Sarrothripus undulanus* (Hübner)
Oak nycteoline moth

This species is associated mainly with oak (*Quercus*) but, at least in mainland Europe, also attacks certain kinds of poplar (*Populus*) and willow (*Salix*), including amenity trees. Moths occur from late summer onwards. They hibernate during the winter months and finally deposit eggs in the spring. Larvae (up to 20 mm long) are mainly green, marked intersegmentally with yellow, and finely clothed in long, whitish hairs; the spiracles are yellowish with black rims. Larval development continues from May to July, fully fed individuals then pupating in light green, boat-shaped cocoons formed on the twigs or on the underside of leaves. Adults are extremely variable in appearance, ranging from greyish white, through brownish to blackish; they are relatively small (13–15 mm wingspan) and superficially similar in appearance to certain members of the family Tortricidae (pp. 254–85).

Orthosia incerta (Hufnagel) (**904–905**)
Clouded drab moth

A generally common pest of trees and shrubs, including ash (*Fraxinus excelsior*), beech (*Fagus sylvatica*), birch (*Betula*), crab-apple (*Malus*), elm (*Ulmus*), flowering cherry (*Prunus*), hawthorn (*Crataegus*), hornbeam (*Carpinus betulus*), lime (*Tilia*), *Populus*, oak (*Quercus*), pussy willow (*Salix caprea*) and rose (*Rosa*); often troublesome in garden trees and on nursery stock. Eurasiatic. Widespread throughout Europe.

DESCRIPTION
Adult: 34–40 mm wingspan; fore wings light grey to reddish brown or purplish brown, with darker markings and a pale submarginal line partly edged with brown; hind wings greyish. **Egg:** 0.7 mm across; dirty creamy white, with a purplish band and micropyle. **Larva:** up to 40 mm long; blackish green, bluish green or light green, dotted with white; a prominent white dorsal stripe down the back, a pair of often indistinct subdorsal lines and a white stripe along each side, the latter edged above with blackish green or yellow; head pale bluish green, brownish green or yellowish green. **Pupa:** 14 mm long; shiny, dark reddish brown; cremaster with two spines.

903 Oak nycteoline moth (*Nycteola revayana*).

904 Clouded drab moth (*Orthosia incerta*).

905 Larva of clouded drab moth (*Orthosia incerta*).

LIFE HISTORY
Adults occur from March to late May or early June, but are most numerous in April and early May. Eggs are laid in groups in cracks and crevices in the bark of host plants, and hatch in 10–14 days. Tiny, very mobile, larvae immediately invade the bursting buds and

unfurling leaves. The larvae feed mainly at night, young individuals hiding during the day in the shelter of young blossom trusses and partially unfurled leaves. Larvae are fully grown in about six weeks, usually in June. They then enter the soil to pupate in flimsy silken cocoons formed a few centimetres below the surface. Adults emerge in the following spring.

DAMAGE
The larvae cause extensive defoliation and also damage buds and flowers. On rose, the larvae feed on the outer tissue and also burrow into the centre of the blooms.

Orthosia gothica (Linnaeus) (**906–907**)
Hebrew character moth
Larvae of this widely distributed and often abundant species are polyphagous on trees and shrubs, including birch (*Betula*), broom (*Cytisus*), hawthorn (*Crataegus*), lime (*Tilia*), oak (*Quercus*), poplar (*Populus*), pussy willow (*Salix caprea*) and *Sorbus*; they also attack herbaceous plants. Although occurring mainly on wild hosts, individuals (which are up to 35 mm long, green,

striped with white or whitish yellow) sometimes cause minor damage to ornamentals. They feed from April to June. The pale purplish-brown to reddish-brown adults (32–34 mm wingspan) are readily distinguished from other species of the genus by the characteristic brownish-black mark on each fore wing. They occur in greatest numbers in March and April.

Orthosia gracilis (Denis & Schiffermüller) (**908–909**)
Powdered quaker moth
Generally common in association with trees, shrubs and herbaceous plants; sometimes a minor pest in gardens and nurseries. Adults (35–40 mm wingspan) have whitish-beige fore wings, each more or less tinged with grey, and marked with a few black dots and a pale yellowish or pinkish submarginal line. They occur in late April and May, rather later than other members of the genus. Eggs are laid in untidy batches on twigs or senescent foliage, and hatch about ten days later. The larvae, which feed from May to July, cause slight defoliation and also injure buds and flowers. Fully

906 Larva of Hebrew character moth (*Orthosia gothica*).

907 Hebrew character moth (*Orthosia gothica*).

908 Powdered quaker moth (*Orthosia gracilis*).

909 Larva of powdered quaker moth (*Orthosia gracilis*).

grown individuals are 40–45 mm long and yellowish green to pinkish green, with several white spots and three pale lines along the back, a yellow stripe (broadly edged above with greenish black) along each side and a yellowish-brown head.

Orthosia miniosa (Denis & Schiffermüller) (910–911)
Blossom underwing moth

A widespread but local species, confined mainly to oak woodlands. The moths occur in March and April, eggs being laid in batches on the shoots of oak (*Quercus*), especially scrub-oaks. At first, larvae feed gregariously, clustering together within the shelter of slight webs spun close to the shoot tips. Such webs are sometimes noted during May on young amenity or nursery trees growing in the vicinity of oak woods. Later, the larvae disperse to feed individually; they may then infest various other trees and shrubs, including beech (*Fagus sylvatica*), birch (*Betula*) and *Sorbus*. Fully grown larvae (35–40 mm long) are mainly blue to bluish grey, with three yellow lines along the back and prominent, black, dorsal spots.

They complete their development in June and then pupate. The mainly pinkish-grey to reddish-grey adults (32–35 mm wingspan) appear in the following spring.

Orthosia munda (Denis & Schiffermüller) (912–913)
Twin-spotted quaker moth

This widely distributed, mainly woodland species is associated with various trees and shrubs, including aspen (*Populus tremula*), crab-apple (*Malus*), elm (*Ulmus*), honeysuckle (*Lonicera*) and oak (*Quercus*). The larvae (up to 45 mm long) are light brown to dark brown, with brownish-yellow dorsal and subdorsal lines, and a pale yellowish-brown spiracular stripe. They feed from April to June, and sometimes cause minor damage to the young foliage of nursery stock and garden trees. The reddish-ochreous to pale greyish-ochreous adults (38–40 mm wingspan) occur mainly in March and April.

910 Larva of blossom underwing moth (*Orthosia miniosa*).

911 Blossom underwing moth (*Orthosia miniosa*).

912 Larva of twin-spot quaker moth (*Orthosia munda*).

913 Twin-spotted quaker moth (*Orthosia munda*).

Orthosia stabilis (Denis & Schiffermüller) (914–915)

Common quaker moth

A generally common species, the larvae occasionally attacking trees and shrubs in gardens, parks and nurseries but more usually associated with forest and hedgerow trees such as beech (*Fagus sylvatica*), birch (*Betula*), elm (*Ulmus*), oak (*Quercus*) and pussy willow (*Salix caprea*). The pale reddish-ochreous to pale greyish-ochreous adults (32–35 mm wingspan) occur in March and April. The larvae feed on foliage from April to June but numbers on cultivated plants are usually small. Individuals are plump, up to 40 mm long, yellowish green and finely dotted with yellow; there are three yellow stripes along the back, a broader one along each side, and a prominent yellow bar across the first thoracic and last abdominal segments; the head is bluish green.

Panolis flammea (Denis & Schiffermüller) (916)

Pine beauty moth

A locally common and widely distributed Eurasiatic pest of Scots pine (*Pinus sylvestris*) and certain other trees, including maritime pine (*P. pinaster*) and Weymouth pine (*P. strobus*). Although mainly of importance as a forestry pest, larvae also cause damage in nurseries and private gardens, primarily in more southerly parts of its range and in the vicinity of pine woodlands. Adults (32–35 mm wingspan) are light greyish red, with greyish and whitish markings on each fore wing. They occur from late March to early May, and often rest on the trunks or branches of pine trees during the daytime. Eggs are laid singly or in rows on the needles of host trees, and hatch about a week later. The dark green, white-striped larvae (up to 40 mm long) browse on the foliage and are usually fully grown by mid-July. They then pupate in silken cocoons spun amongst debris on the ground or in crevices on the bark. The larvae are capable of causing considerable defoliation, and loss of needles has an adverse effect on bud development in the following year. Damage is of particular significance on young trees, but significant attacks are restricted to established plantations rather than to isolated ornamental trees and nursery stock.

Peridroma saucia (Hübner) (917–918)

Pearly underwing moth

A polyphagous pest of agricultural and horticultural crops, including ornamentals such as carnation (*Dianthus caryophyllus*), *Chrysanthemum*, *Geranium* and *Pelargonium*. Cosmopolitan. Widely distributed in America and elsewhere, including the warmest parts of Europe. A migrant to northern Europe (e.g. Britain, Denmark and the Netherlands), arriving annually in greater or lesser numbers.

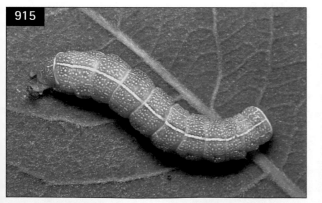

914 Common quaker moth (*Orthosia stabilis*).

915 Larva of common quaker moth (*Orthosia stabilis*).

916 Pine beauty moth (*Panolis flammea*).

917 Pearly underwing moth (*Peridroma saucia*).

918 Larva of pearly underwing moth (*Peridroma saucia*).

919 Angle-shades moth (*Phlogophora meticulosa*).

920 Larva of angle-shades moth (*Phlogophora meticulosa*).

DESCRIPTION

Adult: 45–55 mm wingspan; fore wings light brown to dark brown, with blackish markings; hind wings translucent with a pearly sheen, but dark towards the border and on the veins. **Larva:** up to 45 mm long; mainly ochreous, rusty or purplish brown, with an undulating longitudinal stripe above the spiracles and a dark mark on the eighth abdominal segment.

LIFE HISTORY

Eggs are laid in batches on the foodplant and hatch a week or so later. The cannibalistic larvae then feed for several weeks, their rate of development being strongly dependent on temperature. In sufficiently warm conditions, the pest completes two or more generations annually, but in northern Europe no more than one generation seems possible. Pupae form the overwintering stage, but the pest is unable to survive through northern European winters.

DAMAGE

Larvae burrow into buds and also browse on open flowers and leaves.

Phlogophora meticulosa (Linnaeus) (**919–920**)
syn. *Trigonophora meticulosa* (Linnaeus)
Angle-shades moth

An often common pest of greenhouse-grown herbaceous plants, including *Chrysanthemum*, *Cineraria*, *Geranium*, *Pelargonium*, violet (*Viola*) and various ferns; attacks are most frequent in greenhouses with artificial lighting. Infestations also occur on, for example, *Anemone*, *Dahlia*, *Fuchsia*, hollyhock (*Alcea rosea*), *Iris*, ivy (*Hedera*), primrose (*Primula vulgaris*) and wallflower (*Cheiranthus cheiri*) growing in outdoor flower beds and borders. Eurasiatic. Present throughout Europe.

DESCRIPTION

Adult: 40–50 mm wingspan; fore wings mainly pale pinkish brown, each with a darker base and a large inverted triangular olive-green median mark; hind wings whitish brown. **Egg:** 0.8 mm across; hemispherical and strongly ribbed; pale yellow, with darker mottling. **Larva:** up to 40 mm long; velvety yellowish green or brownish, with a fine, white, interrupted dorsal line, and a series of faint V-shaped marks down the back; spiracles usually white with black rims; body rather plump; early instars are leech-like, bright green and translucent, with indistinct dusky segmental markings. **Pupa:** 18 mm long; reddish brown and plump; cremaster with a pair of long spines.

LIFE HISTORY
Adults occur from May to October, mainly as one generation in May and June and another in the autumn. The moths often rest on greenhouse walls and other structures, with the wings folded lengthwise to resemble a crumpled leaf. If disturbed during the daytime, they fly erratically for a short distance before resettling. Eggs are deposited on leaves, either singly or in small groups. They hatch within a few days at normal greenhouse temperatures. Larvae are most abundant from July to September, but may be found in greater or lesser numbers throughout much of the year, their rate of development being greatly affected by both temperature and humidity. Pupation takes place in the soil within a slight silken cocoon.

DAMAGE
Young larvae 'window' the leaves but older individuals bite right through the leaf blade, causing considerable defoliation; larvae also attack the growing points, flower buds and blossom trusses.

Polychrysia moneta (Fabricius) (**921**)
syn. *Chrysoptera moneta* (Fabricius)
Delphinium moth
A minor garden pest of *Aconitum*, *Delphinium* and globe flower (*Trollius europaeus*). Widely distributed in Europe but most numerous in southern areas.

921 Female delphinium moth (*Polychrysia moneta*).

DESCRIPTION
Adult: 38–45 mm wingspan; fore wings pale gold, with light and dark brown markings and a silver reniform stigma; hind wings brownish black. **Larva:** up to 40 mm long; dark green, with a darker dorsal line, a white line along each side and white pinacula; abdominal segments humped; three pairs of prolegs. **Pupa:** plump; dark reddish brown, with the head and wing pads green.

LIFE HISTORY
Adults occur from mid-June to early July, depositing eggs on open flowers or amongst the buds. Larvae feed from July onwards, attacking the buds, flowers and young leaves, and tying the shoots together with silk; older larvae feed in exposed situations and may also attack the developing seeds. Some larvae complete their development during the summer, descending the foodplant to pupate in yellow cocoons spun amongst the lower leaves. These individuals produce a partial second generation of moths in the autumn. Other larvae, along with those resulting from eggs laid by second-generation adults, overwinter in hollow stems, either above or below ground level, and complete their development in the spring. They pupate in May or June.

DAMAGE
Defoliation, bud damage and loss of flowers may be important locally, affecting the marketability of commercially grown plants.

Scoliopteryx libatrix (Linnaeus) (**922–923**)
Herald moth
A generally common pest of poplar (*Populus*) and willow (*Salix*). Widely distributed in Europe.

DESCRIPTION
Adult: 50–60 mm wingspan; fore wings purplish brown to greyish brown, more or less tinged with reddish orange; hind wings light brown. **Larva:** up to 60 mm long; slender, smooth-bodied and velvety; yellowish green to dark green, with three dark dorsal lines; head green.

LIFE HISTORY
Adults hibernate in barns, hollow trees, and various other situations, reappearing in the following spring. The larvae feed on poplars and willows from May or June onwards, attacking the young leaves and shoots which they often spin together with silk. They complete their development in the summer, and young adults emerge in August. In favourable parts of mainland Europe there are two generations annually.

DAMAGE
Defoliation on young trees and nursery stock can be extensive but attacks on older plants are usually of little or no significance.

Shargacucullia verbasci (Linnaeus) (924–925)
syn. *Cucullia verbasci* (Linnaeus)
Mullein moth

Often common, particularly on light soils, infesting mullein (*Verbascum*) and figwort (*Scrophularia*). Sometimes a pest of cultivated plants; attacks also occur on buddleia (*Buddleja*). Eurasiatic. Widely distributed in Europe.

DESCRIPTION
Adult: 45–55 mm wingspan; fore wings lanceolate, mainly ochreous brown to dark brown; hind wings dark brown, grading to ochreous brown basally. **Larva:** up to 55 mm long; whitish to light green or light blue, each segment marked posteriorly with a yellow dorsal crossband and on the sides with black lines and spots; pinacula black; head yellow, spotted with black.

LIFE HISTORY
Adults appear from mid-April to the end of May, depositing eggs singly on the underside of leaves of host plants. The larvae feed in June and July, each eventually pupating in the soil in a strong, silken cocoon several centimetres below the surface. The period of pupation is protracted, and often persists through four or five winters. This probably accounts for the species being most frequent on light soils and rather scarce in heavy-clay districts.

DAMAGE
Young larvae bite out small holes in the leaves. Older individuals feed ravenously, and often cause significant defoliation. Flower spikes are not only destroyed but also contaminated by masses of expelled frass.

922 Herald moth (*Scoliopteryx libatrix*).

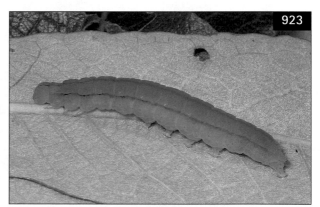

923 Larva of herald moth (*Scoliopteryx libatrix*).

924 Mullein moth (*Shargacucullia verbasci*).

925 Larva of mullein moth (*Shargacucullia verbasci*).

926 Larva of striped lychnis moth (*Shargacucullia lychnitis*).

927 Mediterranean brocade moth (*Spodoptera littoralis*).

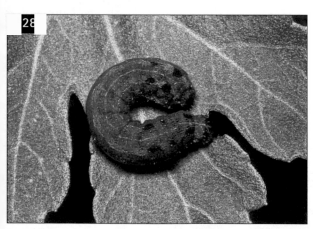

928 Mediterranean climbing cutworm (*Spodoptera littoralis*).

Shargacucullia lychnitis (Rambur) (926)

syn. *Cucullia lychnidis* Rambur
Striped lychnis moth
This very local, generally uncommon and declining Eurasiatic species is very similar in appearance to *S. verbasci*, but associated mainly with dark mullein (*Verbascum nigrum*) and white mullein (*V. lychnitis*). In areas where the moth still survives, larvae (which feed later in the year than those of *S. verbasci*) are sometimes found feeding on the flower spikes of cultivated mulleins.

Spodoptera littoralis (Boisduval) (927–928)

Mediterranean brocade moth
larva = Mediterranean climbing cutworm
A mainly tropical and subtropical species but recently established in southern Europe; sometimes introduced into northern Europe on imported *Chrysanthemum* cuttings and on other ornamentals such as *Hibiscus* and *Kalanchoe*; a potentially serious pest of greenhouse-grown crops, especially chrysanthemums.

DESCRIPTION
Adult: 40 mm wingspan; fore wings blackish brown, with lighter markings, and often with a purplish sheen; hind wings mainly white. **Larva:** up to 45 mm long; light brown to blackish brown, finely speckled with white; dorsal and subdorsal lines orange-brown, the latter interrupted on the first and eighth abdominal segments by black patches and on the other segments by yellow spots, each bordered above with a black patch; yellow spots and black patches are also present on the second and eighth thoracic segments; subspiracular line broad and light reddish brown; dark forms are mainly blackish brown. **Pupa:** 16–20 mm long; reddish brown.

LIFE HISTORY
Eggs are deposited in large groups on the underside of leaves of host plants or on nearby surfaces and then coated with scales from the female's abdomen. The eggs hatch within a few days at normal greenhouse temperatures. Larvae then attack the plants, feeding voraciously and producing noticeably wet faecal pellets. Development is completed in 2–4 weeks; the larvae then pupate in flimsy cocoons, and adults emerge about two weeks later. Breeding is continuous under suitably warm conditions. In northern Europe, the pest is unlikely to survive the winter outdoors or in unheated structures.

DAMAGE
Young larvae 'window' the leaves; older individuals bite right through the leaf tissue, causing considerable defoliation; they also attack the flowers and stems.

Spodoptera exigua (Hübner)
syn. *Laphygma exigua* (Hübner)
Small mottled willow moth
This polyphagous, mainly tropical and subtropical species is an introduced pest in the Netherlands, attacking various greenhouse-grown ornamentals. It also occurs elsewhere in northern Europe (including the British Isles) as a rare immigrant or as an accidentally introduced species on plants such as *Chrysanthemum*. Breeding is continuous under suitable conditions but the pre-adult stages are susceptible to cold and damp, and this reduces the likelihood of their survival outdoors in northern Europe. The moths (28–32 mm wingspan) are yellowish grey, mottled with yellowish brown, the stigmata characteristically ochreous or pinkish, and the hind wings pearly white with a darker border. The larvae (30–35 mm long when fully grown) vary from green or olive-green to purplish black, with a paler or darker dorsal region and the spiracles set in pinkish, yellowish or orange patches. In some parts of the world this pest is an often destructive and notorious pest; the larva is commonly known in America as the 'beet army-worm'.

Xylocampa areola (Esper) (**929–930**)
syn. *Dichonia areola* (Esper)
Early grey moth
Adults of this widely distributed, mainly grey, blackish-marked moth (30–40 mm wingspan) occur from March to early May, and are often found at rest on garden fences, posts and walls. They are active at night, when they are frequent visitors to sallow catkins. Eggs are deposited singly on the stems of honeysuckle (*Lonicera*), and minor infestations sometimes occur on ornamental bushes. The larvae feed in May and June, older individuals remaining flattened against the stem of the foodplant during the daytime and moving onto the leaves to feed at night. Individuals (up to 45 mm long) are elongate, pale yellowish brown, with a brown dorsal stripe and a fine spiracular line; the head is greyish brown marked with grey. Fully fed larvae pupate in the soil, each within a tough white cocoon. There is just one generation annually.

929 Early grey moth (*Xylocampa areola*).

930 Larva of early grey moth (*Xylocampa areola*).

Order **TRICHOPTERA** (caddis flies)

Family **LIMNEPHILIDAE**

A large family of caddis flies, associated mainly with slow-moving water. Antennae of adult about as long as the fore wings and each with a bulbous basal segment. Larval case often very large.

Halesus radiatus (Curtis)

A generally common species, sometimes a pest of aquatic ornamentals such as white water-lily (*Nymphaea alba*) and yellow water-lily (*Nuphar lutea*). Widely distributed in Europe.

DESCRIPTION
Adult: 42 mm wingspan; fore wings broad, with a rounded apex, pale yellow striated with dark grey. **Larva:** up to 22 mm long; head brown; thorax yellowish brown to dark brown, spotted with black; abdomen whitish to dark brown, with gill filaments on the first to seventh segments. **Case:** 30×6 mm; formed from small pieces of plant tissue, with up to three thin sticks cemented along the sides.

LIFE HISTORY
Eggs are laid in the autumn on plants or stones above the water of suitable aquatic habitats, the egg masses being protected by a greenish gelatinous coating. After egg hatch, the larvae enter the water to begin feeding, each constructing an elongated, tube-like case within which to shelter. Further material is added to the case as the larva grows. The insect remains within this habitation throughout its development, merely protruding the anterior end of the body from the case when feeding or collecting case-building material or walking about. Larvae feed throughout the winter, and complete their development in the following summer. Pupation occurs within the sealed-off case, usually after its attachment to a submerged plant or stone, the pupa breaking free of the case and floating to the surface immediately before the appearance of the adult.

DAMAGE
Buds, leaves, stalks and roots of aquatic plants are all attacked, and tissue often becomes extensively tattered and torn.

Limnephilus marmoratus Curtis

A widely distributed and often abundant species, which breeds in most kinds of water and is sometimes damaging to ornamental aquatic plants, including water-lilies. The larvae (up to 28 mm long) inhabit cases measuring *c*. 27×8 mm, each shelter composed of short lengths of plant stalk, arranged either transversely or obliquely. Adults (30–35 mm wingspan) have relatively narrow, mainly yellowish, brownish-marked, parchment-like fore wings; they occur in the autumn.

Order **HYMENOPTERA** (ants, bees, sawflies and wasps)

Family **PAMPHILIIDAE**

A small group of primitive, fast-flying, flattened, broad-bodied sawflies; antennae long and thin, with 18–24 segments. Larvae lack abdominal prolegs, and are often gregarious and web-forming.

Neurotoma saltuum (Linnaeus) (**931**)
syn. *N. flaviventris* (Retzius)
Social pear sawfly
Although mainly a pest of pear fruit trees, infestations of this widespread but local, southerly-distributed pest occur on various other rosaceous plants, including ornamentals such as *Cotoneaster*, flowering cherry (*Prunus*), hawthorn (*Crataegus*) and medlar (*Mespilus germanica*). The unmistakable larvae feed in communal webs in June and July, rapidly stripping the branches of foliage; fully fed individuals are 20–25 mm long, yellowish orange, with a shiny black head and a pair of prominent anal cerci. They overwinter in the soil and pupate in the spring. Adults appear in May and June.

Pamphilius varius (Lepeletier) (**932**)
A minor pest of birch (*Betula*). The solitary, green-bodied larvae (*c.* 20 mm long), which have very small thoracic legs and no abdominal prolegs, each inhabit a rolled leaf. They are sometimes present on young trees during the summer, and cause minor damage to the foliage. Fully grown individuals overwinter in the soil and pupate in the spring. The stout-bodied adults appear in May and June. Species of *Pamphilius* are also associated with other hosts, including aspen (*Populus tremula*), common sallow (*Salix atrocinerea*), oak (*Quercus*) and rose (*Rosa*).

931 Larva of social pear sawfly *(Neurotoma saltuum)*.

932 Larva of *Pamphilius varius.*

933 Larva of barberry sawfly (*Arge berberidis*).

934 Larva of large rose sawfly (*Arge ochropus*).

935 Large rose sawfly (*Arge ochropus*) oviposition scar in shoot of *Rosa*.

Family **ARGIDAE**

Slow-moving, heavily built, moderate-sized sawflies with a short ovipositor; antennae divided into a scape, a pedicel and a fused flagellar segment, the last-mentioned sometimes distinctly bifid like a tuning fork.

Arge berberidis Schrank (**933**)
Barberry sawfly
A well-known pest of barberry (*Berberis*). Infestations also occur on *Mahonia*. Widely distributed in central and southern Europe; also now present in various parts of northern Europe, including southern England.

DESCRIPTION
Adult female: 7–10 mm long; dark metallic blue, with smoky wings. **Larva:** up to 18 mm long, body relatively plump and whitish, marked with yellow patches and numerous black verrucae; head and anal plate black.

LIFE HISTORY
There are typically two or more generations annually, with adult females active from May onwards. Pupation occurs in silken cocoons spun on the host plant or in the ground beneath infested bushes.

DAMAGE
Larvae often cause extensive defoliation; *Berberis thunbergii* is particularly susceptible.

Arge ochropus (Gmelin in Linnaeus) (**934–935**)
Large rose sawfly
A local and often destructive pest of rose (*Rosa*), the larvae occurring on both wild and cultivated bushes. Eurasiatic. Widely distributed in mainland Europe; in Britain generally uncommon (cf. *Arge pagana*).

DESCRIPTION
Adult: 7–10 mm long; head and thorax black but with the pronotum and tegulae yellow; abdomen yellow; legs yellow with the apices of the tibiae and tarsi black; wings yellowish. **Larva:** up to 28 mm long; head black or orange; body bluish green, marked with yellow along the back; body covered with numerous black verrucae, on most segments the latter usually forming one relatively indistinct and two distinct transverse rows (cf. *Arge pagana*); the last two segments bear only small verrucae; anal plate black; five pairs of abdominal prolegs (cf. *A. nigripes*).

LIFE HISTORY

Adults emerge in late May or June and may then be found on or near rose bushes. Eggs are deposited in double rows along the young vegetative shoots or flower stalks, the female penetrating the tissue with her short, stubby ovipositor to form distinctive oviposition scars; up to 20 eggs are inserted into each affected shoot. The eggs hatch a few weeks later. The larvae then attack the expanded foliage. At first they graze on the lower epidermis; later, they bite completely through the leaf tissue, either in the middle or at the edge. Fully fed larvae drop to the ground and eventually pupate in brown, double-walled cocoons formed in the soil. There is often just one generation annually but, if conditions are favourable, a second brood of larvae occurs in the autumn.

DAMAGE

Shoots containing oviposition scars become blackened and distorted, and heavy attacks affect growth and flowering. Larvae cause extensive defoliation, commonly skeletonizing the leaves.

Arge pagana (Panzer) (**936–937**)

syn. *A. stephensii* (Leach)

Variable rose sawfly

A locally common and important pest of rose (*Rosa*). Eurasiatic. Present throughout much of mainland Europe; in Britain most abundant in southern, south-eastern and south-western England.

DESCRIPTION

Adult: 7–9 mm long; head, thorax and legs mainly black; abdomen mainly yellow; wings blackish. **Larva:** up to 25 mm long; head black or orange; body bluish green, suffused above with yellowish green; body bearing numerous, usually very prominent, black verrucae; these verrucae form three distinct transverse rows per segment but with just one transverse row on the last (cf. *Arge ochropus*); anal plate black; five pairs of abdominal prolegs (cf. *Arge nigripes*).

LIFE HISTORY

Adults occur from May to October, depositing eggs in the young shoots as described under *Arge ochropus*. The larvae feed voraciously from June onwards, there being two main broods annually. When fully grown, the larvae enter the soil to pupate in brown, double-walled cocoons.

DAMAGE

Leaf skeletonization is often extensive, seriously affecting shoot growth and spoiling the appearance of bushes.

936 Variable rose sawfly (*Arge pagana*).

937 Larva of variable rose sawfly (*Arge pagana*).

Arge nigripes (Retzius in Degeer) (**938–939**)

Widely distributed on rose (*Rosa*) and sometimes a pest of cultivated bushes. Oviposition occurs in the leaf margins, close to the tips of the main serrations, the position of each egg being indicated by a distinct swelling. The larvae feed gregariously from late May onwards, and often cause noticeable defoliation. Individuals complete their development in about a month, each then pupating on the ground in a double-walled cocoon; there are one or two generations annually. Larvae (up to 30 mm long) are greenish and translucent, the body bearing numerous small, mainly black, verrucae; unlike *Arge ochropus* and *A. pagana*, there are seven pairs of abdominal prolegs. Adults (9–11 mm long) are mainly black; they occur in April and May, with members of a second generation appearing in July or early August.

Arge ustulata (Linnaeus) (**940**)

A widely distributed and often common species on birch (*Betula*) and willow (*Salix*), including ornamentals. Larvae (up to 22 mm long) are dark green and shiny, with a pair of white lines down the back; there are numerous pairs of black verrucae on the body, those on the thorax being largest and most conspicuous. Larvae occur from July to September or October. They feed along the edge of the leaves and cause noticeable, but usually only minor, defoliation. Adults (10–11 mm long) are bronzy black with yellowish wings.

938 *Arge nigripes* egg pouches in leaf of *Rosa*.

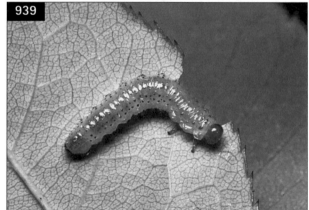

939 Larva of *Arge nigripes*.

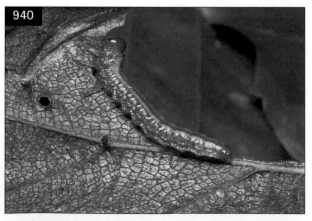

940 Larva of *Arge ustulata*.

Family **CIMBICIDAE**

Medium-sized to large, stout-bodied, fast-flying sawflies, with distinctly clubbed antennae.

Cimbex femoratus (Linnaeus) (**941–942**)
syn. *C. sylvarum* (Fabricius)
Large birch sawfly
A common but minor pest of birch (*Betula*), including, occasionally, nursery and garden trees. Present throughout Europe.

DESCRIPTION
Adult: 20–22 mm long: head, thorax and abdomen black, but sometimes partly, if not entirely, reddish brown or yellow; antennae and tarsi yellowish orange; wings mainly clear, with brownish margins. **Larva:** up to 50 mm long; head pale yellow, with distinct black eyes; body green, marked with yellow and with a distinct black, blue-centred dorsal line; body distinctly wrinkled and covered with white verrucae; spiracles black.

LIFE HISTORY
Adults occur from May to July, eggs then being laid in the stalks of birch leaves. The larvae feed from July to September. When at rest, they remain curled up on the underside of an expanded leaf and, in spite of their size, often escape detection. Fully fed larvae pupate in large (*c.* 10 × 20 mm), barrel-shaped, reddish-brown cocoons spun on the host plant. Adults emerge in the following spring.

DAMAGE
Larvae cause slight defoliation but damage is of little or no significance.

941 Female large birch sawfly (*Cimbex femoratus*).

942 Larva of large birch sawfly (*Cimbex femoratus*).

Family **DIPRIONIDAE**

Slow-moving, stout-bodied sawflies, the males with antennae strongly bipectinate; antennae of females distinctly toothed. Larvae are associated with coniferous plants.

Diprion pini (Linnaeus) (**943–944**)
Pine sawfly
A locally common pest of pine (*Pinus*), sometimes associated with ornamentals. Widely distributed in central and northern Europe.

DESCRIPTION
Adult female: 10 mm long; head and thorax mainly dark brown; abdomen pale yellowish brown, marked with black; antennae serrated. **Adult male:** 8–9 mm long; blackish brown, with partly paler legs. **Larva:** up to 25 mm long; head brown; body pale yellowish to yellowish green, but darker dorsally, with a row of black spots along each side.

LIFE HISTORY
Adults are active in May or June. Females eventually deposit rows of eggs in slits made in the previous year's needles. Larvae emerge 2–3 weeks later and then feed communally on the needles. If disturbed, members of the colony adopt a threatening posture, grasping the leaf edge with the thoracic legs and extending the abdomen upwards in the shape of an 'S'. The larvae are fully fed in July. They then spin cocoons on the host plant or on the ground. A second generation of adults appears in late July or August. Larvae of this generation feed during the later summer; in September or early October, when fully grown, they enter the soil or leaf litter to spin cocoons. Pupation occurs in the following spring.

DAMAGE
Larvae devour the needles and may also browse on the bark of the young shoots. Attacked shoots are often stripped of needles, and small trees may be completely defoliated.

943 Female pine sawfly (*Diprion pini*).

944 Larva of pine sawfly (*Diprion pini*).

Neodiprion sertifer (Geoffroy in Fourcroy) (**945**)
syn. *N. rufus* (Latreille)
Fox-coloured sawfly

A locally common and destructive pest of pine (*Pinus*), often attacking young ornamental and amenity trees. Eurasiatic, widespread in Europe; also present in North America.

DESCRIPTION

Adult female: 8–10 mm long; mainly brownish to reddish yellow, partly marked with black. **Adult male:** 7–9 mm long; mainly black, the underside partly brownish red; legs brownish yellow. **Larva:** up to 25 mm long; head shiny black; body dirty greyish green, with a diffuse, black, lateral stripe and small, black verrucae.

LIFE HISTORY

This species overwinters in the egg stage. The larvae commence feeding in May, forming small groups at the tips of the needles immediately below the opening buds. If disturbed, the larvae jerk the anterior end of the body over their backs, members of the group typically acting in unison. The larvae feed gregariously on the older needles throughout their development, often causing extensive damage. When fully grown, in late June or July, larvae enter the soil to pupate, each in a tough, oval cocoon. Adults appear in September or October, eggs then being laid in rows in the edges of the needles.

DAMAGE

Needles produced in the previous year are chewed down to the basal stalks, leaving unsightly lengths of bare wood.

Family **TENTHREDINIDAE**

The main family of sawflies with, in females, a characteristic, saw-like ovipositor; antennae with 7–15 (but usually 9) segments. Larvae are mainly free-living, often feeding gregariously on the leaves of trees, shrubs and herbaceous plants; they usually possess 6–8 pairs of abdominal prolegs which, unlike those of lepidopterous larvae, lack crotchets.

Allantus cinctus (Linnaeus) (**946**)
syn. *Emphytus cinctus* (Linnaeus)
Banded rose sawfly

An often common pest of rose (*Rosa*). Widely distributed in Europe; also present in North America.

DESCRIPTION

Adult: 7–10 mm long; mainly shiny black, female with a distinct white or creamy-white band on the fifth abdominal segment; body elongate; wings pale yellow, with brown veins. **Larva:** up to 15 mm long; head pale yellowish brown; body greyish green above, with numerous white verrucae on the back, whitish below.

LIFE HISTORY

Adults occur in May or early June, flying strongly in sunny weather. Eggs are laid in leaves of host plants, usually one or two per leaf. The eggs swell considerably after laying, causing conspicuous bulges on the upper surface of the leaves. They hatch about two weeks later. The larvae feed from July onwards, curling into a ball on the underside of a leaf when at rest and dropping to the ground if disturbed. At first they browse on the underside of the leaves, the upper surface remaining intact. Later, they bite right through the leaf blade, usually feeding along the leaf edge. The larvae are fully

945 Larvae of fox-coloured sawfly (*Neodiprion sertifer*).

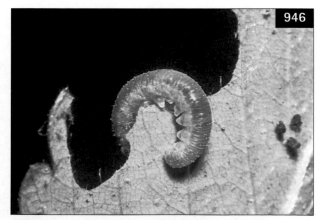

946 Larva of banded rose sawfly (*Allantus cinctus*).

fed in about three weeks; individuals then burrow into decaying wood or the pith of pruned shoots to pupate, each in a flimsy, semitransparent, greenish cocoon. Adults emerge a few weeks later. Larvae of a second generation feed from August to September or October. They then spin cocoons but do not pupate until the following spring.

DAMAGE
Although loss of leaf tissue disfigures host plants, and might cause concern, the extent of damage is usually insufficient to affect plant growth.

947 Larva of *Allantus togatus*.

948 Larva of *Allantus viennensis*.

Allantus togatus (Panzer) (947)

Widely but locally distributed on oak (*Quercus*); less frequently found on birch (*Betula*) and broad-leaved willows such as grey willow (*Salix cinerea*) and pussy willow (*S. caprea*). Larvae feed on the leaves in the summer and early autumn; they are sometimes found on young trees but, owing to their late appearance, do not cause significant damage. Fully grown individuals (*c.* 15 mm long) are dirty greenish to greyish white, with several small, white verrucae on the back and an orange-yellow head; when at rest they curl into a ball on the underside of the leaves, dropping to the ground if disturbed. Fully fed individuals overwinter as prepupae in rotting wood. Adults (8–9 mm long) are mainly black, marked with yellowish white, and possess partly cloudy wings; they occur from June to August.

Allantus viennensis (Schrank) (948)

This species occurs in mainland Europe but not in the British Isles, the larvae infesting wild and cultivated rose (*Rosa*). Eggs are laid in the young shoots, usually some distance from the tip. The young larvae feed gregariously on the youngest leaves, forming distinctive circular holes in the leaf blades. In their final feeding stage they move downwards (then tending to occur singly), often invading other branches. They then typically devour tissue around the leaf margins. Larvae are fully fed in 2–3 weeks. They then pupate in snags or broken branches on the host plant. Individuals of the final generation overwinter. Most adults appear in the spring, but the emergence of some is delayed until the summer. The larvae are relatively large (up to 20 mm long), with an orange-yellow head and a green to bluish-green body, and the first, second and fourth fold of each segment marked with prominent white verrucae. Larvae may be found from June onwards with, in favourable situations, up to three generations annually.

Amauronematus leucolaenus (Zaddach) (949–950)
syn. *A. saarineni* Lindqvist

Locally common on willows, including common sallow (*Salix atrocinerea*), eared willow (*S. aurita*) and grey willow (*S. cinerea*). Eurasiatic. Widespread in central and northern Europe.

DESCRIPTION
Adult: 5.5–6.5 mm long; mainly black, with pale legs.
Larva: up to 18 mm long; elongate, green and shiny, with small black verrucae; tracheae conspicuous.
Prepupa: green to yellowish green, with small, black verrucae.

949 Adult of *Amauronematus leucolaenus.*

950 Larva of *Amauronematus leucolaenus.*

LIFE HISTORY

Adults of the genus *Amauronematus* are active early in the spring, those of *A. leucolaenus* occurring from March to April or early May. Eggs are then deposited in the leaves of willow. Larvae feed mainly in April and early May. They lie stretched out on the leaves, blending well with their surroundings, often several on each infested shoot. Fully grown larvae moult to a non-feeding prepupal stage and enter the soil. Here they overwinter, each in a dark brown cocoon. Pupation occurs shortly before the adults emerge.

DAMAGE

Larvae cause some defoliation. However, as this occurs relatively early in the season, plants are well able to compensate and growth is not affected.

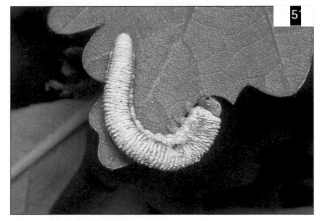

951 Larva of *Apethymus filiformis.*

Apethymus filiformis (Klug) (**951**)

syn. *A. abdominalis* (Lepeletier); *A. autumnalis* Forsius

A minor, locally common pest of oak (*Quercus*). Widely distributed in central and northern Europe.

DESCRIPTION

Adult: 9–11 mm long; shiny black, with a mainly yellow abdomen; antennae black; legs yellow. **Larva:** up to 19 mm long; head dark grey above but yellow below, covered in white, waxy powder; body light green, and mealy coated. **Prepupa:** yellowish green and very shiny, with a mainly yellow head.

LIFE HISTORY

Unlike all other known European members of the family Tenthredinidae, adults of the genus *Apethymus* occur only in the late summer or autumn, individuals of *A. filiformis* occurring from September to November.

Eggs laid in the autumn on host plants hatch in the following May or early June. Larvae are then common on young oak leaves in June, often curling into a ball when at rest. They usually complete their development in July, fully grown individuals moulting to an active non-feeding prepupal stage and entering the soil to pupate but without forming a cocoon. Adults emerge a few weeks later.

DAMAGE

Larvae browse the foliage but damage is not significant.

Apethymus serotinus (Müller) (952)

syn. *A. braccatus* (Gmelin in Linnaeus);
A. tibialis (Panzer)

This widely distributed but local species is associated with oak (*Quercus*). Larvae feed on the foliage from May to July, and commonly rest fully exposed on the upper surface of expanded leaves. Fully grown individuals are 15–18 mm long, with a shiny black head and the much wrinkled body dark grey above and light grey to yellowish grey at the sides and below. Adults (8–10 mm long) are mainly black, with the hind tibiae white basally and the sixth to eighth antennal segments also usually white. The adults occur from August to October.

952 Larva of *Apethymus serotinus.*

Ardis brunniventris (Hartig) (953–954)

syn. *A. bipunctata* (Cameron)
Rose shoot sawfly

A locally common and often destructive pest of rose (*Rosa*). Widely distributed in Europe; also present in North America.

DESCRIPTION

Adult: 5.5–6.5 mm long; black, with pronotum, tegulae and legs partly yellowish white. **Larva:** up to 12 mm long; brownish white, with a light brown head; legs poorly developed.

LIFE HISTORY

Adults occur from May onwards, the period of emergence being very protracted (often giving the impression of two generations) and also varying considerably from year to year. Eggs are laid in leaf tissue on the young terminal buds, and hatch a few days later. Each young larva feeds briefly on the leaf tissue but soon bores into the shoot tip, typically one per infested shoot. Feeding continues within the shoot for about three weeks, the larva burrowing downwards for a few centimetres and expelling considerable quantities of wet, black frass through the original entry hole. When fully grown the larva vacates the shoot and drops to the ground, leaving behind a characteristic exit hole at the base of the feeding gallery. Larvae overwinter in earthen cells and pupate in the spring.

DAMAGE

Infested shoot tips wilt and die, affected tissue turning black. Growth of young bushes is greatly affected, and loss of terminal shoots results in the development of lateral shoots. Damage is most severe on nursery rootstock.

Blennocampa pusilla (Klug) (955–958)

Leaf-rolling rose sawfly

Generally common on wild and cultivated rose (*Rosa*); often an important pest in nurseries and gardens. Present throughout Europe.

DESCRIPTION

Adult: 3.0–4.5 mm long; mainly black, with whitish knees, tibiae and tarsi. **Larva:** up to 10 mm long; head brown; body light green, with short, spiny hairs on the back.

LIFE HISTORY

Adults are active during May and June, females depositing eggs singly in the underside of rose leaves (leaflets), close to the leaf margin. Females also use their ovipositor to probe the tissue on either side of the midrib; this causes the leaf blade to roll tightly inwards to form a protective tube in which the larva eventually feeds; leaves in which eggs have not been laid also become rolled if they have been probed by the female. The larvae feed from late May or June to July or August. They then enter the soil to overwinter in cocoons. Pupation occurs in the spring. There is just one generation annually.

DAMAGE

The presence of rolled, drooping leaves spoils the appearance of bushes and the quality of nursery stock; severe attacks also reduce plant vigour. Certain cultivars, including various climbing roses, are attacked more frequently than others. Infestations are uncommon on standard-grown bushes.

953 Larva of *Ardis brunniventris*.

954 *Ardis brunniventris* larval exit hole in shoot of *Rosa*.

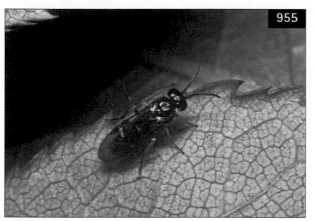

955 Leaf-rolling rose sawfly (*Blennocampa pusilla*).

956 Larva of leaf-rolling rose sawfly (*Blennocampa pusilla*).

957 Larval habitations of leaf-rolling rose sawfly (*Blennocampa pusilla*) on *Rosa*.

958 Leaf-rolling rose sawfly (*Blennocampa pusilla*) damage to foliage of *Rosa*.

959 Female oak slug sawfly (*Caliroa annulipes*).

960 Oak slugworm (*Caliroa annulipes*).

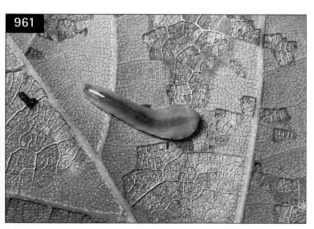

961 Young oak slugworm (*Caliroa annulipes*).

Caliroa annulipes (Klug) (959–961)

Oak slug sawfly

larva = oak slugworm

A generally common pest of oak (*Quercus*); infestations also occur on lime (*Tilia*) and, less frequently, birch (*Betula*), beech (*Fagus sylvatica*), common sallow (*Salix atrocinerea*), grey willow (*S. cinerea*) and other broad-leaved willows. Widespread in central and northern Europe.

DESCRIPTION

Adult: 7–8 mm long; body black; wings blackish with an iridescent sheen, but distinctly subhyaline apically; females with base of the hind tibiae and basitarsi white. **Larva:** up to 12 mm long; head blackish brown, with the anterior third yellowish brown; body pear-shaped, pale yellowish, covered in yellowish slime, shiny and translucent; the gut contents clearly visible; young larva whitish and translucent, with a black head.

LIFE HISTORY

Adults appear in the spring, the females depositing eggs in the leaves of oak and other host plants. The larvae feed in groups on the underside of the leaves during May and June. Eventually, they enter the soil and pupate in tough, blackish cocoons. Larvae of a second generation feed in late July and August. In favourable situations there may be a further generation in the autumn.

DAMAGE

Foliage is grazed from below, the upper epidermis remaining intact but turning brownish or whitish. Although leaves are not distorted and attacks have little or no effect on plant growth, the presence of damaged leaves on nursery or ornamental trees may be undesirable.

Caliroa cerasi (Linnaeus) (962–963)

syn. *C. limacina* (Retzius)

Pear slug sawfly

larva = pear & cherry slugworm

A locally common pest of rosaceous trees and shrubs, including, almond (*Prunus dulcis*), flowering cherry (*Prunus*), hawthorn (*Crataegus*), japonica (*Chaenomeles speciosa*), ornamental pear (*Pyrus calleryana* 'Chanticleer') and rowan (*Sorbus aucuparia*). Eurasiatic. Present throughout Europe; accidentally introduced into many other parts of the world, including Africa, Australasia, North and South America.

DESCRIPTION

Adult: 4–6 mm long; black and shiny. **Larva:** up to 10 mm long; greenish yellow to yellowish orange but covered in olive-black, shiny slime; body pear-shaped, tapering towards the hind end; head and legs inconspicuous.

LIFE HISTORY

Adults appear in late May and June. Eggs are then laid in small slits cut into the underside of leaves, often several on the same leaf. The larvae rest and feed whilst fully exposed on the upper surface of expanded leaves, browsing away the epidermis but without biting through the entire blade. Larvae are fully grown in July. They then pupate in small, black cocoons formed in the soil about 10 cm below the surface. Adults emerge a week or two later. There are usually two, sometimes three, generations each year, the larvae occurring throughout the summer and early autumn. Reproduction is parthenogenetic, males being very rare.

DAMAGE

Leaf browsing is often extensive, and very disfiguring to ornamentals. Severe infestations cause premature leaf fall and affect the growth of plants in the following season.

Caliroa cinxia (Klug) (964)

This species occurs on oak (*Quercus*) in various parts of mainland Europe; also present in central and southern England. The larvae browse on the underside of the leaves, damage being identical to that caused by *Caliroa annulipes*. *C. cinxia* is single brooded, infestations being most evident in the autumn from September to October. Adults are mainly black, and distinguished from those of *C. annulipes* by their less cloudy wings and (in females) by the less extensive area of white on the hind legs; the larvae of both species are also similar but those of *C. cinxia* have a uniformly reddish-brown head.

962 Pear & cherry slugworm (*Caliroa cerasi*).

963 Pear & cherry slugworm (*Caliroa cerasi*) damage to leaves of *Sorbus aucuparia*.

964 Young larvae of *Caliroa cinxia*.

Cladius difformis (Panzer) (965–967)

Lesser antler sawfly

Widespread and common on wild and cultivated rose (*Rosa*), the larvae also feeding on marsh cinquefoil (*Potentilla palustris*), meadowsweet (*Filipendula ulmaria*) and strawberry (*Fragaria*). Present throughout Europe; also occurs in North America, but probably introduced.

DESCRIPTION

Adult: 5–7 mm long; body black; legs yellowish white; antennae of male with characteristic, long projections on the two basal segments and shorter projections on the third and fourth segments; female with slight projections on the first and second antennal segments. **Larva:** up to 12 mm long; somewhat flattened and distinctly hairy; head yellowish brown; body translucent, yellowish to light green, with darker subdorsal stripes; young larva paler, with a blackish head.

965 Female lesser antler sawfly (*Cladius difformis*).

LIFE HISTORY

Adults occur in May, depositing eggs singly in the leaf stalks. After egg hatch the larvae browse on the underside of leaves, most often attacking fully expanded leaflets. Larvae are fully grown in four or five weeks and then pupate in thin, double-walled, brownish cocoons formed on the leaves or amongst debris on the ground. Adults emerge two or three weeks later, usually in late July or August. Larvae of a second generation feed during August and September, eventually overwintering in cocoons and pupating in the spring.

DAMAGE

At first, larvae 'window' the leaves, the upper epidermis remaining intact; older larvae make holes right through the leaf blade and also browse on the leaf edges. Such damage is disfiguring but of significance only if extensive.

Cladius pectinicornis (Geoffroy in Fourcroy)

Antler sawfly

This abundant sawfly occurs widely in Europe and parts of Asia, the larvae often attacking cultivated rose (*Rosa*). Adults of *Cladius pectinicornis* are distinguished from those of *C. difformis* (with difficulty in females) by the longer and increased number of antennal projections. Both species have a similar lifecycle and cause similar damage.

Claremontia waldheimii (Gimmerthal)

syn. *C. subcana* (Zaddach); *C. subserrata* (Thomson)

Larvae of this widely distributed species infest cultivated *Geum* during the summer months. They are very similar in appearance to those of the more frequently encountered species *Monophadnoides geniculatus* (p. 377) but slightly larger (up to 15 mm long). Adults (5.5–6.5 mm long) are shiny black with mainly white tibiae. They occur in May and June.

966 Larva of lesser antler sawfly (*Cladius difformis*).

967 Young larva of lesser antler sawfly (*Cladius difformis*).

968 Hazel sawfly (*Croesus septentrionalis*).

969 Larva of hazel sawfly (*Croesus septentrionalis*).

Croesus septentrionalis (Linnaeus) (**968–969**)
Hazel sawfly

A common and often serious pest of ornamental trees and shrubs, especially alder (*Alnus*), birch (*Betula*), hazel (*Corylus*), poplar (*Populus*), rowan (*Sorbus aucuparia*) and willow (*Salix*). Eurasiatic. Widespread in Europe; also present in parts of Asia.

DESCRIPTION

Adult: 8–10 mm long; head and thorax black; abdomen with the basal two and the apical two or three segments black, the rest reddish brown; hind basitarsus and tip of hind tibia greatly expanded; wings mainly clear but apex of each fore wing cloudy. **Larva:** up to 22 mm long; head shiny black; body yellowish to bluish green, variably marked with orange-yellow, and with prominent black patches along the sides.

LIFE HISTORY

Adults appear in May and June. Eggs are deposited in the leaf veins of host plants and hatch about two weeks later. The larvae then feed gregariously along the leaf edge, grasping on with their thoracic legs and arching the body over the head. If disturbed, they thrash their bodies violently in the air. Fully fed larvae pupate in the soil, each spinning an elongate, brown cocoon a short distance below the surface. Larvae occur from mid-June onwards, those of the second generation, in late summer and autumn, being most numerous.

DAMAGE

The larvae are voracious feeders and cause considerable defoliation, often completely stripping the foliage from small trees.

Croesus latipes (Villaret) (**970**)

A widespread but usually uncommon species on birch (*Betula*). Infestations occur from May or June onwards, the larvae feeding gregariously and causing noticeable defoliation; damage sometimes occurs on nursery trees. Fully grown larvae are 15–20 mm long, and shiny black to shiny brownish black, with the legs and underside of the last few abdominal segments pale yellowish. Adults are similar to those of *Croesus septentrionalis* (see above) but slightly smaller (7.5–10 mm long), with the cloudy apex of the fore wings less extensive and the hind femora of females reddish below. They occur from May to June and, in favourable districts where there are two generations annually, also from July to August.

970 Larva of *Croesus latipes*.

Dineura stilata (Klug) (**971**)

A generally common but minor pest of hawthorn (*Crataegus*), sometimes present on nursery stock and garden hedges. Widely distributed throughout central and northern Europe.

DESCRIPTION
Adult: 5–6 mm long; head black; thorax black, with reddish marks; abdomen black and yellow, and relatively narrow; legs yellowish. **Larva:** up to 14 mm long; head yellowish green; body green, semi-translucent, slightly flattened and distinctly tapered from front to rear; head and body bear short, whitish hairs.

971 Larva of *Dineura stilata*.

LIFE HISTORY
The larvae feed on hawthorn from August to October, lying pressed flat against the leaf surface. They often occur in small groups, typically browsing away the upper epidermis and leaving the lower surface intact. When fully fed, larvae enter the soil, to spin single-walled cocoons; individuals pupate in the following spring. Adults emerge in late May or June.

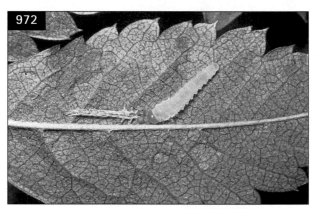

972 Recently moulted larva of *Dineura testaceipes*.

DAMAGE
Infested leaves become noticeably discoloured where the upper epidermis is removed but damage has no effect on plant growth.

Dineura testaceipes (Klug) (**972**)

A widespread pest of rowan (*Sorbus aucuparia*), the light green larvae feeding on the leaves during the summer months. Minor infestations sometimes occur on ornamental trees, the leaves being disfigured by pale patches, but the insect is more numerous on wild hosts.

Dineura viridorsata (Retsius in Degeer) (**973**)

Although most abundant in birch woods, this widely distributed sawfly is also a common pest of ornamental and amenity birch (*Betula*) trees. The larvae (*c.* 16 mm long when fully grown) are greenish white, tapered and flattened, with a brownish head. They rest during the daytime on the underside of leaves, where the cast skins of earlier instars are often present. Feeding takes place at night, the larvae removing the upper epidermis of the leaves to produce distinct whitish blotches; such damage is often extensive but is of little or no significance. When fully grown, larvae enter the soil to overwinter in black, thin-walled cocoons. Larvae pupate in the spring. Adults occur in late May or June. They are 5–8 mm long and mainly black with a yellow abdomen.

973 Larva of *Dineura viridorsata*.

974 Larva of rose slug sawfly (*Endelomyia aethiops*).

975 Rose slug sawfly (*Endelomyia aethiops*) damage to leaf of *Rosa*.

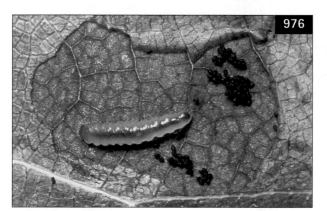

976 Larva of *Fenusa dohrnii*.

977 Mines of *Fenusa dohrnii* in leaf of *Alnus*.

Endelomyia aethiops (Fabricius) (974–975)
syn. *E. nigricolle* (Cameron); *E. rosae* (Harris)
Rose slug sawfly
An often common pest of rose (*Rosa*). Widespread throughout Europe; also present in North America.

DESCRIPTION
Adult: 4–5 mm long; black, with smoky wings; legs mainly black, with the tibiae and knees of the fore and middle legs yellowish white. **Larva:** up to 15 mm long; head yellowish brown; body yellow and translucent, the gut contents imparting a greenish tinge.

LIFE HISTORY
This species is mainly parthenogenetic. Adult females occur from May to June, depositing eggs on the underside of rose leaves in slits cut close to the edges. The eggs hatch in 1–2 weeks and the larvae then feed on the upper surface, leaving the lower surface more or less intact. Larvae are fully fed in 3–4 weeks; they then enter the soil, usually in late June or early July, to construct silken cocoons in which to overwinter. Individuals pupate in the spring, shortly before the emergence of

adults, but development is sometimes delayed for a further year.

DAMAGE
Infested leaves become extensively blanched, the damaged areas eventually turning brown and shrivelling up. Severely infested bushes appear scorched and their growth is checked.

Fenusa dohrnii (Tischbein) (976–977)
syn. *F. melanopoda* (Cameron); *F. westwoodi* (Cameron)
A generally common pest of alder (*Alnus*). Holarctic. Widespread in central and northern Europe.

DESCRIPTION
Adult: 3–4 mm long; body black and shiny, the abdomen stumpy; wings blackish. **Larva:** 8–10 mm long; whitish, with a light brown head and prothoracic plate; a large black plate present between the first pair of legs, and a small black spot sited between the second and third pairs, these markings absent after the last moult.

978 Larva of *Fenusa pusilla*.

979 Composite mine of *Fenusa pusilla* in leaf of *Betula pendula* 'Dalecarlica'.

LIFE HISTORY

The larvae occur from July to October. They mine within the leaves of alder, the galleries frequently commencing close to the midrib (cf. *Heterarthrus vagans*, p. 373); there are often several mines in the same leaf. There are usually two generations annually, adults appearing in May or June and in August; in favourable conditions there may be a third generation. Reproduction is parthenogenetic.

DAMAGE

The feeding galleries disfigure the foliage and, if numerous, cause significant defoliation.

Fenusa pusilla (Lepeletier) (**978–979**)

syn. *F. pumila* (Klug)

A locally common and sometimes important pest of birch (*Betula*), especially downy birch (*B. pubescens*). Holarctic, but probably introduced to North America. Present throughout Europe.

DESCRIPTION

Adult: 2.5–3.5 mm long; mainly shiny black, the legs with brownish tibiae and tarsi; antennae short. **Larva:** up to 8 mm long; flattened and tapered from front to rear; head brownish; body white and translucent, with a greenish tinge, finally becoming yellowish white; body before final moult with a black dumbbell-shaped ventral mark on the first thoracic segment and blackish or brownish dots on the following two, three or four segments; prolegs are lacking on the anal segment.

LIFE HISTORY

Adults occur mainly in May and June, and in July and August. Eggs are laid in leaves of birch, usually towards the midrib in the tissue between two major lateral veins, the females tending to select young leaves on the new

shoots. The larvae feed in kidney-shaped blotches, infested leaves commonly containing several mines which either remain separated throughout their development or eventually unite into a common gallery which may occupy the whole leaf. Fully grown larvae pupate in the soil without forming cocoons. Occupied mines occur from June to October, there being two or more generations per year.

DAMAGE

Infested leaves are disfigured by the larval mines. Heavily infested foliage turns completely brown and drops prematurely, reducing the vigour of affected shoots.

Fenusa ulmi Sundewall (**980**)

Elm leaf-mining sawfly

A generally common pest of elm (*Ulmus*). Holarctic but probably introduced into North America. Widely distributed in Europe.

DESCRIPTION

Adult: 3–4 mm long; black and shiny, with short antennae and slightly smoky wings. **Larva:** up to 10 mm long; white to yellowish white, with a light brown head; legs banded with brown; a ventral plate on the first and a black dot on several of the following segments.

LIFE HISTORY

This insect is mainly parthenogenetic, adult females occurring from late April to June and in August. The larvae feed in blister mines formed in the leaves of elm from June to July, and in September. The mines are brown and tend to extend between the major veins without crossing the midrib; there are usually several in each infested leaf.

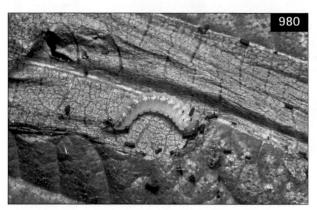

980 Larva of elm leaf-mining sawfly (*Fenusa ulmi*).

981 Larva of *Heterarthrus vagans*.

DAMAGE

The mines cause slight distortion of the leaf blade but, even when numerous, have little or no effect on growth.

Hemichroa crocea (Geoffroy in Fourcroy)

A sporadically abundant pest of alder (*Alnus*) and birch (*Betula*). Widely distributed in Europe; also occurs in North America.

DESCRIPTION

Adult female: 5–8 mm long; reddish yellow, with antennae, part of thorax and legs black. **Larva:** up to 20 mm long; head light brown to black; body mainly greyish green, with the first thoracic segment yellowish orange, and with a black longitudinal line above the spiracles and two lines of black spots above the legs.

982 Mines of *Heterarthrus vagans* in leaf of *Alnus*.

LIFE HISTORY

Adults of this mainly parthenogenetic species appear from May onwards. Eggs are laid in alder or birch leaves, along either side of a leaf stalk. After egg hatch, the larvae move onto the leaf blade to feed. Older larvae feed in company along the leaf edges but they become less gregarious as they mature. When fully grown, the larvae enter the soil to pupate, each in a dark brown, single-walled cocoon. This species is usually double brooded but, in favourable conditions, there may be a third generation.

DAMAGE

Young larvae form distinctive sigmoid-shaped holes in the leaf blades. Older larvae devour most of the leaf tissue, apart form the major veins, and are capable of causing considerable defoliation; greatest damage is caused to alder, especially red alder (*Alnus rubra*).

Heterarthrus vagans (Fallén) (**981–982**)

An often abundant pest of alder (*Alnus*), including ornamentals and trees planted as windbreaks. Widespread in central and northern Europe.

DESCRIPTION

Adult: 3–5 mm long; mainly blackish, the abdomen orange-yellow below; legs yellowish; wings smoky. **Larva:** up to 9 mm long; head small, brown and pointed anteriorly; body flattened and tapering back from a relatively broad thoracic region; whitish and translucent, with the green gut contents clearly visible; distinctive black markings present between the thoracic legs (except in final instar); legs very small, the anal pair of abdominal prolegs fused.

983 Mines of *Heterarthrus aceris* in leaf of *Acer*.

984 Larva of *Heterarthrus microcephalus*.

985 Mine of *Heterarthrus microcephalus* in leaf of *Salix*.

LIFE HISTORY

Adults occur in the spring, depositing eggs in the leaves of alder. The larvae then feed within distinctive brownish blotch mines in the leaves. There are commonly two or three mines on each infested leaf; these commence their development separately close to the leaf edge (cf. *Fenusa dohrnii*, pp. 371–2), but may eventually coalesce to form one large chamber. When feeding, the larvae remove all the tissue between the upper and lower leaf surfaces, and eject frass through a small slit made in the wall of the gallery; occupied mines occur in June and July, and the larvae are clearly visible if a mined leaf is held up to the light. Pupation occurs within the mine in a brown, flat, disc-like cocoon. A second generation of adults appears in the summer, with a second and usually larger brood of larvae feeding in the autumn.

DAMAGE

Infested leaves are extensively disfigured, spoiling the appearance of specimen trees. Heavy attacks in the autumn lead to premature leaf fall.

Heterarthrus aceris (Kaltenbach) (**983**)

Locally common on field maple (*Acer campestre*) and sycamore (*A. pseudoplatanus*). Adults (3.5–4.5 mm long) are mainly black, with white legs; they occur in May and June, eggs being laid singly in the tips of the leaf lobes. The whitish larvae mine the leaves during the summer, each forming a large, brown blotch. When fully fed, each larva cuts out a circular section from the upper cuticle of its feeding gallery, to which it attaches a flat, disc-like cocoon. These habitations drop to the ground before leaf fall, the occupants overwintering and eventually pupating shortly before the emergence of the adults. By flexing their bodies, individuals are capable of propelling their cocoons; hence the colloquial name 'jerking disc sawfly'. Although usually of little significance as a pest, severe outbreaks of this mainly parthenogenetic sawfly have occurred, with extensive damage reported on ornamental maples growing as amenity trees.

Heterarthrus microcephalus (Klug) (**984–985**)

A generally common species, associated with various kinds of willow (*Salix*) including, occasionally, nursery and ornamental trees. The whitish larvae feed mainly from July or August to September or October, forming brown blotch mines on the leaves. Adults (3–5 mm long) are black and shiny, with pale yellow legs; they occur from May to July.

Macrophya punctumalbum (Linnaeus) (986–988)

In continental Europe, an important pest of ash (*Fraxinus excelsior*) growing as shade trees; infestations also occur on privet (*Ligustrum vulgare*) and lilac (*Syringa*), and are sometimes of significance in nurseries. Present throughout Europe; locally common in Britain, particularly in southern England. An introduced pest in North America.

DESCRIPTION

Adult female: 7–8 mm long; mainly black with two white marks on the prothorax, a white scutellum, and white marks on the sides of the abdomen; legs mainly black with partly white tibiae and bright red hind femora. **Larva:** up to 16 mm long; dull green with a yellowish-brown head; young larva yellowish green to whitish green, with a green head.

LIFE HISTORY

Adults occur in May and June, but males are extremely rare and reproduction is normally parthenogenetic. Eggs are deposited in the leaves of host plants from June onwards, larvae feeding throughout the summer and becoming fully grown by the autumn. Individuals then enter the soil and eventually pupate in dark brown cocoons. Adults appear in the following spring.

DAMAGE

Larvae cause considerable defoliation, attacked leaves becoming extensively holed and often totally destroyed. Adults are also responsible for damage, macerating leaves with their jaws.

986 Adult female of *Macrophya punctumalbum.*

987 Larva of *Macrophya punctumalbum.*

988 *Macrophya punctumalbum* damage to leaves of *Syringa.*

Messa hortulana (Klug) (**989–990**)

A locally distributed, leaf-mining and mainly parthenogenetic sawfly associated with black poplar (*Populus nigra*), infestations occurring occasionally on ornamental trees. The larvae (up to 10 mm long) are whitish and shiny, with a brownish-black head, a blackish prothoracic plate and black plates between the thoracic legs. They occur from May or June to July, each feeding within a brown, blister-like, frass-filled blotch. Mines develop from the leaf margin, and cause slight distortion of the leaf blade. If attacks are heavy, several eggs may be deposited in the same leaf, mines then eventually merging to occupy most if not all of the available tissue. Larvae are fully fed in about four weeks. They then vacate their mines to pupate in cocoons formed in the ground. The adult females are 4.0–4.5 mm long, stout-bodied and mainly black, with white tegulae and yellowish-white knees, tibiae and tarsi. They usually emerge in May, there being just one generation annually.

Messa nana (Klug) (**991**)

syn. *M. quercus* (Cameron)

Infestations of this widely distributed, mainly parthenogenetic species occur on birch (*Betula*), especially downy birch (*B. pubescens*). The larvae (up to 8 mm long) are whitish to yellowish white, with a brown head and prothoracic plate, and black plates between the thoracic legs. They form brown, frass-filled blotches on the leaves, feeding from July or August to September. Adults are similar to those of the previous species; they appear in May and June.

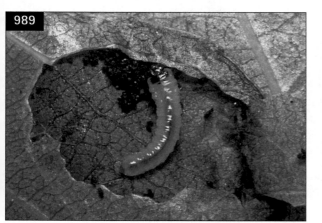

989 Larva of *Messa hortulana.*

990 Mine of *Messa hortulana* in leaf of *Populus.*

991 Mine of *Messa nana* in leaf of *Betula.*

Metallus gei (Brischke) (**992–993**)
Geum leaf-mining sawfly

Generally common on wild and cultivated *Geum*, and frequently a pest in gardens and nurseries. Widely distributed in Europe.

DESCRIPTION
Adult female: 3.5–4.5 mm long; body mainly black but the abdomen partly brown; hind legs mainly white. **Larva:** pale greenish white, with a dark head, prothoracic plate and anal plate.

LIFE HISTORY
Adult females of this parthenogenetic species occur in May or early June, with a second generation in July and August. The larvae are leaf miners, each forming an expansive, light brown blotch on the upper side of a fully expanded leaf of geum. The larvae feed from June to July and from August to September or October, fully grown individuals eventually pupating in cocoons in the soil. In contrast with many other leaf miners, larval frass does not accumulate in the mine but is ejected through a small slit in the wall of the gallery.

DAMAGE
Infested plants are disfigured by the conspicuous mines. There are often several mines in each infested leaf and, if attacks are severe, plants are seriously weakened.

Monophadnoides geniculatus (Hartig) (**994**)
syn. *Blennocampa geniculata* (Hartig)

Geum sawfly

A generally common pest of *Filipendula* and *Geum*; also attacks *Rubus*. Eurasiatic. Widely distributed in Europe.

DESCRIPTION
Adult: 5–6 mm long; stout-bodied and mainly black. **Larva:** up to 14 mm long; head greenish yellow; body green to dark green, with numerous white, branched spines.

LIFE HISTORY
Adults occur in May and early June, females depositing eggs in the underside of leaves of host plants. The larvae feed from late May to early July. They then enter the soil and spin cocoons, individuals eventually pupating and adults emerging in the followings spring. There is just one generation annually.

DAMAGE
Larvae bite out large, irregular holes in the foliage; such damage is often extensive, and heavily infested plants are seriously disfigured and weakened.

992 Mine of geum leaf-mining sawfly (*Metallus gei*).

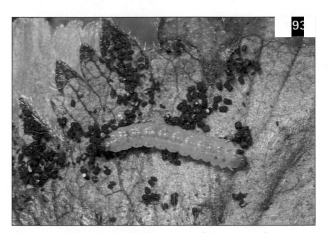

993 Larva of geum leaf-mining sawfly (*Metallus gei*).

994 Larva of geum sawfly (*Monophadnoides geniculatus*).

Nematus melanaspis Hartig (**995**)
syn. *N. maculiger* Cameron
Gregarious poplar sawfly

A generally common pest of poplar (*Populus*); infestations also occur on birch (*Betula*) and willow (*Salix*). Widely distributed throughout central and northern Europe.

DESCRIPTION
Adult: 6–8 mm long; pale yellowish white, the thorax and upper side of the abdomen mainly black; head brownish, with a dark facial mark; antennae short; legs mainly pale, the hind legs partly black. **Larva:** up to 21 mm long; head shiny black; body mainly light green to dull green, with the first thoracic segment orange; three black stripes along the back, a double series of shiny, black verrucae along the sides and another series of verrucae above the legs; anal plate black.

LIFE HISTORY
Adults occur from May to June and from July to August. Eggs are then deposited along the major veins on the underside of leaves of poplar. The larvae feed gregariously from June onwards, members of the second generation completing their development in the autumn. Fully grown larvae pupate in the soil in tough, brown cocoons.

DAMAGE
Infested leaves are reduced to a network of veins, heavy infestations affecting the appearance and vigour of host plants.

Nematus nigricornis Lepeletier (**996**)
A minor pest of poplar (*Populus*), especially black poplar (*P. nigra*), and willow (*Salix*). Widely distributed in central and northern Europe.

DESCRIPTION
Adult: 6–8 mm long; head black and hairy; thorax black, marked with red; abdomen mainly yellow, with a black back; legs reddish. **Larva:** up to 18 mm long; head black; body relatively plump and mainly whitish green to pale yellowish green, with several shiny black verrucae along the back and sides; thoracic legs mainly black.

LIFE HISTORY
Adults occur in two main generations, from May to June and from July to August. Eggs are deposited in the leaf stalks of poplar or willow. The larvae feed gregariously along the edges of leaves, typically lying with the top of the abdomen curled downwards on one side of the leaf blade. They occur from June to September, fully grown individuals eventually pupating in cocoons formed in the soil.

DAMAGE
Larvae devour all but the major leaf veins, and infested branches are significantly defoliated.

995 Larva of gregarious poplar sawfly (*Nematus melanaspis*).

996 Larva of *Nematus nigricornis.*

Nematus pavidus Lepeletier (**997**)

Lesser willow sawfly

An abundant and often destructive pest of willow (*Salix*); infestations also occur on alder (*Alnus*) and poplar (*Populus*). Widely distributed in Europe.

DESCRIPTION

Adult: 6–7 mm long; head and thorax black; abdomen mainly yellow (male usually with black crossbands); antennae black; legs mainly yellowish; wings hyaline with blackish veins, the pterostigma on each fore wing brownish black. **Larva:** up to 20 mm long; head shiny black; body green, with the first thoracic and last three abdominal segments mainly orange; three black stripes along the back, a single series of shiny black verrucae along the sides and another series above the legs.

LIFE HISTORY

Adult sawflies occur in two main generations, from April to June and from August to September. Eggs are deposited in dense groups on the underside of willow leaves. Each female lays either fertilized or unfertilized eggs. The former eventually give rise to females and the latter develop parthenogenetically to produce males. After eggs have hatched, the larvae feed along the edge of the leaves, with the hind part of the body arched upwards. The larvae occur at any time from May to October, and are voracious feeders. Large numbers typically occur together on the same leaf or branch. Fully grown larvae eventually pupate in the soil in tough, brownish cocoons.

DAMAGE

Larvae rapidly reduce leaves to a mere network of veins, heavy infestations affecting the appearance and vigour of plants. Attacks are often very damaging in osier beds.

Nematus salicis (Linnaeus) (**998**)

Willow sawfly

A local pest of willow (*Salix*), especially crack willow (*S. fragilis*) and white willow (*S. alba*). Widely distributed in central and northern Europe.

DESCRIPTION

Adult female: 8–10 mm long; head and thorax black; abdomen and legs mainly yellow; wings hyaline with black veins. **Larva:** up to 30 mm long; head shiny black; body bluish green, with the thoracic and last three abdominal segments brownish orange; numerous black verrucae arranged in rows along the body, and (predominantly on the green segments) five rows of black marks along the back.

LIFE HISTORY

Adults occur most commonly from May to June and from August to September, depositing eggs on the underside of willow leaves. Young larvae feed gregariously but as development progresses they eventually disperse, older individuals usually occurring singly or in pairs. When fully grown, the larvae enter the soil to pupate in brownish-black, double-walled cocoons.

DAMAGE

Larvae devour the tissue of fully expanded leaves but cause far less damage than those of *Nematus pavidus*.

997 Larva of lesser willow sawfly (*Nematus pavidus*).

998 Larva of willow sawfly (*Nematus salicis*).

999 Aruncus sawfly (*Nematus spiraeae*).

1000 Larva of aruncus sawfly (*Nematus spiraeae*).

Nematus spiraeae Zaddach (**999–1000**)
Aruncus sawfly

A common and destructive pest of goat's beard (*Aruncus dioicus*). Widely distributed in central and northern Europe; in Britain, first recorded in 1924 and now often an abundant garden pest.

DESCRIPTION
Adult: 5–6 mm long; yellowish brown, darker above and with head, thorax and antennae brownish black; tegulae pale; wings clear with brown veins. **Egg:** 1 mm long; whitish and capsule-shaped. **Larva:** up to 20 mm long; head brownish green to brown; body green and translucent, with short, pale hairs arising from pale, inconspicuous verrucae.

LIFE HISTORY
Adults of this parthenogenetic sawfly emerge in late April or May, eventually depositing eggs on the underside of leaves of host plants. Eggs hatch in about a week. The larvae feed in groups on the leaves during May and June, rapidly devouring the tissue between the major veins. Individuals are fully grown in 4–5 weeks; they then enter the soil to pupate in silken cocoons. A second generation of adults appears in late July or August, and second-brood larvae feed in August and September; in some seasons there may be a partial third generation. Larvae of the final brood overwinter in silken cocoons, and pupate early in the spring.

DAMAGE
Defoliation is often extensive, the leaves of plants becoming skeletonized with only the major veins remaining.

Nematus tibialis Newman (**1001–1003**)
False acacia sawfly

A minor pest of false acacia (*Robinia pseudoacacia*). Widely distributed in mainland Europe and in England, having been introduced from North America along with its foodplant in the early nineteenth century.

DESCRIPTION
Adult: 6–7 mm long; yellow, marked above with black; antennae black; legs mainly yellow, the hind tibiae and tarsi black. **Larva:** up to 12 mm long; head brownish green marked with black; body green and shiny.

LIFE HISTORY
Adults occur in May and June, depositing eggs in association with the young growth of false acacia. Following egg hatch, the young larvae feed on expanded leaves, each forming a small hole through the leaf blade and resting along the cut edge; later, the larvae devour more of the tissue. Larvae are fully fed in two or three weeks. They then enter the soil and pupate in tough, dark brown cocoons. Adults emerge shortly afterwards. A second generation occurs in the late summer with, in favourable seasons, a partial third developing in the autumn.

DAMAGE
Holed or partially devoured leaves attract attention on small ornamental trees and nursery stock but damage is rarely of any significance.

1001 False acacia sawfly (*Nematus tibialis*).

1002 Larva of false acacia sawfly (*Nematus tibialis*).

1003 Young larva of false acacia sawfly (*Nematus tibialis*).

1004 Larva of *Nematus bergmanni*.

Nematus bergmanni Dahlbom (**1004**)
syn. *N. curtispina* (Thomson)

A generally common species, the larvae feeding on the leaves of willow (*Salix*), especially crack willow (*S. fragilis*) and white willow (*S. alba*). Infestations occur from May to October, but the larvae are rarely numerous and do not cause extensive damage. Individuals (up to 18 mm long) are green, with a broad, translucent-white or pinkish-white dorsal stripe and a pair of darker subdorsal stripes; the head is pale yellowish brown, marked with brownish black, and the tapered body terminates in a pair of reddish cerci. Adults (6–8 mm long) are mainly black, with the lighter parts of the body green; they occur from April onwards.

1005 Larva of *Nematus umbratus.*

1006 Adult of *Nematus umbratus.*

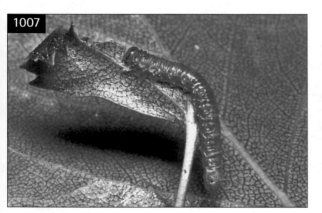

1007 Larva of *Nematus viridescens.*

1008 Adult of *Nematus viridescens.*

Nematus umbratus Thomson (**1005–1006**)

syn. *N. collinus* Cameron; *N. similis* (Forsius)

A gregarious species, associated with birch (*Betula*); sometimes damaging to young nursery or amenity trees but usually uncommon. The larvae (up to 25 mm long) are blackish green to whitish green, shiny and translucent. They feed gregariously from May onwards, causing considerable damage to the foliage. There are two generations annually, the mainly yellowish-orange adults (5.5–7.5 mm long) occurring from May to June and from late July to early September.

Nematus viridescens Cameron (**1007–1008**)

A generally common species, associated with birch (*Betula*); sometimes present on nursery trees and young ornamentals but not an important pest. Adults occur from April onwards. The mainly green larvae feed on the leaves during the summer and autumn. There are two or more generations annually. Adults (6–8 mm long) are mainly green.

Periclista lineolata (Klug) (**1009–1011**)

A locally common but minor pest of oak (*Quercus*), especially English oak (*Q. robur*). Widely distributed in central and northern Europe.

DESCRIPTION

Adult: 5–6 mm long; mainly shiny black with a short greyish or whitish pubescence; tegulae, edge of pronotum, knees and tibiae whitish; wings hyaline and iridescent, with black veins and each fore wing with a black pterostigma. **Larva:** up to 17 mm long; head black; body green with numerous black, mainly Y-shaped spines along the back and sides, and a series of smaller spines (some Y-shaped and some simple) above the legs.

LIFE HISTORY

Adults occur in April and May, eggs then being laid close to the veins of unfurling leaves. The larvae feed on the underside of oak leaves from mid-May to mid-June, making large holes in the leaves. When fully grown they moult to a mobile prepupal stage (which is devoid of

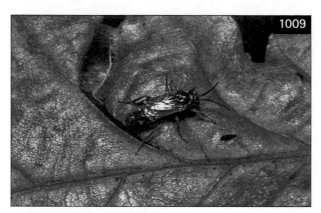

1009 Adult of *Periclista lineolata.*

1010 Larva of *Periclista lineolata.*

1011 *Periclista lineolata* egg pouches in young leaves of *Quercus.*

1012 Adult of *Periclista albida.*

spines), and enter the soil to pupate in a silken cocoon. There is just one generation annually.

DAMAGE
Infestations are most common on young trees, but damage caused is of little or no significance.

Periclista albida (Klug) (**1012–1013**)
syn. *P. melanocephala* (Fabricius)

A widely distributed species in association with oak (*Quercus*). The adult females (5–7 mm long) have a mainly reddish-yellow thorax, a yellowish, brown-spotted abdomen and a black head and black antennae; although the wing veins are black, the pterostigma on each fore wing is yellow. The larvae are green with a black head, the body bearing numerous, mainly black, Y-shaped, thick-stemmed spines. Adults occur in April and May, and the larvae feed on the underside of the expanded leaves from May to June. Pupation takes place in the ground in silken cocoons. Individuals overwinter as prepupae and pupate in the spring.

1013 Larva of *Periclista albida.*

1014 Larva of *Phyllocolpa leucapsis*.

1015 Larval habitation of *Phyllocolpa leucapsis* on leaf of *Salix*.

1016 Larval habitation of *Phyllocolpa leucosticta* on leaf of *Salix*.

Phyllocolpa leucapsis (Tischbein) (**1014–1015**)

A generally common but minor pest of common sallow (*Salix atrocinerea*), eared willow (*S. aurita*), grey willow (*S. cinerea*) and pussy willow (*S. caprea*). Present throughout central and northern Europe.

DESCRIPTION

Adult: 3.5–5.0 mm long; mainly black, with a short, broad abdomen; legs brownish, marked with white or yellow. **Larva:** up to 10 mm long; head reddish brown; body bluish green to greyish, with a pair of small, black cerci; young larva mainly yellowish.

LIFE HISTORY

Adults occur from May onwards, and deposit eggs in the leaves of various broad-leaved willows. The larvae develop during June and July, each within a folded leaf-edge which forms a flattened pouch on the lower surface; a second brood appears later in the season. Fully grown individuals construct brown cocoons in the soil and then pupate, those of the second generation overwintering within their cocoons and producing adults in the spring.

DAMAGE

Heavily infested bushes are disfigured but growth is not impaired.

Phyllocolpa leucosticta (Hartig) (**1016**)

syn. *P. sharpi* (Cameron)

Infestations of this species occur widely on common sallow (*Salix atrocinerea*), grey willow (*S. cinerea*) and other broad-leaved willows, the larvae feeding within elongate folds along the leaf edges. The young larva is whitish and later becomes greenish white with a blackish to brownish head. Larvae are slightly larger than those of the previous species; adults (4.5–5.5 mm long) are distinguished by their entirely reddish-yellow hind femora.

Phymatocera aterrima (Klug) (**1017–1019**)

syn. *P. robinsoni* (Curtis)

Solomon's seal sawfly

A destructive pest of Solomon's seals (*Polygonatum*), especially *P. multiflorum*. Widely distributed in central and southern Europe. In Britain, first reported in London in 1846 and now common throughout much of England and Wales.

DESCRIPTION

Adult: 8–9 mm long; stout-bodied, black and shiny, with moderately long antennae and smoky, iridescent wings. **Larva:** up to 20 mm long; head black, but

1017 Larva of Solomon's seal sawfly (*Phymatocera aterrima*).

1018 Young larvae of Solomon's seal sawfly (*Phymatocera aterrima*).

inconspicuous; body greyish white and wrinkled, bearing numerous small, black, spinose verrucae; thoracic legs black; young larva greenish yellow, with a prominent black head.

LIFE HISTORY
Adults occur in May and June. Eggs are then deposited in the leaf stalks of host plants and hatch in June. Larvae then feed gregariously on the underside of the expanded leaves, forming elongate holes between the major veins or devouring the leaf edges. Fully grown larvae enter the soil to overwinter in silken cocoons. Pupation takes place in the spring.

DAMAGE
The larvae cause considerable defoliation, the foliage of heavily infested plants being totally destroyed.

1019 Solomon's seal sawfly (*Phymatocera aterrima*) damage to foliage of *Polygonatum*.

Platycampus luridiventris (Fallén) (**1020**)

The unusual, extremely flattened, woodlouse-like larvae of this locally common species are occasionally noted on alder (*Alnus*) but are not important pests; the larvae also occur on birch (*Betula*) and hazel (*Corylus*). They feed on the underside of leaves from July to October, producing irregular holes through the leaf blade between the major veins. When at rest, they lie flat against the lower surface of a leaf, blending well with their surroundings. Fully grown larvae are 10–12 mm long and yellowish green, with pairs of small black spots towards the sides of most abdominal segments; the lateral borders of each segment are rounded, and fringed with white hairs; the head is yellowish green, marked with a pair of yellowish-brown patches which rise sharply upwards from behind and slope gently downwards towards the mouthparts. The mainly black-bodied, orange-legged adults (5–6 mm long) occur in May and June.

1020 Larva of *Platycampus luridiventris*.

Pontania proxima (Lepeletier) (**1021**)

syn. *P. flavipes* (Cameron); *P. gallicola* (Cameron)

Willow bean-gall sawfly

An often abundant pest of crack willow (*Salix fragilis*) and white willow (*S. alba*). Widespread throughout central and northern Europe.

DESCRIPTION

Adult: 3.0–45 mm long; black and shiny, with whitish-yellow legs. **Larva:** up to 10 mm long; head brownish black and shiny; body yellowish green, the legs whitish, with brown claws.

LIFE HISTORY

Adults first appear in May, eggs being deposited in the leaf buds. As attacked leaves unfurl, galls develop on the leaf blades, each housing a single larva. The mature galls, which project equally from both sides of the leaf, are red above and yellowish green below, with the surface roughened by numerous small protuberances and ridges. During its development, the larva expels frass through a small hole formed in the lower wall of the gall. Fully fed larvae pupate in cocoons spun in the ground or in bark crevices. Second-generation adults emerge in July. These deposit eggs in the expanded leaves, giving rise to gall-inhabiting larvae which usually complete their development in October.

DAMAGE

Galls spoil the appearance of infested plants but, although often numerous, have little or no effect on tree growth.

Pontania bridgmanii (Cameron)

Sallow bean-gall sawfly

A widely distributed and generally common sawfly, the larvae developing in relatively flat, dark green, bean-shaped galls on the leaves of broad-leaved willows such as common sallow (*Salix atrocinerea*), grey willow (*S. cinerea*) and pussy willow (*S. caprea*). The galls are similar in appearance to those of *Pontania proxima* but usually expanded more above the leaf than below; also, their surface is smooth and somewhat pubescent, particularly below; galls on grey sallow are noticeably larger and less hairy than those on other hosts. The lifecycle is similar to that of *P. proxima*.

Pontania pedunculi (Hartig) (**1022–1023**)

syn. *P. baccarum* (Cameron); *P. bellus* (Zaddach)

Sallow pea-gall sawfly

This common and widely distributed species forms light greenish, densely hairy, pea-shaped galls, attached to the midrib on the underside of leaves of broad-leaved willows, including common sallow (*Salix atrocinerea*), grey willow (*S. cinerea*) and pussy willow (*S. caprea*) (cf. *Pontania viminalis*). Each gall measures about 7 mm across and encloses a single larva. Tenanted galls occur during two main generations, from late May or early June onwards. Fully grown larvae (10–12 mm long) are whitish, each with a brown head. They pupate

1021 Galls of willow bean-gall sawfly (*Pontania proxima*).

1022 Gall of sallow pea-gall sawfly (*Pontania pedunculi*).

1023 Larva of sallow pea-gall sawfly (*Pontania pedunculi*).

in cocoons spun amongst debris on the ground. Adults are 3.5–4.5 mm long, and mainly black with white legs; they occur from April to June and from July to August.

Pontania triandrae Benson (**1024–1025**)

This species is essentially similar to *Pontania proxima* but infests the leaves of almond willow (*Salix triandra*). The characteristically smooth, glabrous galls are at first light green but soon become red above and pale yellowish green below. In common with those of *P. proxima*, they are developed equally on both sides of the leaf blade; also, there are two generations annually. The galls are often associated with plants cultivated in willow beds but they do not normally cause damage. However, if present in large numbers on the top-most growth of tall canes, the extra weight of the galls can result in shoots becoming bent.

Pontania viminalis (Linnaeus) (**1026–1027**)
syn. *P. saliciscinereae* (Retzius); *P. vollenhoveni* (Cameron)
Willow pea-gall sawfly

A generally common, double-brooded species, forming pea-shaped galls on the leaves of purple willow (*Salix purpurea*) and, less commonly, crack willow (*S. fragilis*) and osier (*S. viminalis*); the galls are pinkish to orange-yellow, with a glabrous, somewhat warty surface (cf. *Pontania pedunculi*). The adults (4–5 mm long) occur in May and June, and in July and August, the galls appearing on the underside of the leaves from late May onwards. The larvae (up to 13 mm long) are whitish green, each with a light greenish-brown head.

1024 Galls of *Pontania triandrae*, viewed from above.

1025 Galls of *Pontania triandrae*, viewed from below.

1026 Galls of willow pea-gall sawfly (*Pontania viminalis*).

1027 Larva of willow pea-gall sawfly (*Pontania viminalis*).

Priophorus morio (Lepeletier) (**1028**)
syn. *P. brullei* Dahlbom; *P. tener* (Zaddach)
Small raspberry sawfly

A widespread and locally common species, sometimes infesting rowan (*Sorbus aucuparia*), including ornamental trees in gardens, but associated mainly with raspberry and other cane fruits. The larvae feed on the expanded leaves from late May or early June onwards but do not cause significant damage. Fully grown individuals (*c*. 12 mm long) are whitish and translucent, but mainly black dorsally, with numerous whitish hairs arising from pale verrucae; the head is black and shiny. Adults are about 5–7 mm long and mainly black with pale legs. There are two or more generations annually.

Pristiphora aquilegiae (Vollenhoven) (**1029)**
syn. *P. alnivora* (Hartig)
Columbine sawfly

A common and often important pest of columbine (*Aquilegia vulgaris*). Widely distributed in central Europe; now well established in southern England, where it was first reported in 1946.

1028 Larva of small raspberry sawfly (*Priophorus morio*).

1029 Columbine sawfly (*Pristiphora aquilegiae*).

DESCRIPTION
Adult: 4.5–5.5 mm long; mainly black, with light brown legs. **Larva:** up to 10 mm long; head greenish yellow to blackish; body green and shiny.

LIFE HISTORY
Adults occur from April or May onwards, there being three or more generations annually. The larvae feed on the leaves of host plants, biting out large holes in the edges of leaves whilst lying with the abdomen curled beneath the leaf blade. They are fully fed within a few weeks, and then drop from the host plant to pupate in brownish-orange cocoons. During the summer months, pupal development is rapid and adults emerge within about two weeks.

DAMAGE
Defoliation is often very extensive, affecting both the appearance and vigour of infested plants.

Pristiphora testacea (Jurine) (**1030–1031**)
syn. *P. betulae* (Retzius)

A locally common pest of birch (*Betula*). Widely distributed in central and northern Europe but absent from more northerly areas.

DESCRIPTION
Adult: 5–7 mm long; mainly black, with orange-yellow tegulae and an orange-yellow abdomen; wings hyaline, the fore wings each with a black pterostigma. **Larva:** up to 16 mm long; head black; body yellowish green and shiny, with large orange-yellow lateral patches on the second and third thoracic segments; body also with a small, orange-yellow mark on each side of the first abdominal segment, and with large lateral patches on the second to eighth abdominal segments; legs light green, with black claws.

LIFE HISTORY
Adults occur from late April or May to June, and in July and August. Eggs are deposited in birch leaves, each close to a major vein. The larvae feed gregariously on the leaf edges during the summer months, a second generation appearing in the autumn. Pupation occurs in the soil in brownish cocoons.

DAMAGE
Larvae bite out large pieces from the leaves but attacks are rarely of significance, except on small ornamental or nursery trees.

1030 Adult female of *Pristiphora testacea.*

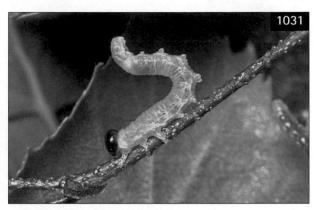

1031 Larva of *Pristiphora testacea.*

1032 Adult female of *Pristiphora punctifrons.*

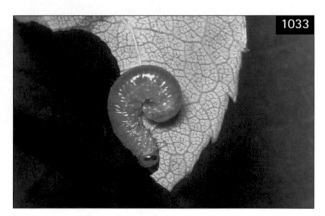

1033 Larva of *Pristiphora punctifrons.*

Pristiphora abietina (Christ)
syn. *P. pini* (Retzius)
Gregarious spruce sawfly
Widely distributed and locally common on fir (*Abies*) and spruce (*Picea*). Although regarded mainly as a forest pest, damage also occurs on Norway spruces (*P. abies*) intended for the Christmas tree market, particularly when these are being raised near established spruce plantations. The larvae (up to 15 mm long) are green with a yellowish or brownish head. They occur in May and June, and feed gregariously on the needles of the new shoots. Infestations occur mainly at the top of plants, sometimes causing distortion and shoot death, and thereby reducing the quality of nursery trees.

Pristiphora geniculata (Hartig)
Associated with rowan (*Sorbus aucuparia*) and locally important as a pest of forest and ornamental trees; widespread in mainland Europe, from Italy northwards, but uncommon in the British Isles. Also present in North America. The larvae feed gregariously from June

onwards, and sometimes cause noticeable defoliation. Fully grown larvae (15–18 mm long) are mainly yellow, marked with black. There are up to two generations annually. Adults (6.5–7.5 mm long) are mainly black; they occur in May and June, and in July and August.

Pristiphora punctifrons (Thomson) (**1032–1033**)
syn. *P. viridana* Konow
Minor infestations of this widespread species occur occasionally on cultivated rose (*Rosa*), causing slight damage to the foliage. The mainly black to dull greyish-yellow adults (4–5 mm long) occur from April to May or June. The larvae (up to 11 mm long) are green and shiny, with a few small blackish plates dorsally on the thoracic segments, pale yellowish thoracic legs and a dirty-yellow, blackish-marked head; they feed on the edges of the leaves in May and June, typically resting with the tip of the abdomen curled downwards. Fully fed individuals enter the soil, where they eventually pupate. New adults appear in the spring.

Pristiphora wesmaeli (Tischbein) (**1034**)
Larch sawfly

A widely distributed pest of larch (*Larix*), particularly young trees; of greatest importance in forestry plantations but sometimes also present on ornamentals. The mainly yellow, black-marked adults (5.0–6.5 mm long) occur from May to June or early July, eggs being deposited in young needles at the tips of new shoots. The larvae (up to 18 mm long) are green, with a yellowish-green head. They feed during the summer, often stripping the needles from the terminal shoots.

Profenusa pygmaea (Klug) (**1035**)
Oak leaf-mining sawfly

This widely distributed, locally common leaf miner is associated with English oak (*Quercus robur*) and sessile oak (*Q. petraea*), including young trees. The larvae mine the upper side of the leaves in the summer and early autumn, forming prominent, brown blotches. These mines are disfiguring but they do not affect plant growth. When fully fed the larvae vacate the mines to overwinter in the soil. Adult females (3–4 mm long) are black, with white tegulae, knees, tibiae and tarsi. They occur from May to July. Males are unknown, and reproduction is entirely parthenogenetic. The larvae (up to 8 mm long) are white, with a light brown head, a black prothoracic plate and black plates between the thoracic legs; abdominal prolegs are absent.

Protemphytus carpini (Hartig) (**1036–1037**)
syn. *P. glottianus* Cameron; *Ametastegia carpini* (Hartig)
Geranium sawfly

This widely distributed species is associated with *Geranium*, including cultivated forms, on which the larvae cause noticeable leaf damage. Larvae (up to 12 mm long) are mainly olive-green to greyish black, but paler below. There are two generations annually. Adults (6–8 mm long) are mainly black and shiny; they occur from May to June and from July to August.

Protemphytus pallipes (Spinola)
syn. *P. grossulariae* (Klug); *Ametastegia pallipes* (Spinola)
Viola sawfly

An extremely widespread pest of *Viola*, including pansy (*V. tricolor*) and sweet violet (*V. odorata*); often present on cultivated plants. Adults (6–8 mm long) are mainly black; the larvae (up to 10 mm long) are greenish grey, but paler below, with a brown head. Adults occur from April to September and larvae from May onwards, there being three or more generations annually.

Rhadinoceraea micans (Klug) (**1038**)
Iris sawfly

A locally distributed pest of mainly waterside irises, occurring on wild yellow flag (*Iris pseudacorus*) and also cultivated species such as *I. laevigata* and butterfly iris (*I. spuria*). Most common in central Europe; in Britain found mainly in the southern half of England.

DESCRIPTION
Adult: 7–8 mm long; body black and stout; wings slightly smoky. **Larva:** up to 20 mm long; head black; body bluish grey to dirty greenish yellow, with pale verrucae on the back and sides.

LIFE HISTORY
Adults occur in May and June, depositing eggs on the leaves of irises. The larvae feed from late May onwards, most completing their development by the end of July. They then enter the soil and spin silken cocoons in which to pupate. There is just one generation annually.

DAMAGE
Young larvae bite out V-shaped wedges along the leaf edges; older larvae graze away longitudinal sections of the leaf margins, feeding from the tips downwards and eventually reducing the foliage to ragged stumps. Larvae also attack flower buds.

Scolioneura betuleti (Klug) (**1039**)
syn. *S. betulae* (Zaddach)

A generally common pest of birch (*Betula*), and often injurious to young amenity trees and nursery stock. Widely distributed in central and northern Europe.

DESCRIPTION
Adult: 4–5 mm long; shiny black and slender bodied, with mainly reddish-yellow legs. **Larva:** up to 10 mm long; white with a brown head, a large prothoracic plate and a large plate ventrally on the first thoracic segment; distinct black markings usually present ventrally on the remaining two thoracic and first (sometimes also second) abdominal segments, plus a pair of marks ventrally on the penultimate abdominal segment; black markings also present along the sides; legs banded with black.

LIFE HISTORY
Adults occur in May and June, and from July to September. Eggs are laid at the edge of the leaves of birch, the larvae subsequently mining and forming brownish, semitransparent blotches. There may be one or several mines per leaf, the galleries often uniting and

1034 Larva of larch sawfly (*Pristiphora wesmaeli*).

1035 Mine of oak leaf-mining sawfly (*Profenusa pygmaea*) on *Quercus*.

1036 Larva of geranium sawfly (*Protemphytus carpini*).

1037 Geranium sawfly (*Protemphytus carpini*) damage to foliage of *Geranium*.

1038 Larva of iris sawfly (*Rhadinoceraea micans*).

1039 Larva of *Scolioneura betuleti*.

the larvae then feeding gregariously. Fully grown larvae eventually vacate the leaf mine to pupate in the soil. Occupied mines occur in June and early July, and again in the autumn.

DAMAGE
Attacked leaves may contain one or several mines which, if numerous, destroy much of the leaf tissue; heavily infested trees appear brown.

1040 Poplar sawfly (*Trichiocampus viminalis*).

1041 Larva of poplar sawfly (*Trichiocampus viminalis*).

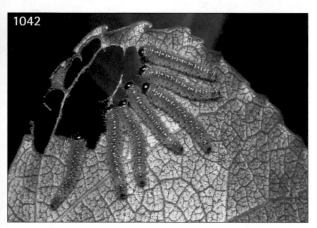

1042 Young larvae of poplar sawfly (*Trichiocampus viminalis*).

Trichiocampus viminalis (Fallén) (**1040–1042**)
syn. *T. luteicornis* (Stephens)
Poplar sawfly
An often common pest of poplar (*Populus*); sometimes also attacking willow (*Salix*). Present throughout the whole of Europe; also occurs in North America.

DESCRIPTION
Adult: 7–9 mm long; body mainly yellow, with the head and dorsal part of thorax black; antennae relatively long, black above and yellowish below; wings subhyaline, the fore wings each with a brown pterostigma. **Larva:** up to 20 mm long; head black and shiny; body varying from glassy-green to greenish yellow or orange, each side bearing a row of prominent black patches; body also somewhat flattened and hairy, the hairs arising from pale verrucae.

LIFE HISTORY
Adults occur from May to August, eggs being deposited in rows within the stalks of poplar leaves. After egg hatch, the oviposition sites appear as characteristic depressions running along the length of the leaf stalks, and these often betray the presence nearby of young larvae. The larvae occur in groups, sheltering beneath the leaves during the daytime; they also typically feed whilst lying alongside one another. The larvae continue to feed in groups until the final stages of their development. Individuals then wander over the trunks of host trees in search of suitable pupation sites. They eventually spin double-walled cocoons in bark crevices or amongst debris on the ground. Adults emerge shortly afterwards. Larvae occur throughout the summer and early autumn, there being two or more generations annually.

DAMAGE
Very young larvae browse on the lower epidermis, the upper side of the leaf turning brown; older individuals bite through the entire leaf, and sometimes cause significant defoliation.

Family **CYNIPIDAE** (gall wasps)

A group of small to tiny, mainly gall-forming wasps, often with highly complex lifecycles. Many species are associated with ornamental trees and shrubs.

Andricus kollari (Hartig) (**1043–1044**)
Marble gall wasp
A common and often abundant pest of oak (*Quercus*), especially English oak (*Q. robur*) and sessile oak (*Q. petraea*); often of importance on young trees. Widespread in Europe.

DESCRIPTION
Adult (asexual) female: 4–5 mm long; reddish yellow, with gaster partly black; legs pale; wings hyaline but tinged with red. **Larva:** plump, whitish.

LIFE HISTORY
This cynipid forms large (up to 28 mm in diameter), smooth, green to reddish, marble-like galls on young oak trees. Each gall, which develops from the base of a bud from late spring onwards, surrounds a small central cavity and contains a single larva. The galls mature in August or September. Adult (asexual) female wasps either emerge during the autumn or early in the following spring. These wasps deposit eggs in the axillary buds of Turkish oak (*Quercus cerris*), to produce an often overlooked bisexual generation of larvae. Mated adult females from this early-spring generation then initiate the familiar unisexual, summer generation of marble galls. (In areas where Turkish oak does not exist, the insect is represented merely by the asexual generation.) Deserted marble galls, which soon become woody, may persist for several years and are often a common sight on scrub-oaks. The galls are commonly invaded by parasitoids and inquilines, which generally emerge in the spring through a series of small exit holes; parasitized galls may also remain small and become prematurely woody.

DAMAGE
Infestations cause considerable distortion on young trees and can reduce the marketability of nursery stock. The galls are also disfiguring, particularly when numerous.

1043 Galls of marble gall wasp (*Andricus kollari*).

1044 Old galls of marble gall wasp (*Andricus kollari*).

Andricus curvator Hartig (**1045**)

Oak bud collared-gall cynipid

Infestations of this species occur abundantly on oak (*Quercus*). Female wasps of the asexual generation occur in February or March, initiating galls which eventually develop on the major veins or stalks of the young leaves. The galls cause considerable distortion of the foliage, and often attract attention when present on small trees. Each gall, which contains a single larva in a large chamber, is light green to light brown and measures 5–8 mm across. The larvae feed during April and May, developing into male or female wasps which emerge in June. After mating, females of the bisexual generation deposit eggs between the scales of the leaf buds, where small, inconspicuous 'collared galls' are eventually formed. Such galls are commonly initiated in buds already hosting the 'larch-cone gall' generation of *Andricus fecundator* (see below). Development within the galls commences in the summer and extends throughout the following year, the complete lifecycle thus occupying two years.

Andricus fecundator (Hartig) (**1046–1047**)

syn. *A. foecundatrix* (Hartig)

Larch-cone gall cynipid

This species is responsible for transforming the buds of oak (*Quercus*) into artichoke-like larch-cone galls. Each gall, which begins its development in June, is up to 20 mm long at maturity, and contains a single larva. Adult female wasps emerge from the galls in the following spring but their emergence may be delayed for up to three years. These wasps lay eggs in male flower buds on oak, a bisexual generation then developing in May to June inside hairy, catkin galls. The resulting adults mate, females then depositing fertilized eggs in leaf buds and thereby initiating the next round of larch-cone galls. Attacks are most common on scrub-oaks but also occur on nursery trees. The dead remains of old larch-cone galls often persist on host plants for several years.

Andricus quercuscalicis (Burgsdorf) (**1048–1050**)

Acorn cup gall cynipid

A widespread and common pest in mainland Europe and, having appeared in southern England in the late 1950s, now well established in England and Wales; the pest reached Scotland in 1995. Adult females appear in the early spring, depositing eggs on the flower initials of Turkish oak (*Quercus cerris*), a bisexual generation then developing in galls on the catkins. Later, after the production of males and females, eggs are deposited in the acorn primordia of English oak (*Q. robur*). Characteristic 'knopper' galls (usually one per infested acorn) then develop from the acorn-cup tissue, each containing a single, whitish larva. During the summer the galled tissue expands into irregular, green, sticky outgrowths; by the autumn these become brown and woody; any acorns surviving within these infested cups are also malformed. Mature galls fall from the tree in October, still associated with the peduncle, the occupants pupating before the onset of winter, each in a small pupal cell within the shelter of the gall. Infestations of this wasp are sometimes heavy and totally prevent acorn production on some trees; however, there appears to be no real effect on tree growth.

Andricus quercustozae (Bosc)

A southern European, gall-forming pest of cork oak (*Quercus suber*), downy oak (*Q. pubescens*), English oak (*Q. robur*), Hungarian oak (*Q. frainetto*), Pyrenean oak (*Q. pyrenaica*) and sessile oak (*Q. petraea*), with a single asexual generation. The females appear in early spring and lay eggs singly in the buds of host plants. Larvae then develop separately in unilocular, more or less spherical, hard-walled, chestnut-brown galls, each 20–40 mm in diameter and with a central ring of hump-like or thorn-like projections. Development from egg to adult takes up to two years.

1045 Galls of oak bud collared-gall cynipid (*Andricus curvator*).

1046 Gall of larch-cone gall cynipid (*Andricus fecundator*) on *Quercus.*

1047 Old galls of larch-cone gall cynipid (*Andricus fecundator*) on *Quercus.*

1048 Asexual female of acorn cup gall cynipid (*Andricus quercuscalicis*).

1049 Young knopper galls of acorn cup gall cynipid (*Andricus quercuscalicis*).

1050 Mature knopper galls of acorn cup gall cynipid (*Andricus quercuscalicis*).

Biorhiza pallida (Olivier) (**1051**)
syn. *B. aptera* (Fabricius)
Oak-apple gall wasp

Galls caused by this widespread cynipid are often common on scrub-oaks (*Quercus*). The galls develop in spring, following egg laying in the base of axillary and terminal buds by newly emerged wingless females. Affected buds swell rapidly into smooth, slightly irregular, whitish to brownish-yellow, soft, spongy galls about 25–40 mm in diameter. These galls, which contain several larvae, each within its own internal chamber, become extensively tinged with pinkish red as maturity is reached. Fully winged adults of both sexes appear in June or July. Vacated galls eventually darken and their remains often persist on trees long after the emergence of the original occupants. After mating, females of the summer generation enter the soil and give rise to a unisexual brood which develops inside root galls. Larvae of this generation complete their development in about 16 months. The wingless females then appear and ascend the trunks to initiate the next crop of oak-apple galls.

Cynips divisa Hartig (**1052**)
syn. *C. verrucosa* (Schlechtendal)
Oak bud red-gall cynipid

A locally distributed cynipid, forming smooth, whitish-yellow to bright red, woody, thick-walled galls on the underside of the leaves of oak (*Quercus*). Each gall arises from a major vein, and encloses a small cavity within which a small, whitish larva develops. The galls represent the asexual generation and occur from July onwards; adults appear in October. Although sometimes numerous on young trees, the galls do not distort the foliage and are often overlooked. A sexual generation occurs in the spring, larvae developing in small insignificant galls associated with the young leaves or terminal buds.

Cynips longiventris Hartig (**1053**)
syn. *C. substituta* Kinsey
Oak leaf striped-gall cynipid

Asexual galls formed by this locally common species also occur on the underside of the leaves of oak (*Quercus*). They are irregular in shape, 7–8 mm across and slightly flattened, with a hard, roughened wall; each encloses a small larval cavity. The galls are whitish yellow, marked with red, and often appear striped. They develop during the summer months, reaching maturity in October. The sexual generation develops in the spring in the adventitious buds on old oak trees.

Cynips quercusfolii Linnaeus (**1054–1055**)
syn. *C. taschenbergi* (Schlechtendal)
Oak leaf cherry-gall cynipid

A common and widely distributed cynipid, inducing the formation of cherry-like galls on the underside of the leaves of oak (*Quercus*), mainly English oak (*Q. robur*) and sessile oak (*Q. petraea*). The smooth, rounded galls, each 15–20 mm in diameter, are at first green or yellowish green but eventually become yellowish brown, flushed with red. The galls arise from the major veins but, in spite of their size, they do not cause distortion, even when several occur on the same leaf. On young trees, the galls are sometimes present in considerable numbers. Each gall contains a single larva, which develops within a small central cavity surrounded by spongy tissue. The galls reach maturity by October, but remain attached to the leaves after leaf fall. The adult wasps, although fully formed, delay their escape until mid-winter. These wasps initiate a bisexual generation in dormant buds, adult males and females eventually appearing in June. Fertilized eggs are then deposited in leaf veins, initiating the next unisexual cherry-gall generation.

1051 Young gall of oak-apple gall wasp (*Biorhiza pallida*).

1052 Galls of oak bud red-gall cynipid (*Cynips divisa*).

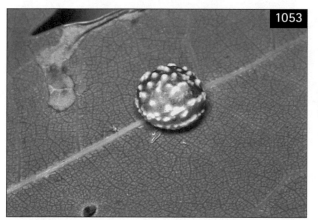

1053 Gall of oak leaf striped-gall cynipid (*Cynips longiventris*).

1054 Galls of oak leaf cherry-gall cynipid (*Cynips quercusfolii*).

1055 Asexual female of oak leaf cherry-gall cynipid (*Cynips quercusfolii*).

1056 Bedeguar gall wasp (*Diplolepis rosae*).

1057 Larva of bedeguar gall wasp (*Diplolepis rosae*).

1058 Pupa of bedeguar gall wasp (*Diplolepis rosae*).

1059 Young gall of bedeguar gall wasp (*Diplolepis rosae*) on *Rosa*.

1060 Old gall of bedeguar gall wasp (*Diplolepis rosae*) on *Rosa*.

Diplolepis rosae (Linnaeus) (**1056–1060**)
Bedeguar gall wasp
Generally common on wild rose (*Rosa*) and sometimes present on cultivated bushes. Present throughout Europe; also present in North America.

DESCRIPTION
Adult: 3.5–4.5 mm long; black with a mainly red gaster; tip of body with a distinct spine; legs partly reddish; fore wings mainly dark, each with a darker subapical patch. **Larva:** up to 6 mm long; whitish, with a pale yellowish-brown head. **Pupa:** 5 mm long; white, with purplish-white eyes.

1061　Galls of rose smooth pea-gall cynipid (*Diplolepis eglanteriae*).

LIFE HISTORY
Adult wasps emerge in the spring from May onwards, depositing eggs in unopened rose buds. Affected tissue begins to swell into a large, compact gall containing up to 60 hard, wooden, cherry-stone-like cells, each containing a single larva. The galls are surrounded by a dense, sticky mass of branched, moss-like filaments and measure up to 10 cm across; they change in colour from green, through pink and bright red to reddish brown, and darken further during the winter months. Larvae develop throughout the summer, eventually pupating in their cells in the following spring. Adult wasps appear shortly afterwards. Males are unknown, reproduction being entirely parthenogenetic. Old weather-worn galls remain attached to host plants long after the emergence of the adults, and are most obvious during the winter and early spring.

1062　Gall of rose spiked pea-gall cynipid (*Diplolepis nervosa*).

DAMAGE
The galls have little effect on bush growth, but are disfiguring if present in large numbers.

Diplolepis eglanteriae (Hartig) (**1061**)
Rose smooth pea-gall cynipid
A local but often overlooked species, forming smooth, pea-like galls (4–5 mm in diameter) on the underside of the leaves of rose (*Rosa*). The galls, which sometimes bear slight surface depressions or small tubercles, also occur occasionally on the upper surface. They appear in July and are at first light green. They mature in September or October, by which time they have turned rose-red; each gall, which encloses a single larva, then drops to the ground. The occupant pupates within its chamber, and the adult emerges in the following May or June. Reproduction is mainly parthenogenetic, males being very rare. Infestations are associated mainly with wild, rather than cultivated, bushes and are unimportant.

Diplolepis nervosa (Curtis) (**1062**)
Rose spiked pea-gall cynipid
This generally common species is essentially similar to *Diplolepis eglanteriae* but usually produces pea-like leaf galls characterized by the presence of one or more long, thorn-like spines. Young galls are yellowish green; they later become flushed with pink, and finally turn brown.

1063 Spangle galls of oak leaf spangle-gall cynipid (*Neuroterus quercusbaccarum*).

1064 Currant galls of oak leaf spangle-gall cynipid (*Neuroterus quercusbaccarum*).

Neuroterus quercusbaccarum (Linnaeus) (**1063–1065**)

syn. *N. lenticularis* (Olivier)
Oak leaf spangle-gall cynipid
A generally abundant gall wasp, associated with oak (*Quercus*) and often present on young trees. Widely distributed in Europe.

DESCRIPTION
Adult (asexual) female: 2–3 mm long; black and shiny, with brownish-yellow legs; wings mainly hyaline; ovipositor long and curved. **Adult (sexual) female:** 1.5–2.0 mm long; black with yellowish legs; wings hyaline to subhyaline; ovipositor short. **Adult male:** similar to sexual female, but with a long petiolus and no ovipositor.

1065 Oak leaf spangle-gall cynipid (*Neuroterus quercusbaccarum*) damage to leaf of *Quercus*.

LIFE HISTORY
In summer, mated female wasps deposit large numbers of eggs in the tissue on the underside of expanded oak leaves, inducing the formation of characteristic spangle galls. The disc-like galls are slightly hairy (the hairs stellate) and yellowish to yellowish white. Each gall contains a central chamber within which a single larva develops. The galls, often 100 or more on a leaf, occur from July onwards; they mature in October, when about 6 mm in diameter. They then fall to the ground and swell as they take up moisture. The larvae, which represent the unisexual generation of the species, overwinter within the galls and pupate in the spring. Asexual female wasps then emerge, depositing unfertilized eggs in male catkins. Characteristic currant-like galls (currant galls) then develop in strings on the catkins, each changing from green through pink to red; they measure about 4 mm in diameter when mature.

Larvae, destined to become either males or females, occur singly within these galls, completing their development in June. Adult wasps eventually emerge and, after mating, the females initiate the familiar generation of spangle galls.

DAMAGE
Spangle galls often occur in vast numbers, and cause spotting of the foliage, visible from above. However, infested trees are seldom if ever harmed. The currant galls are also unimportant.

1066 Spangle galls of oak leaf smooth-gall cynipid (*Neuroterus albipes*).

1067 Galls of oak leaf blister-gall cynipid (*Neuroterus numismalis*).

Neuroterus albipes (Schenck) (**1066**)
syn. *N. laeviusculus* Schenck
Oak leaf smooth-gall cynipid
Spangle galls formed on the leaves of oak (*Quercus*) by this cynipid wasp are smooth, irregularly saucer-shaped with a slight central knob, and vary from creamy white to reddish. They develop from July to October, and finally drop to the ground when about 4 mm in diameter. The galls are usually present in relatively small numbers, and sometimes occur in company with those of the more abundant species *N. quercusbaccarum*. Larvae of the bisexual generation develop within so-called Schenck's galls, which occur on the margins of leaves in May and June. These galls are oval (*c.* 2 × 1 mm), green and smooth, although slightly hairy when young.

Neuroterus numismalis (Geoffroy in Fourcroy) (**1067**)
syn. *N. politus* Hartig; *N. vesicator* (Schlechtendal)
Oak leaf blister-gall cynipid
The very characteristic golden-brown galls formed by this species, develop on the underside of the leaves of oak (*Quercus*) from August onwards. The galls (known as spangle galls) measure up to 3 mm across, and have a distinct central pit and a dense silky covering of short hairs. They often occur in very large numbers, sometimes more than 1,000 appearing on a single leaf. The galls also cause a distinct discoloration of the upper surface of infested leaves. Larvae of the bisexual generation occur in May and June, within greenish or greyish, hemispherical, partly ribbed blister galls (*c.* 3 mm across); these are formed on either side of young leaves.

Neuroterus tricolor (Hartig)
syn. *N. fumipennis* Hartig
Oak leaf cup-gall cynipid
This wasp forms spangle galls on the underside of the leaves of oak (*Quercus*) from July onwards. The galls are yellowish green, 3 mm in diameter and button-like, each with a slightly raised rim bearing reddish hairs, and a slight but noticeable central elevation. The bisexual generation develops in shiny, whitish, yellowish or greenish, pea-like galls, each with a temporary coating of reddish hairs; the hairs drop off when about 5 mm long. These galls, each up to 6 mm in diameter, arise from the midrib or major veins on the underside of leaves during May and June; they often occur singly, but may coalesce and cause noticeable distortion of infested leaves.

Family **EULOPHIDAE**

A group of tiny insects, most of which are parasitoids.

Leptocybe invasa Fisher & La Salle
Blue gum chalcid
A major new, invasive pest of *Eucalyptus*. First discovered (as recently as the year 2000) in the Middle East, but probably of Australian origin. Now present in various parts of Africa, Asia, North America and southern Europe, including France, Greece, Italy, Portugal and Spain.

DESCRIPTION
Adult female: 1.2 mm long; body dark brown, with a metallic blue or green sheen; legs and antennae pale.

LIFE HISTORY
This devastating species breeds parthenogenetically, the females depositing eggs in the upper side of young leaves, on either side of the midrib. Eggs are also laid in leaf stalks and shoots. Affected tissue turns corky and swells, and also changes colour from green to pink, and eventually becomes light brown or red. Adult exit holes also become evident in the walls of mature galls. There are two or three overlapping generations annually.

DAMAGE
Galls cause stunting of growth and loss of vigour, and the impact on young trees and seedlings is considerable. Premature leaf fall, dieback of shoots and death of severely damaged trees are also recorded.

Ophelimus maskelli (Ashmead)
Eucalyptus gall wasp
In recent years this Australian pest has become well established in the Mediterranean Basin (including France, Greece, Italy and Spain) on *Eucalyptus*, upon which it forms tiny (*c.* 1 mm diameter) pimple-like galls on the upper surface of young expanded leaves. The pest has also spread to Portugal, and it or a close relative is believed to have been accidentally introduced to England. Although leaves may be peppered with galls, and infestations may result in premature leaf fall, the pest is not considered as devastating as *Leptocybe invasa* (see above). There are up to three generations annually.

Family **EURYTOMIDAE**

A varied group of tiny chalcid wasps, including phytophagous species and parasitoids.

Eurytoma orchidearum (Westwood)
Cattleya 'fly'
A tropical Central and South American pest of orchids, especially *Cattleya*. Accidentally introduced along with host plants to many other parts of the world, including Europe.

DESCRIPTION
Adult: 4–5 mm long; head and thorax dull black; abdomen black and shiny. **Larva:** up to 5 mm long; creamy white.

LIFE HISTORY
Adult female wasps insert eggs singly in the rhizome of host orchids, often depositing several in one and the same plant. The eggs hatch 1–2 weeks later. Larvae then feed on the internal tissue, each forming a large cavity. Feeding galleries often coalesce, and larvae then develop gregariously. Larvae are fully grown in 2–3 months. They then pupate internally, and adults emerge shortly afterwards. There are several generations annually.

DAMAGE
Infested rhizomes are deformed and weakened. As a result, vegetative growth is interrupted, and attacked plants usually fail to flower. Damaged rhizomes are also liable to be invaded by bacterial and fungal pathogens.

Family **FORMICIDAE** (ants)

Ants, such as the generally abundant and well-known common black ant (*Lasius niger*), are sometimes harmful to ornamental plants but are of only minor importance. They often ascend plants to collect nectar from flowers, occasionally also damaging young buds of trees and shrubs in their quest for moisture. More frequently, the presence of ants on plants is an indication that the leaves, shoots or branches are infested by honeydew-excreting pests such as aphids and scale insects; ants sometimes construct earthen shelters on plants to protect colonies of aphids, notably conifer aphids (genus *Cinara*) (p. 87–90) and root aphids such as *Maculolachnus submacula* (p. 91). Subterranean activities by ants are also harmful, damage most often being restricted to the accidental disturbance of seedlings and established plants as soil around the roots becomes loosened; seriously affected plants wilt and die, damage being most severe in hot, dry conditions. Ants are sometimes troublesome on lawns, particularly on those freshly constructed in light, sandy soils inhabited by the yellow meadow ant (*Lasius flavus*), this usually being the damaging species. Ants also remove seeds from pots, seed boxes and seed trays, but losses are rarely significant.

Family **VESPIDAE** (1068–1069)

Social wasps (*Vespula* spp.) are unimportant pests of ornamental plants but they are sometimes a nuisance in gardens and nurseries, commonly establishing their colonies in banks, buildings, hollow trees and walls. Such wasps frequently visit fences, sheds, shrubs and trees to scrape off pieces of dead wood which are formed into a papery material used in the construction of their nests; they also gnaw tissue from the woody stems of ornamentals such as *Dahlia*; severely injured plants then collapse. Although in this way sometimes injurious, social wasps are also beneficial, particularly during the spring and early summer when they collect large numbers of harmful caterpillars and other insect pests which they then feed to their developing brood. Nest-building hornets (*Vespa crabro*) also sometimes cause damage to ornamental trees and shrubs, including ash (*Fraxinus excelsior*), birch (*Betula*), *Dahlia*, elder (*Sambucus*), lilac (*Syringa*), poplar (*Populus*) and willow (*Salix*). Adult hornets are readily distinguished from social wasps by their larger size (queens up to 35 mm long; workers up to 24 mm long), their mainly reddish-brown head and thorax, and by the position of the ocelli which lie anterior to the hind margin of the compound eyes at some distance from the back of the head.

Family **ANDRENIDAE**

A relatively large group of often very hairy, solitary, burrowing bees; the tongue short, ovate and pointed; the abdomen somewhat flattened dorsoventrally.

Andrena fulva (Müller in Allioni)
syn. *A. armata* (Gmelin in Linnaeus)
Tawny burrowing bee

An often abundant species, nesting in the soil and sometimes causing concern when its burrows are formed in lawns. Widely distributed in Europe.

DESCRIPTION
Adult female: 12–14 mm long; thorax and abdomen clothed in bright brown hairs, those on the abdomen distinctly reddish; hairs on the face, legs, at tip of abdomen and on the underside of the body black. **Adult male:** 12–14 mm long; body clothed in mainly brown hairs.

1068 Worker social wasp (*Vespula* sp.).

1069 Queen hornet (*Vespa crabro*).

LIFE HISTORY

Adults appear in March and April. They are active in warm, sunny weather, and then forage on various spring flowers. Mated females excavate deep (15–30 cm long) burrows in the soil, the excavated soil forming volcano-like mounds on the surface. Each burrow consists of a main tunnel, with a cell at the bottom, and a series of short, lateral branches, each also ending in a single cell. Into each cell the female places a quantity of nectar and pollen, upon which she then lays an egg; when all the cells have been provisioned, she seals the burrow with soil and flies away. After the eggs have hatched, the entombed larvae feed on their supplies of food; fully fed larvae eventually pupate, and adult bees emerge in the following spring.

DAMAGE

Heaps of excavated soil are a nuisance when fine lawns are invaded by large numbers of bees attracted by particularly favourable nesting conditions.

Family **MEGACHILIDAE**

Megachile centuncularis (Linnaeus) (**1070**)

Common leaf-cutter bee

Generally common, the adult females sometimes damaging the leaves of rose (*Rosa*) and other ornamental plants, including *Laburnum*, lilac (*Syringa*), privet (*Ligustrum vulgare*), *Rhododendron* and snowy mespilus (*Amelanchier laevis*); petals of *Geranium* plants are also damaged. Widely distributed in Europe.

DESCRIPTION

Adult female: 10–12 mm long; black bodied; head and thorax clothed in golden-brown hairs; abdomen clothed above with black hairs but banded with pale hairs, particularly basally; pollen-collecting hairs on underside of abdomen orange-red and projecting beyond the sides to form an apparent fringe; legs with pale hairs; wings smoky.

LIFE HISTORY

Adults are active in June and July. They occur commonly in gardens, foraging during the daytime on various flowers from which they collect both nectar and pollen. When ready to breed, the females form elongate burrows in decaying wood, soft brickwork or light soil; each then collects several fresh leaf fragments from nearby rose bushes or other suitable plants, and carries these into her burrow to form a series of thimble-like brood cells. When completed, each cell is provisioned with a mixture of nectar and pollen, and an egg deposited on the surface. The cell is then capped with a leaf fragment and another cell constructed above it. A fully furnished nest usually contains a series of about six cells placed end to end; the burrow is then sealed with wood-pulp or soil and abandoned. Larvae feed on their store of food from late summer onwards. They complete their development in the following spring and then pupate.

DAMAGE

Attacked leaves have large, regular, oblong or semicircular pieces removed from the blade. Such damage often causes concern. However, although plants are disfigured, growth is not affected.

1070 Common leaf-cutter bee (*Megachile centuncularis*) damage to leaves of *Rosa*.

Chapter 3
Mites

Family **PHYTOPTIDAE** (gall mites)

Phytoptus avellanae Nalepa (**1071**)
syn. *Phyllocoptella avellanae* (Nalepa)
Filbert bud mite

Associated with hazel (*Corylus*), on which it induces the formation of greatly enlarged buds ('big buds'). The mites breed within the shelter of the galled buds throughout the autumn and winter. In early spring, female mites migrate from the galls to invade the leaves, where further eggs are laid. The eggs produce active protonymphs that feed and eventually moult into more or less sedentary deutonymphs; these, unlike adults and protonymphs, are distinctly flattened and have few abdominal tergites. In late June or July the summer deutonymphs moult into 'normal' adults. These invade new terminal buds. These buds then swell and become very noticeable from September onwards. Such galls are often abundant in hedgerows. They are also often numerous on ornamentals and nursery stock, affecting both the appearance and potential structure of young plants. The causal mites are whitish and elongate (*c.* 0.3 mm long), with numerous abdominal tergites and sternites, and a pair of short, anterior (frontal) setae and a pair of longer posterior (dorsal) setae on the prodorsal shield.

Phytoptus tetratrichus Nalepa (**1072**)
This species forms tight upward leaf-roll galls along the edge of the leaves of various kinds of lime (*Tilia*). The galls cause slight distortion but are less noticeable than those formed on lime by the midge *Dasineura tiliamvolvens* (p. 180).

1071 Gall of filbert bud mite (*Phytoptus avellanae*).

1072 Galls of *Phytoptus tetratrichus* on leaf of *Tilia*.

Family **ERIOPHYIDAE** (gall mites and rust mites)

Many different species of eriophyid mite are associated with ornamental plants, some forming galls or bronzing the leaves but others existing merely as inquilines or as leaf vagrants. Most species mentioned in the following account are indigenous to northern Europe but some are relatively new arrivals, having been introduced along with their host plants from other parts of the world. The nomenclature for eriophyid mites is very confusing, with many species referred to in the literature under two or more specific if not subspecific names. Confusion at generic level is also widespread. As a result it is often difficult, if not impossible, to assign correctly an observed plant symptom to any particular species, subspecies or race of mite. These problems are exacerbated by the occurrence in eriophyid galls of inquilines, which are not the gall-forming species. Further difficulties arise in those eriophyids exhibiting deuterogeny (two or more structural female forms – a winter form or 'deutogyne' and a summer form or 'protogyne'). In this book, to reduce confusion and to aid correct cross-reference to names used in other publications, species are catalogued under their main host plants rather than in any taxonomic order. Apart from occasional reference to colour, size and form, taxonomic descriptions of the mites have been excluded. Although most, if not all, of the more important pest species are included, the following account is far from exhaustive, there being many others on ornamental plants, especially shrubs and trees. Also, various galls are known that cannot be attributed with certainty to any particular species of mite.

ACER

Aculops acericola Nalepa
Sycamore gall mite

Although sometimes considered a gall-forming species, this mite is merely an inquiline in erinea produced on sycamore (*Acer pseudoplatanus*) by *Eriophyes psilomerus* (p. 408). The mites are *c*. 0.12 mm long with about 23 abdominal tergites; microscopically, therefore, the two species are readily distinguishable.

Artacris cephaloneus (Nalepa) (**1073–1074**)
Maple bead-gall mite

A generally abundant species, responsible for the development of pimple-like bead galls on the leaves of maple (*Acer*) and sycamore (*A. pseudoplatanus*). Female mites overwinter in sheltered situations on the shoots and become active in the early spring. They then invade the expanding foliage, each initiating the development of numerous galls. These galls occur from April onwards, and change from green to red as they mature. The galls are often present in very large numbers, particularly on young trees, but are generally harmless. Breeding continues in the galls throughout the summer, although many galls will be found on examination to be empty. The causal mites are *c*. 0.16 mm long and slender-bodied, with about 65 abdominal tergites and sternites, and a pair of backwardly directed setae arising from tubercles on the hind margin of the prodorsal shield.

1073 Young galls of maple bead-gall mite (*Artacris cephaloneus*) on leaves of *Acer campestre*.

1074 Mature galls of maple bead-gall mite (*Artacris cephaloneus*) on leaf of *Acer campestre*.

NOTE
The specific name '*cephaloneus*' (sometimes cited as '*macrorhynchus cephaloneus*') is often restricted to the mites producing bead galls on maple; the name '*macrorhynchus*' (see below) (sometimes cited as '*macrorhynchus aceribus*') is then applied to the mites on sycamore, without recognizing differences between the two types of gall found on that host.

Artacris macrorhynchus (Nalepa) (**1075**)
Sycamore leaf gall mite
Widespread and generally common, inducing the development of elongate (2–4 mm long), dark red galls on the upper surface of the leaves of sycamore (*Acer pseudoplatanus*). The galls, which occur throughout the spring and summer, are distinctly longer than those of the previous species.

Eriophyes eriobius Nalepa (**1076**)
Infestations of this species induce the development of erinea on the underside of the leaves of maple (*Acer*). The galls commence as whitish patches which later become flushed with purple. The upper surface of affected leaves is slightly discoloured but damage caused is unimportant. The causal mites (*c.* 0.2 mm long) are pale yellowish, with about 68 abdominal tergites and sternites.

Eriophyes eriobius pseudoplatani Nalepa (**1077**)
This subspecies is associated with sycamore (*Acer pseudoplatanus*), inhabiting yellowish to brownish erinea on the underside of the leaves. The galls occur from May onwards, and tend to be concentrated alongside and at the junctions of the major veins (cf. galls formed by *Eriophyes psilomerus*). Although infested leaves are often extensively galled, the foliage is not distorted.

1075 Galls of sycamore leaf gall mite (*Artacris macrorhynchus*) on *Acer pseudoplatanus*.

1076 Galls of *Eriophyes eriobius* on underside of leaf of *Acer campestre*.

1077 Galls of *Eriophyes eriobius pseudoplatani* on underside of leaf of *Acer pseudoplatanus*.

Eriophyes macrochelus (Nalepa) (**1078–1079**)
Maple leaf gall mite

An often abundant mite on ornamental maple (*Acer*) trees, producing distinctive pouch galls on the leaves. The galls appear from May onwards, commencing as greenish warts but later becoming more or less red. They frequently arise at the junction of the major veins; they are often very brightly coloured but eventually turn black. On heavily infested leaves, the galls coalesce and cause considerable distortion. Mites inhabiting the galls are *c*. 0.16 mm long and cylindrical, with about 60 abdominal tergites and sternites, and a relatively small pair of backwardly directed setae arising from tubercles on the hind margin of the prodorsal shield.

Eriophyes psilomerus Nalepa (**1080**)
The leaves of sycamore (*Acer pseudoplatanus*) are often disfigured by large, irregular erinea induced by this generally common gall mite. Each gall appears as a pale green to brownish blister on the upper surface of the leaf, with the underside densely clothed in whitish to purplish hairs. Such galls occur from May onwards, and gradually darken as they mature. Although infested foliage looks unsightly, tree growth is not affected. The causal mites are *c*. 0.18 mm long, with about 65 abdominal tergites and sternites (cf. *Aculops acericola*, p. 406).

AESCULUS

Tegonotus carinatus Nalepa
Horse chestnut rust mite

This widespread and generally common species is free-living on the underside of the leaves of horse chestnut (*Aesculus hippocastanum*). When present in large numbers, the mites cause extensive bronzing and premature leaf fall. Infestations are sometimes of importance on nursery trees. The mites are *c*. 0.19 mm long and rather flattened, with few abdominal tergites; there are distinct overwintering and summer forms: deutogynes and protogynes, respectively. The latter are thought to aestivate within bark crevices on the previous year's shoots when leaves harden during the early summer.

ALNUS

Acalitus brevitarsus (Fockeu) (**1081–1082**)
Alder erineum mite

A generally common pest of various kinds of alder (*Alnus*), infested leaves becoming distorted by large, irregular, blister-like erinea. The upper surface of each gall is pale, somewhat warty and shiny; below, the galled tissue is densely coated with whitish to pale brown, multi-headed hairs. The galls develop from June to October, changing in colour from light green, through pale yellow, to reddish brown. The inhabitants are *c*. 0.16 mm long, with about 60 abdominal tergites and

1078 Young galls of maple leaf gall mite (*Eriophyes macrochelus*).

1079 Maple leaf gall mite (*Eriophyes macrochelus*) damage to foliage of *Acer*.

1080 Galls of *Eriophyes psilomerus* on underside of leaf of *Acer pseudoplatanus*.

1081 Galls of alder erineum mite (*Acalitus brevitarsus*) viewed from above.

1082 Galls of alder erineum mite (*Acalitus brevitarsus*) viewed from below.

sternites, and a pair of backwardly directed setae arising from tubercles close to the hind margin of the prodorsal shield. This species is deuterogenous.

Acaricalus paralobus Keifer
Alder leaf rust mite
One of several free-living eriophyid mites responsible for bronzing the foliage of alder (*Alnus*). Affected leaves become dull and noticeably discoloured; heavy infestations reduce the vigour of young trees, including nursery stock.

1083 Galls of *Eriophyes inangulis* on leaf of *Alnus*.

Eriophyes inangulis (Nalepa) (**1083**)
This mite induces the development of prominent swellings in the angles between the midrib and the major veins of leaves of alder (*Alnus*). The position of each gall is demarcated above by a discoloured, shiny swelling and below by a small patch of whitish to reddish-brown hairs. The galls develop from May onwards, changing from green, through yellow and red, to brown.

Eriophyes laevis (Nalepa) (**1084**)
Alder bead-gall mite
A deuterogenous species, forming small, compact, pimple-like galls on the upper surface of the leaves of alder (*Alnus*). The galls often occur in vast numbers, sometimes several hundred on a leaf, and cause significant distortion. Attacks on established trees are of little importance but damage to young nursery stock affects plant vigour. The galls develop from June to October, and vary in colour from green, through yellow, to purplish brown. The causal mites are relatively large (*c.* 0.28 mm long), with about 65 abdominal tergites and sternites, and a pair of short, backwardly directed setae arising from tubercles in front of the hind margin of the prodorsal shield.

1084 Galls of alder bead-gall mite (*Eriophyes laevis*).

1085 Cone gall of birch witches' broom mite (*Acalitus rudis*).

1086 Witches' broom galls on *Betula*.

BETULA

Acalitus rudis (Canestrini) (**1085–1086**)
syn. A. rudis calycophthirus (Nalepa)
Birch witches' broom mite
Abundant on birch (*Betula*), in association with various symptoms. In some cases, buds are invaded. These then fail to open and, instead, become greatly enlarged and cone-like. Shoot growth on affected branches is then disrupted but, because infestations are most frequently established on mature trees, damage caused is insignificant. The mites are *c*. 0.19 mm long, whitish and slender-bodied, with about 65 abdominal tergites and sternites, and a pair of backwardly directed setae arising from tubercles close to the hind margin of the prodorsal shield. The mites also inhabit witches' broom galls on birch trees, and were once considered to be their initiators; however, it is now generally accepted that they are merely inquilines, the growths being induced by fungi (*Taphrina* spp.). Species of *Taphrina* are also responsible for the development of witches' broom galls on certain other trees. Such fungi also induce leaf-curl galls, including the well-known peach leaf curl (caused by *T. deformans*); these galls are sometimes mistaken for pest-induced deformities, particularly if invertebrates (such as mites or small insects) are sheltering within them.

Aceria leionotus (Nalepa)
syn. Eriophyes laevis lionotus Nalepa
Birch bead-gall mite
An often common species, inhabiting small (*c*. 1 mm diameter), red, pimple-like galls on the upper side of the leaves of birch (*Betula*). Although sometimes numerous the galls cause little or no distortion.

Aceria longisetosus (Nalepa) (**1087**)
syn. A. rudis longisetosus (Nalepa)
This mite inhabits whitish, often reddish-tinged erinea on the underside of the leaves of birch (*Betula*). The galls also develop on the upper surface of the foliage, and are capable of causing noticeable distortion.

CARPINUS

Eriophyes macrotrichus (Nalepa) (**1088**)
Hornbeam leaf gall mite
A southerly distributed species which induces an interveinal furrowing on the leaves of hornbeam (*Carpinus betulus*), the affected foliage becoming distinctly crinkled and discoloured. Although damage is disfiguring, and sometimes appears of some significance, plant growth is not affected. The causal mites are *c*. 0.16 mm long and relatively plump.

CHRYSANTHEMUM

Epitrimerus alinae Liro
Chrysanthemum leaf rust mite
Free-living on greenhouse-grown *Chrysanthemum*, the stems of which become russeted, notably around the uppermost leaf stalks; such damage causes the leaves to wilt and may result in premature leaf fall. The adult mites are 0.16 mm long, with about 45 abdominal tergites and a greater number of sternites.

CRATAEGUS

Epitrimerus piri (Nalepa)
Pear rust mite
The foliage of hawthorn (*Crataegus*) is sometimes bronzed following the development of infestations of this generally common mite. The pest sometimes occurs on nursery stock but is more important in pear orchards, russeting both the foliage and the developing fruits.

Eriophyes pyri crataegi
See under *Sorbus*, p. 419

Phyllocoptes goniothorax (Nalepa) (1089)
Hawthorn leaf gall mite
This mite causes a tight downward leaf-edge rolling on hawthorn (*Crataegus*), and is often present on cultivated plants. The rolled edges vary in colour from light green to yellowish green, often with a reddish tinge, and sometimes result in noticeable distortion of the leaf blade. Galling is apparent from spring to autumn and is often extensive, but not of significance, on unclipped hedges. The causal mites are whitish, *c.* 0.18 mm long, with about 54 abdominal tergites and sternites, and a pair of small forwardly directed setae arising from tubercles in front of the hind margin of the prodorsal shield. Forms of this mite also occur in whitish erinea on the underside of the leaves. These are often regarded as subspecies, and include *Phyllocoptes goniothorax malinus* and *P. goniothorax sorbeus* on crab-apple (*Malus*) and rowan (*Sorbus aucuparia*), respectively.

1087 Galls of *Aceria longisetosus* on underside of leaf of *Betula*.

1088 Hornbeam leaf gall mite (*Eriophyes macrotrichus*) damage to leaf of *Carpinus*.

1089 Galls of hawthorn leaf gall mite (*Phyllocoptes goniothorax*).

CYTISUS

Eriophyes genistae (Nalepa) (1090)
Broom gall mite

A locally common pest in parks and gardens, breeding throughout the spring and summer in galled buds of broom (*Cytisus*) and greenweed (*Genista*). Young buds are invaded in the spring, each then developing into a tight cluster of fleshy, pale green lobes, with a downy coating of whitish hairs. The galls measure about 20–30 mm across at maturity, and often persist on infested plants for several years. Infested shoots are stunted and, if attacks persist, bushes are severely disfigured. Mites inhabiting the galls are brownish, *c*. 0.13 mm long, with about 70 abdominal tergites and sternites, and a pair of backwardly directed setae arising from tubercles on the hind margin of the prodorsal shield.

EUONYMUS

Eriophyes convolvens (Nalepa) (1091)
Spindle leaf-roll gall mite

This mite causes an upward leaf-edge rolling on spindle (*Euonymus*). Attacks on wild European spindle (*E. europaeus*) are often extensive, and galling is sometimes also noted on cultivated bushes. Although unsightly, galls do not affect plant growth. The mites are whitish and relatively small (*c*. 0.11 mm long), with about 60 abdominal tergites and sternites, and a pair of small, forwardly directed, convergent, setae arising from tubercles in front of the hind margin of the prodorsal shield.

FAGUS

Aceria stenaspis (Nalepa) (1092)
syn. *A. stenaspis plicans* (Nalepa)

A locally common pest of beech (*Fagus sylvatica*), the mites causing considerable malformation and stunting of young leaves, and death of opening buds. The mites, which breed within the shelter of the galled tissue throughout the spring and summer, are whitish and *c*. 0.14 mm long, with about 75 abdominal tergites and sternites. Damage frequently occurs on garden hedges but is often overlooked. A species of *Acaricalus* is also associated with leaf deformation on beech, but may be merely an inquiline in buds damaged by *Aceria stenaspis*.

Aceria stenaspis stenaspis (Nalepa) (1093)

This subspecies breeds in marginal leaf-roll galls formed on the extreme edge of infested beech (*Fagus sylvatica*) leaves. The tissue rolls over the upper surface, to form a tight, hair-lined tube 1–2 mm in diameter; distortion and discoloration may also spread onto the leaf blades. The galls occur from late April or May onwards, changing from green to brown as they mature.

Eriophyes nervisequus (Canestrini) (1094)
Beech leaf-vein gall mite

A common species, forming hairy, white to brownish ridges along the major lateral veins on the upper surface of leaves of beech (*Fagus sylvatica*). The galls are most often noticed in June, during the early stages of development before the hairs darken. Mites inhabiting these galls are *c*. 0.13 mm long, with about 60 abdominal tergites and sternites, and a pair of backwardly directed setae arising from tubercles on the hind margin of the prodorsal shield.

Eriophyes nervisequus fagineus Nalepa (1095)
Beech erineum gall mite

This locally common subspecies is responsible for the formation of conspicuous erinea between the major veins on the underside of beech (*Fagus sylvatica*) leaves. The galls develop throughout the summer months, commencing as pale patches of enlarged, club-shaped hairs, amongst which various stages of the mite may be found; the erinea soon turn red and finally brown. Heavy infestations cause young leaves to curl at the edges.

1090 Gall of broom gall mite (*Eriophyes genistae*).

1091 Galls of spindle leaf-roll gall mite (*Eriophyes convolvens*).

1092 *Aceria stenaspis* damage to young shoots of *Fagus*.

1093 Galls of *Aceria stenaspis stenaspis* on leaf of *Fagus*.

1094 Galls of beech leaf-vein gall mite (*Eriophyes nervisequus*).

1095 Galls of beech erineum gall mite (*Eriophyes nervisequus fagineus*).

FRAXINUS

Aculus epiphyllus (Nalepa) (**1096–1097**)
Ash rust mite

This generally common mite causes extensive bronzing of the foliage of ash (*Fraxinus excelsior*) and is often present on nursery trees. In severe cases, the foliage becomes brittle and distorted, and the tips of new shoots turn black and die. Loss of terminal growth often results in the forking of young trees, and this is a serious problem in nurseries. The pale, yellowish, pear-shaped mites are *c*. 0.15 mm long, with about 30 abdominal tergites, and a pair of very short, backwardly directed setae arising from widely spaced tubercles on the hind margin of the prodorsal shield (cf. *Tegonotus collaris*, below). When present in large numbers, the mites are clearly visible against the darker background of infested leaves or bud scales.

1096 Ash rust mite (*Aculus epiphyllus*) damage to leaves of *Fraxinus*.

1097 Ash rust mite (*Aculus epiphyllus*) damage to shoots of

Eriophyes fraxinivorus Nalepa (**1098**)
Ash inflorescence gall mite

A widespread species, causing a distinctive galling of the inflorescences of ash (*Fraxinus excelsior*). The mites overwinter in bark crevices. In the spring, they invade the emerging inflorescences, depositing eggs and producing a succession of overlapping generations throughout the summer. The mites cause considerable distortion, the flower clusters swelling into a series of brownish lumps up to 20 mm across; galled flower stalks may also coalesce. The galled inflorescences remain attached to trees throughout the year and are particularly obvious after leaf fall. Infestations are often heavy on mature trees but any effect on vegetative growth appears to be slight. Adult mites are *c*. 0.18 mm long, with about 65 abdominal tergites and sternites, and a pair of backwardly directed setae arising from tubercles on the hind margin of the prodorsal shield.

Tegonotus collaris Nalepa
Often present on bronzed foliage of ash (*Fraxinus excelsior*), usually in company with *Aculus epiphyllus* (q.v.). The mites are *c*. 0.16 mm long and rather stumpy, with about 13 broad, roof-like abdominal tergites, and a pair of posteriorly directed setae arising from tubercles on the hind margin of the prodorsal shield.

FUCHSIA

Aculops fuchsiae Keifer
Fuchsia gall mite

This important pest of *Fuchsia* was first discovered in 1972 in Brazil. It has since been introduced into North America and, more recently, France. In 2006, following its arrival in Europe, the pest was found in the Channel Islands (both Guernsey and Jersey). In 2007 it was also discovered on fuchsia plants in southern England. The

1098 Galls of ash inflorescence gall mite (*Eriophyes*

mites cause severe galling of infested tissue, particularly at the shoot tips. Leaves at first become light green and felt-like, and eventually tinged with red. Infested plants, including the flowers, are also greatly deformed. There are several generations annually, and development from egg to adult takes about three weeks at a temperature of 18°C. The mites (*c.* 0.20–0.25 mm long) are pale yellow to whitish, with about 50 abdominal tergites and 70 sternites.

JUGLANS

Aceria erinea (Nalepa) (1099)
Walnut leaf erineum mite
A generally abundant mite, responsible for the formation of erinea on the underside of the leaves of walnut (*Juglans*). The galls appear as large, reddish-tinged blisters on the upper surface of the expanded leaves; the lower surface is coated with a felt-like mass of whitish hairs, within which the mites breed. Galling is often extensive on both nursery stock and mature trees but is of little or no importance. The causal mites are *c.* 0.22 mm long and whitish, with a pair of moderately long, posteriorly directed setae arising from tubercles on the hind margin of the prodorsal shield.

Aceria tristriatus (Nalepa) (1100)
syn. *Eriophyes tristriata* (Nalepa)
Walnut leaf gall mite
A locally common pest of walnut (*Juglans*), forming small (1–2 mm diameter) hard, dark red, pimple-like galls on the upper surface of leaves, each with a tiny opening through the lower surface. Galls are often located alongside the midrib and other major veins. Heavily infested leaves may be distorted, but there appears to be little or no impact on tree growth. Adult females (0.20–0.24 mm long) are pinkish to reddish and elongate, with a pair of moderately long, posteriorly directed setae arising from tubercles located on the hind margin of the prodorsal shield.

LAURUS

Several European species of eriophyid mite attack bay laurel (*Laurus nobilis*). The following two species are of particular significance.

Calepitrimerus russoi Di Stefano
Bay rust mite
This brownish-orange, pear-shaped mite (0.15–0.19 mm long) is a pest of bay laurel (Laurus nobilis) in various parts of central and southern Europe. The pest has also been found on greenhouse-grown plants in

Belgium, England and the Netherlands. The mites are free-living and cause noticeable bronzing of infested foliage. Heavy infestations also result in the development of black lesions on the underside of leaves, followed by desiccation and premature leaf fall.

Cecidophyopsis malpighianus (Canestrini & Massalongo)
Bay big bud mite
This little-known Mediterranean species feeds and breeds in the flower buds of bay laurel (*Laurus nobilis*). Infested buds swell considerably, and also become noticeably discoloured and distorted. The worm-like mites (*c.* 0.27 mm long) are whitish or yellowish in colour, and occur in vast numbers within infested buds. The pest was originally discovered in Italy, but has since also been found in certain other countries, including Belgium, England, the former Yugoslavia and the USA.

1099 Galls of walnut leaf erineum mite (*Aceria erinea*).

1100 Galls of walnut leaf gall mite (*Aceria tristriatus*).

MALUS

Aculus schlechtendali (Nalepa)
Apple rust mite

Although more frequently reported as a pest in apple orchards, infestations of this generally common species also occur on crab-apple (*Malus*) and are occasionally of significance on young trees. The mites are *c*. 0.17 mm long and yellowish brown; they are deuterogenous, breeding throughout the spring and summer on the underside of leaves. Deutogynes shelter during the winter in bark crevices and beneath bud scales. Heavy infestations lead to bronzing and shrivelling of the foliage, and may also affect the growth of new shoots.

Eriophyes pyri
Pear leaf blister mite
See under *Sorbus*, p. 419

Phyllocoptes goniothorax malinus
Apple leaf erineum mite
See under *Crataegus*, p. 411

POPULUS

Aceria dispar (Nalepa)
Associated with aspen (*Populus tremula*), the mites infesting the young, lateral shoots. Affected leaves are rolled and crinkled, and the internodes of heavily infested shoots become noticeably shortened, to form brush-like clumps of small, deformed leaves.

Phyllocoptes populi (Nalepa) (1101)
Poplar erineum mite
Associated with aspen (*Populus tremula*) and black poplar (*P. nigra*), inhabiting erinea on the underside of the leaves. The upper surface of affected leaves is disfigured by the development of pallid, blister-like areas. These galls should not be confused with similar distortions caused on poplar by the generally common fungal disease *Taphrina populina*, which produces bright yellow blisters on the underside of the leaves; in the case of the disease the upper surface of an infected leaf, although noticeably distorted, remains green.

PRUNUS

Aculus fockeui (Nalepa & Trouessart) (1102)
Plum rust mite
Although mainly a pest in damson and plum orchards, infestations of this generally common species also occur on various other kinds of *Prunus*, including ornamentals. Affected foliage becomes distinctly bronzed and may develop a characteristic yellowish flecking; small blotches sometimes also appear on the new wood of the young shoots. Damage of greatest significance is caused in the early spring by the overwintered mites as they feed on the newly developing tissue. This species is deuterogenous, with distinct winter and summer forms. Deutogynes are *c*. 0.16 mm long, with about 32 abdominal tergites and sternites; protogynes are *c*. 0.17 mm long, with about 30 abdominal tergites and 50 sternites.

Eriophyes padi (Nalepa) (1103)
Plum leaf gall mite
The erect, dark red leaf galls produced by this locally common species occur on blackthorn (*Prunus spinosa*) and certain other species of *Prunus*, and are sometimes of significance on nursery stock. The galls, which are very noticeable, tend to occur in tight clusters towards the middle of the base of the leaf blade. However, they cause only slight distortion, and do not affect tree growth. Occupants of the galls are *c*. 0.22 mm long, with about 55 abdominal tergites and sternites.

Eriophyes similis (Nalepa) (1104)
Plum pouch-gall mite
A widespread and locally common pest, often established in the wild on blackthorn (*Prunus spinosa*). Infestations also occur on ornamental *Prunus*, including nursery trees. The pale greenish to yellowish or reddish, pouch-like galls tend to occur around the periphery of the leaves, sometimes resulting in considerable distortion. The causal mites are *c*. 0.2 mm long, with about 50 abdominal tergites and sternites.

RHODODENDRON

Aculus atlantazaleae (Keifer) (1105–1106)
Azalea bud & rust mite
Although formerly restricted to North America, this pest is now established in various parts of Europe, including southern England and the Netherlands. Infestations occur on certain kinds of azalea (*Rhododendron*), especially Mollis hybrids. The pale yellowish-brown mites (*c*. 0.17 mm long) feed on the leaf bases and buds, and cause considerable distortion. They also occur on the expanding or expanded leaves, and such infestations result in noticeable bronzing of the foliage.

1101 Galls of poplar erineum mite (*Phyllocoptes populi*).

1102 Plum rust mite (*Aculus fockeui*) damage to leaf of *Prunus*.

1103 Galls of plum leaf gall mite (*Eriophyes padi*).

1104 Galls of plum pouch-gall mite (*Eriophyes similis*).

1105 Azalea bud & rust mite (*Aculus atlantazaleae*) damage to young shoot of *Rhododendron* 'Mollis'.

1106 Azalea bud & rust mite (*Aculus atlantazaleae*) damage to leaves of *Rhododendron* 'Mollis'.

ROBINIA

Vasates allotrichus (Nalepa) (**1107**)

False acacia rust mite

Infestations of this mite occur on false acacia (*Robinia pseudoacacia*), puckering and rolling the leaves; heavily infested foliage also becomes blackened and brittle. The mites are often responsible for extensive damage to nursery and specimen trees, affected shoots or branches developing a dull, sickly appearance. The pale yellowish adults are *c.* 0.15 mm long and stumpy, with about 45 abdominal tergites and a slender pair of backwardly directed setae arising from tubercles on the hind margin of the prodorsal shield (cf. *Vasates robiniae*).

Vasates robiniae (Nalepa) (**1108**)

This mite inhabits false acacia (*Robinia pseudoacacia*), causing discoloration and a marginal leaf rolling. Adults are similar in appearance to those of *Vasates allotrichus*, but have fewer (*c.* 25) abdominal tergites and stronger setae on the prodorsal shield.

SALIX

Aculops tetanothrix (Nalepa) (**1109**)

Willow leaf gall mite

Galls formed by this locally common mite occur on willow (*Salix*), especially crack willow (*S. fragilis*). They develop on the leaves from May onwards, and are often present in considerable numbers as irregular, hairy, yellowish-green to red swellings, each 2–4 mm in diameter.

Aculus truncatus (Nalepa) (**1110**)

This mite inhabits distinctive leaf-edge galls on the upper surface of the leaves of purple willow (*Salix purpurea*). On suitable hosts, galling is often extensive and may affect a considerable proportion of the leaves. Each gall becomes distinctly reddened throughout its length, and contains numerous whitish mites. Similar galls are formed on osier (*S. viminalis*) by the midge *Rabdophaga marginemtorquens* (p. 188).

1107 False acacia rust mite (*Vasates allotrichus*) damage to leaf of *Robinia*.

1108 *Vasates robiniae* damage to leaves of *Robinia*.

1109 Galls of willow leaf gall mite (*Aculops tetanothrix*).

1110 Galls of *Aculus truncatus* on leaves of *Salix purpurea*.

Eriophyes triradiatus (Nalepa) (**1111**)

Willow witches' broom gall mite

This widely distributed and generally common species occurs in witches' broom galls on various kinds of willow (*Salix*), including weeping willow (*S. vitellina* var. *pendula*). Although usually abundant in the galls, and breeding within them throughout the summer, the mites are not involved in their initiation or development. Adult females are *c*. 0.17 mm long, with about 80 abdominal tergites and sternites, and a pair of upwardly directed setae on the prodorsal shield.

SAMBUCUS

Epitrimerus trilobus (Nalepa) (**1112**)

Elder leaf mite

Widely distributed in association with elder (*Sambucus*), causing considerable distortion and discoloration of affected foliage. Infestations often occur on ornamental bushes, including nursery stock, but are most common on wild hosts.

SORBUS

Aculus aucuparia Liro

A free-living species on rowan (*Sorbus aucuparia*). The mites cause distinct bronzing of the foliage, and infestations are sometimes noted on nursery stock.

Eriophyes pyri (Pagenstecher) (**1113**)

Pear leaf blister mite

Although most important as a pest of pear fruit trees, this species (or species complex) is also associated with other rosaceous hosts, including ornamentals. The elongate (*c*. 0.22 mm long), brownish mites overwinter under bud scales. They become active in the early spring. Small, light green to yellowish, blister-like galls are then formed on the unfurling foliage, each with a

tiny aperture on its lower surface. Mites breed within these chambers throughout the summer. Colony development ends in the autumn, following the production of overwintering forms. The galls, which appear as pale blisters and gradually darken as they mature, are often common on rowan (*Sorbus aucuparia*), whitebeam (*S. aria*) and wild service tree (*S. torminalis*); galling also occurs on crab-apple (*Malus*), hawthorn (*Crataegus*) and ornamental pear (*Pyrus calleryana* 'Chanticleer'). The pest is usually of minor importance, but heavy attacks on young ornamentals are troublesome. On certain hosts the mites are afforded subspecific status: e.g. *Eriophyes pyri sorbi* on rowan; *E. pyri arianus* on whitebeam; *E. pyri torminalis* on wild service tree. The mites on hawthorn are often cited as *E. pyri crataegi* or treated as a distinct species (namely, *E. crataegi*), and those on *Sorbus* are sometimes collectively described as *E. sorbi*.

1111 Young gall of willow witches' broom gall mite (*Eriophyes triradiatus*) on *Salix*.

1112 Elder leaf mite (*Epitrimerus trilobus*) damage to leaves

1113 Pear leaf blister mite (*Eriophyes pyri*) damage to leaves

Phyllocoptes goniothorax sorbeus
See under *Crataegus*, p. 411.

SYRINGA

Eriophyes löwi Nalepa (**1114**)
Lilac bud mite
A widely distributed pest of lilac (*Syringa*). The mites infest the buds, causing desiccation and a proliferation of dwarfed lateral shoots. Expanded leaves are also invaded, and affected foliage becomes discoloured and the leaf margins distorted. The mites are c. 0.17 mm long, with about 60 tergites and sternites. Infestations also occur on privet (*Ligustrum vulgare*).

1114 Lilac bud mite (*Eriophyes löwi*) damage to leaves of *Syringa*.

TAXUS

Cecidophyopsis psilaspis (Nalepa) (**1115–1116**)
Yew gall mite
A widely distributed and generally common pest of yew (*Taxus*); particularly abundant on nursery stock and on regularly clipped bushes or hedges that produce an abundance of new growth. The mites breed throughout the year, and appear in large numbers in the spring. They then invade terminal and lateral buds on the young shoots. Infested buds fail to open, and instead expand into bloated mite-laden galls ('big buds'). Attacked buds and surrounding tissues also become blackened. Although damage on mature hosts is of little or no consequence, infestations on young plants are very disfiguring (often reminiscent of herbicide damage). The causal mites (*c*. 0.16 mm long) are whitish, with about 75 abdominal tergites and sternites; they lack setae on the prodorsal shield.

TILIA

Aceria exilis (Nalepa)
syn. *Eriophyes tiliae exilis* Nalepa
Leaf galls formed by this mite at the junction of two major veins may be found on lime (*Tilia*) from May or June onwards. The galls develop on the upper surface of the leaf as small, hairy, greenish-yellow to brownish pimples, the position of each being marked below by pale brownish hairs; such hairs also line the inner surface of the galls. The mites are similar in appearance to *Eriophyes tiliae* (q.v.). but have fewer (*c*. 60) abdominal tergites.

1115 Gall of yew gall mite (*Cecidophyopsis psilaspis*).

1116 Yew gall mite (*Cecidophyopsis psilaspis*) damage to young shoot of *Taxus*.

Aculus ballei (Nalepa) (**1117**)

An often common, free-living species that sometimes causes significant leaf bronzing on lime (*Tilia*) trees. Affected foliage appears dull and sickly. Heavy infestations affect the vigour of young trees, including nursery stock.

Eriophyes leiosoma (Nalepa) (**1118**)

syn. *E. tiliae liosoma* (Nalepa)

Lime leaf erineum mite

A generally common species, inhabiting large, irregular, white, downy patches on the underside of leaves of lime (*Tilia*) trees. The galls are formed from May onwards, their upper surface appearing light green; such galls are well developed by mid-summer, and affected tissue eventually turns brown. Although a considerable proportion of the leaf area may be affected, the galls are relatively shallow and usually cause little or no distortion of the leaf blade. However, attacks are sometimes heavy on small trees, resulting in considerable disfigurement.

Eriophyes tiliae (Nalepa) (**1119**)

syn. *E. gallarumtiliae* (Turpin); *E. tiliae* (Pagenstecher)

Lime nail-gall mite

An often abundant species, responsible for the elongated, tack-like galls which often occur in vast numbers on the upper surface of the leaves of large-leafed lime (*Tilia platyphyllos*). The galls are up to 15 mm long, and vary from light greenish to red or brown. They develop from May or June onwards but, although disfiguring the foliage of ornamental trees, appear to have little or no effect on plant growth. The causal mites are elongate, with about 75 abdominal tergites and sternites.

1117 *Aculus ballei* damage to leaves of *Tilia.*

1118 Galls of *Eriophyes leiosoma* on leaf of *Tilia.*

1119 Galls of lime nail gall mite (*Eriophyes tiliae*) on leaf of *Tilia platyphyllos.*

1120 Galls of *Eriophyes tiliae lateannulatus* on leaf of *Tilia cordata*.

1121 Galls of *Aculus ulmicola* on leaves of *Ulmus*.

1122 Galls of *Eriophyes filiformis* on leaf of *Ulmus*.

Eriophyes tiliae lateannulatus Schulze (**1120**)
syn. *E. tiliae rudis* Nalepa

This subspecies is associated with small-leafed lime (*Tilia cordata*), producing nail galls that are much smaller (up to 5 mm long) than those formed on large-leafed lime by *Eriophyes tiliae*; the mites are sometimes regarded as a distinct species.

ULMUS

Aculus ulmicola (Nalepa) (**1121**)
Elm bead-gall mite

An often abundant species, responsible for the development of tiny (*c.* 1 mm diameter) bead-like galls on the upper surface of the leaves of elm (*Ulmus*). Adult females overwinter under the bud scales, invading the underside of the unfurling leaves in the spring and inducing the formation of the distinctive galls. Breeding continues within the galls throughout the late spring and early summer; at the height of development each gall contains up to 200 or more mites. Infestations are often very extensive, and vast numbers of galls often occur on each infested leaf. As a result, the foliage of affected branches develops a distinctly roughened appearance. Heavy infestations have a detrimental effect on host plants, reducing their resistance to severe weather conditions and leaving them more susceptible to other disorders. Mites inhabiting these galls are *c.* 0.17 mm long, with about 55 abdominal tergites and characteristic, two-branched feather-claws.

Eriophyes filiformis (Nalepa) (**1122**)
Elm leaf blister mite

This southerly-distributed species causes the development of pouch-like galls on the leaves of elm (*Ulmus*), including amenity trees. The galls are much larger than those formed by the previous species but are usually less numerous; they cause slight distortion of the foliage but shoot growth is not affected. The mites are *c.* 0.17 mm long, with about 90 abdominal tergites.

YUCCA

Cecidophyopsis hendersonii (Keifer)

A North American species, introduced into Europe on imported *Yucca* plants. In Europe, the pest was first reported on greenhouse-grown blue-stem yucca (*Y. guatamalensis*) in Scandinavia. The upper surface of affected fronds appears dusted with whitish powder. Older infested leaves develop brown markings and eventually turn yellow and die. The causal mites are relatively stumpy and lack setae on the prodorsal shield.

Family **TARSONEMIDAE**

Small, elliptical, light brown to whitish mites with a distinct head-like gnathosoma, short, needle-like chelicerae and pronounced sexual dimorphism; the hind legs of females are clawless and those of males broadly elaborated and often inwardly flanged.

Hemitarsonemus tepidariorum (Warburton)
Fern mite

Recorded in Costa Rica, England and the USA as a pest of ferns, including hen & chick fern (*Asplenium bulbiferum*), Polystichum and *Pteris* in greenhouses.

DESCRIPTION
Adult female: 0.23 mm long; pale yellowish brown; body elongate-oval, the gnathosoma longer than broad and with the palps directed forwards. **Adult male:** 0.15–0.16 mm long; pale yellowish brown; hind leg with a broad triangular tooth on the inner margin of the tibia and a prominent tarsal claw; tibia and tarsus both longer than femur. **Egg:** 0.11–0.12 mm long; oval and whitish. **Larva:** similar to adult, but smaller and 6-legged.

LIFE HISTORY
The mites breed continuously throughout the year, with a succession of overlapping generations, and infestations are encouraged by warm, dark and humid conditions. Eggs are laid singly or in small groups close to the tips of the fronds (within the shelter of the furled leaflets or pinnae) or at the top of the rhizome between the scales. The eggs hatch within a few days at normal greenhouse temperatures but their development is greatly protracted in cool conditions. The larvae feed for 1–2 weeks before entering the quiescent nymphal stage. Adults appear 3–4 days later. Adults and larvae feed on the youngest tissue, imbibing sap from the surface cells which then collapse. All developmental stages and both sexes occur in abundance throughout the summer but winter populations consist mainly of adult females and eggs.

DAMAGE
Infested leaves are speckled with brown, attacked fronds becoming distorted and discoloured. The growth of heavily infested plants is severely checked.

Phytonemus pallidus (Banks) (**1123***)
syn. *Tarsonemus pallidus* Banks
Cyclamen mite

A major pest of greenhouse ornamentals, including African violet (*Saintpaulia hybrida*), azalea (*Rhododendron*), *Begonia*, busy lizzie (*Impatiens*), *Cyclamen, Gerbera, Gloxinia*, ivy (*Hedera*), Japanese aralia (*Fatsia japonica*), *Pelargonium, Petunia* and *Verbena*. In favourable situations infestations also survive on outdoor plants. Distinct biological races are associated with Michaelmas daisy (*Aster*) and strawberry – *Phytonemus pallidus asteris* and *P. pallidus fragariae*, respectively. Virtually cosmopolitan. Widely distributed in Europe.

DESCRIPTION
Adult female: 0.25 mm long; light brown and translucent; body oval-elongate and somewhat barrel-shaped; gnathosoma longer than broad, with the palps directed forwards; hind legs very thin, each bearing a long, whip-like seta. **Adult male:** 0.2 mm long; light brown and oval-bodied; hind legs broad, each with a very large femur bearing a rounded inner flange and terminating in a strong claw. **Egg:** 0.125 × 0.075 mm; elliptical, semitransparent and whitish. **Larva:** whitish, with hind part of body triangular; 6-legged.

1123 Cyclamen mite (*Phytonemus pallidus*) damage to leaf of *Fatsia*.

LIFE HISTORY

In greenhouses this species is active throughout the year, breeding continuously whilst conditions remain favourable. The mites are light-shy and tend to occur on the young, succulent tissue of host plants. All stages (eggs, larvae, quiescent nymphs and adults) shelter within leaf folds, amongst leaf hairs and between bud scales. As the tissue ages and hardens, the mites move to younger, more suitable feeding and breeding sites, commonly invading the still-furled leaves and unopened flower buds. The mites may also spread from plant to plant, particularly if leaves or shoots or adjacent hosts overlap, but they rarely if ever wander over the soil or the greenhouse staging. There are several overlapping generations annually, mites passing from egg to adult in about 2–3 weeks at temperatures of 20–25°C; the rate of development is much reduced at lower temperatures, the egg stage becoming particularly protracted. Although males occur, typically during the summer months, they are usually greatly outnumbered by females and reproduction is mainly parthenogenetic. There are also several overlapping generations annually on outdoor plants, populations reaching a peak in August or September, but breeding usually ceases completely during the winter months.

DAMAGE

Infested foliage become brittle, discoloured and crinkled, the margins of young leaves often rolling tightly inwards; flower buds are also affected. Attacked plants are stunted and young growth significantly distorted; when infestations are severe, leaves, flower buds or complete plants may be killed. Mites on Michaelmas daisy plants, especially *Aster novi-belgii*, cause severe scarring of flower stems, affected 'flowers' being converted into rosettes of small, green leaves.

Polyphagotarsonemus latus Banks (1124–1126)
Broad mite

A tropical and subtropical pest which, in temperate countries, infests a wide variety of greenhouse plants, including ornamentals such as African violet (*Saintpaulia hybrida*), *Begonia*, busy lizzie (*Impatiens*), *Chrysanthemum*, *Cyclamen*, *Dahlia*, *Gloxinia*, *Fuchsia*, *Gerbera*, ivy (*Hedera*), *Hibiscus*, Japanese aralia (*Fatsia japonica*) and stock (*Matthiola*). Well established in Europe, but less frequently reported than formerly.

DESCRIPTION

Adult female: 0.14–0.2 mm long; whitish and translucent, but often greenish or yellowish; body very broad and oval (cf. *Phytonemus pallidus*). **Adult male:** 0.11–0.17 mm long; whitish and translucent; body short and broad, but tapered posteriorly; legs long, the hind pair relatively stout. **Egg:** 0.11 × 0.07 mm; flattened, smooth ventrally; several rows of large, white, mushroom-like tubercles dorsally. **Larva:** similar to adult but smaller and 6-legged.

LIFE HISTORY

The mites feed mainly on the underside of leaves but also invade unopened or unfurling buds and other plant tissue. Female mites, that greatly outnumber males, normally live for about ten days, each depositing up to 50 eggs. Breeding is rapid in warm conditions, with eggs hatching in 2–3 days and larvae feeding for four days at normal greenhouse temperatures. The quiescent nymphal stage is passed entirely within the bloated larval skin. Adult males often carry female nymphs around, holding them aloft with a genital sucker.

DAMAGE

New growth of infested plants becomes stunted and discoloured, and is often shiny, brittle and distorted. On some hosts, including cyclamen, flowers are malformed and unopened buds may drop off. Heavily infested plants may be killed.

Steneotarsonemus laticeps (Halbert) (1127)
syn. *Tarsonemus approximatus* Banks
Bulb scale mite

An important pest of *Hippeastrum* and forced *Narcissus*; also associated with other members of the Amaryllidaceae, including *Eucharis*, Scarborough lily (*Vallota purpurea*) and *Sprekelia*. Present in several parts of Europe, including England, Germany, Ireland, the Netherlands, Poland and Sweden; also found in North America.

DESCRIPTION

Adult female: 0.2 mm long; light brown and translucent; gnathosoma broader than long; palps directed inwards; hind legs thin, each terminating in a long seta. **Adult male:** similar to female but smaller, and with robust hind legs. **Egg:** 0.1 mm long; oval, whitish and translucent. **Larva:** similar to adult but smaller and 6-legged.

1124 Broad mite (*Polyphagotarsonemus latus*) damage to leaves of *Gloxinia*.

1125 Broad mite (*Polyphagotarsonemus latus*) damage to leaf of *Fatsia*.

1126 Broad mite (*Polyphagotarsonemus latus*) damage to flowers of *Cyclamen*.

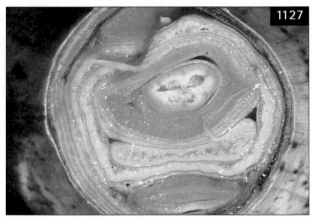

1127 Bulb scale mite (*Steneotarsonemus laticeps*) damage to bulb of *Narcissus*.

LIFE HISTORY

Bulbs are invaded in August and September. The mites enter the spaces between the shrinking scales to feed on the surface of the tissue, particularly around the neck region. Breeding is continuous whilst conditions remain suitable, the lifecycle (which includes egg, larval and quiescent nymphal stages) being completed in about two weeks at bulb-forcing temperatures. Mite development is greatly protracted on infested narcissus bulbs planted out in the autumn, but increases in response to warmer conditions in the following spring and early summer. At this stage the mites occur both within the bulbs and on aerial parts of the plants. However, by the time of lifting in mid-summer most will have moved back into the neck region.

DAMAGE

Hippeastrum: vegetative growth from infested bulbs becomes spotted, and streaked or scarred with red; flowers are malformed and may wither. **Narcissus:** infested bulbs lack vigour, producing weakened, distorted, often sickle-shaped leaves and small, malformed flowers. Emerging foliage tends to be bright green, and lacks the normal greyish bloom. Later, the leaves become streaked with yellow; they are also scarred, and the leaf edges appear saw-like. Heavy attacks result in a marked reduction in both crop yield and flower quality; they also cause early senescence of foliage and, sometimes, death of bulbs. Bulbs in store become very dry and, particularly in the neck region, display brown streaks of dead tissue if sliced across.

Family **TETRANYCHIDAE** (spider mites)

Spider-like mites with long, needle-like chelicerae and a thumb-claw on each palp. They develop from egg to adult through larval, protonymphal and deutonymphal stages.

Bryobia kissophila van Eyndhoven (**1128**)
Ivy bryobia mite
Common in the wild on European ivy (*Hedera helix*) and often a troublesome pest on ornamental cultivars. Present throughout Europe.

DESCRIPTION
Adult female: 0.7 mm long; dark reddish brown or red; body oval and rather flat, with spatulate dorsal setae; first pair of legs very long. **Egg:** 0.2 mm across; dark red and more or less spherical. Larva: bright reddish-orange; 6-legged. **Nymph:** dark red, brown or dark green; 8-legged.

LIFE HISTORY
Adult and juvenile mites are present on the upper surface of ivy foliage throughout much of the year. However, populations often decline during the summer. The mites then occur on clover (*Trifolium*) and usually return to ivy in August. Eggs tend to be deposited on supporting stakes and walls rather than on host plants. Breeding is continuous throughout the year, with about 5–8 overlapping generations, the duration of each generation varying according to ambient temperatures. Although development of the mites is greatly protracted during cold winter weather, there is no diapause stage in the lifecycle. Males are unknown and reproduction is entirely parthenogenetic. Bryobia mites do not produce webbing.

1128 Ivy bryobia mite (*Bryobia kissophila*) damage to leaves of *Hedera*.

DAMAGE
Infested foliage becomes pallid and silvery, and may turn brown.

Bryobia cristata (Dugés)
Grass/pear bryobia mite
A polyphagous species, occurring throughout the year on various grasses and herbaceous plants. Infestations often occur on ornamentals such as bell flower (*Campanula*), *Cyclamen*, *Dianthus*, gentian (*Gentiana*), *Iris*, *Polyanthus* and saxifrage (*Saxifraga*). The mites feed mainly on the upper surface of the leaves, and cause a mottling and general silvering of tissue. In common with *Bryobia kissophila*, breeding is continuous throughout the year, with many generations annually. During May the mites often disperse from their normal hosts to various trees and shrubs, including fruit trees and ornamentals such as flowering cherry (*Prunus*), hawthorn (*Crataegus*) and rose (*Rosa*), where two summer generations occur before a return migration to herbaceous plants. The mites are distinguished from *Bryobia kissophila* by the narrower dorsal setae and by other microscopic features.

Bryobia praetiosa Koch
Clover bryobia mite
During the spring vast numbers of this mite occur on sunny walls, particularly on those of new or recently renovated buildings. Fully grown nymphs and adults of this and the previous species also invade buildings through doors and windows, to moult or to lay eggs in cracks and crevices. Eggs deposited in the spring produce a summer generation of mites which feed on grasses and herbaceous plants (including certain ornamentals) and mature by the autumn. Eggs laid in late summer or autumn produce mites which develop throughout the winter and reach maturity in the following spring. The mites are structurally similar to *Bryobia kissophila*.

Bryobia rubioculus (Scheuten)
Apple & pear bryobia mite
This species overwinters in the egg stage and usually completes no more than three generations annually. It occurs mainly on rosaceous fruit trees, but also attacks related ornamentals such as crab-apple (*Malus*) and flowering cherry (*Prunus*). Although the mites feed on the upper surface of leaves, they often congregate on the trunks and branches. Such aggregations occur mainly in late May and June, and in August and September. The mites also cluster beneath the shoots and main branches whilst moulting from one growth stage to the next, and masses of greyish-white cast skins (which are a

1129 *Eotetranychus carpini* damage to leaves of *Carpinus.*

1130 Lime mite (*Eotetranychus tiliarum*) damage to leaves of *Tilia.*

characteristic sign of an infestation) soon accumulate on the trees. The foliage of infested trees is often malformed and also becomes pale, silvery and brittle; damaged leaves eventually turn brown and may fall prematurely.

NOTE
Oribatid mites (order Cryptostigmata), usually the cherry beetle mite (*Humerobates rostrolamellatus*) (a shiny, dark red to blackish, egg-like and short-legged species about 1 mm long), often cluster in considerable numbers on the bark of trees; they are harmless, feeding mainly on algae and lichens, and should not be mistaken for bryobia mites.

Eotetranychus carpini (Oudemans) (**1129**)
Locally common on various trees, including maple (*Acer*), alder (*Alnus*), common hazel (*Corylus avellana*), hornbeam (*Carpinus betulus*), oak (*Quercus*) and willow (*Salix*), and occasionally noted on cultivated plants. In Europe most often reported in England, Germany and the Netherlands; also present in North and Central America.

DESCRIPTION
Adult female: 0.4 mm long; light green or greenish yellow, with red eyes. **Egg:** 0.1 mm across; light green and globular.

LIFE HISTORY
Adult females hibernate on host plants in bark crevices and other suitable shelter, reappearing in the following spring. They then invade the underside of leaves to form small, compact colonies. These are sheltered by often dense, silken webs. Breeding continues from April to October, and there are about six generations in a season.

DAMAGE
Infested leaves become pallid, usually visible only from below. However, heavy infestations may cause the upper surface of leaves to become speckled with yellow. Plant growth is not noticeably affected.

Eotetranychus fagi (Zacher)
Beech red spider mite
This species occurs on beech (*Fagus sylvatica*) in various parts of mainland Europe, including Belgium, Germany, Italy, Poland and Switzerland; recently, the mite has also been found in England. Infestations cause extensive bronzing of foliage, and damage to nursery stock and garden hedges is of particular concern.

Eotetranychus tiliarum (Hermann) (**1130**)
Lime mite
A sporadically important pest of lime (*Tilia*), particularly on established trees. Widely distributed in Europe; also present in the eastern USA.

DESCRIPTION
Adult female: 0.4 mm long; yellowish to orange-red, with red eyes and long, narrow dorsal setae. **Egg:** 0.1 mm across; yellowish white and globular.

LIFE HISTORY
This species overwinters in the adult stage. In the following spring the mites become active, invading the newly developing leaves upon which eggs are laid. The mites feed on the underside of the leaves, particularly alongside the veins, young individuals passing through larval, protonymphal and deutonymphal stages before becoming adults. There are several overlapping generations each year, populations reaching a peak from late summer onwards. In the early autumn, female mites

1131 *Eotetranychus populi* damage to leaf of *Salix*.

1132 Conifer spinning mite (*Oligonychus ununguis*) damage to needles of *Picea*.

sometimes aggregate in vast numbers amongst webbing on the trunks and main branches of host trees.

DAMAGE
The mites cause noticeable bronzing of leaves, spoiling the appearance of specimen trees; they also cause premature defoliation, heavily infested leaves shrivelling and dying. Host trees are sometimes disfigured by glistening sheets of polythene-like webbing.

Eotetranychus populi (Koch) (**1131**)
Associated with broad-leaved willows such as grey willow (*Salix cinerea*) and various species of poplar (*Populus*), including aspen (*P. tremula*). Present as a pest in various parts of Europe, including England; also found in North America. The adults overwinter under loose bark or within bark crevices on the trunks and branches of host trees, and emerge in the spring. They then invade the underside of the young leaves and soon deposit eggs. The mites often occur on young sucker growth, individuals sheltering amongst the leaf hairs and beneath silken webs. There are several generations annually, and breeding continues so long as conditions remain favourable. Infested leaves become noticeably discoloured and also tend to curl, but damage is of significance only on young trees.

Oligonychus ununguis (Jacobi) (**1132**)
Conifer spinning mite
A generally common and virtually worldwide pest of conifers; most important on young spruce (*Picea*) trees. Present throughout Europe.

DESCRIPTION
Adult: 0.2–0.5 mm long; dark green or orange to brownish or blackish. Egg: greenish brown to orange-red; spherical, with a dorsal spine (stipe). **Larva:** pinkish but soon turning greenish; 6-legged. **Nymph:** greenish; 8-legged.

LIFE HISTORY
Eggs overwintering on the shoots hatch from April or early May onwards. Mites then feed for 2–3 weeks, passing through a larval and two nymphal stages before becoming adults. There are several generations annually, usually up to five. The mites produce considerable quantities of webbing. Summer eggs are laid on the shoots and needles; winter eggs are usually deposited close to the base of the needles.

DAMAGE
Infested needles are noticeably discoloured, becoming mottled and yellowish and eventually turning brown. Affected needles may also fall prematurely, checking shoot growth. Heavily affected shoots become curved and develop with shortened internodes. Spruce seedlings and transplants may be killed, particularly if being raised in dry soil conditions.

1133 Fruit tree red spider mite (*Panonychus ulmi*) damage to leaves of *Sorbus aucuparia*.

1134 *Schizotetranychus schizopus* damage to leaf of *Salix*.

Panonychus citri (McGregor)
Citrus red spider mite

Although found mainly on citrus, this pest also occurs on many other hosts, including evergreen ornamentals such as Californian bay laurel (*Umbellularia californica*), cherry laurel (*Prunus laurocerasus*), *Elaeagnus*, Japanese holly (*Ilex crenata*), Japanese yew (*Taxus cuspidata*) and Mexican orange (*Choisya ternata*). Widely distributed in tropical and subtropical regions, including parts of southern Europe; sometimes also reported in greenhouses in more northerly regions.

DESCRIPTION
Adult female: 0.5–0.6 mm long; body dark purplish red to purplish red, and distinctly rounded, with strong setae arising from pale tubercles. **Adult male:** similar to female but smaller, narrower-bodied and tubercles less noticeable. **Egg:** bright red and more or less spherical. **Larva:** dark red; 6-legged. **Nymph:** brick-red; 8-legged.

LIFE HISTORY
Eggs are laid on the leaves of host plants, each held to the surface by several radiating strands of silk. They hatch within 1–3 weeks, depending upon temperature. The immature stages feed for up to 2 weeks before moulting into adults. Breeding is continuous throughout the year and reaches an optimum at temperatures of about 25°C.

DAMAGE
The mites cause a noticeable silvering, yellowing or speckling of infested leaves; especially under drought conditions, they may also cause premature leaf fall. Heavy infestations weaken host plants and, in severe cases, young shoots may be killed.

Panonychus ulmi (Koch) (**1133**)
Fruit tree red spider mite

This widespread and generally common mite is an important pest of apple and various other fruit crops in many parts of the world; it also occurs on ornamental trees and shrubs such as almond (*Prunus dulcis*), Cotoneaster, crab-apple (*Malus*), flowering cherry (*Prunus*), flowering currant (*Ribes sanguineum*), hawthorn (*Crataegus*), Japanese quince (*Chaenomeles japonica*) and rowan (*Sorbus aucuparia*). Eggs overwinter on the spurs and smaller branches, and hatch from April to mid-June. There are then several overlapping generations throughout the summer months. Mite numbers decline from September onwards as breeding ceases and winter eggs are laid. Unlike *Tetranychus urticae* (p. 430) the mites do not inhabit silken webs. However, damage caused by both species is similar, and heavy infestations lead to significant leaf bronzing and premature leaf fall. Adult females are 0.4 mm long and dark red, with short, pale legs. They are most commonly found (along with adult males, nymphs, 6-legged larvae and summer eggs) on the underside of the leaves.

Schizotetranychus schizopus (Zacher) (**1134**)
Infestations of this widely distributed and often common mite occur mainly on crack willow (*Salix fragilis*), white willow (*S. alba*) and other narrow-leaved willows. The mites feed on the underside of the leaves, particularly along either side of the midrib, causing the foliage to become discoloured. Damage is often severe, affecting the growth and appearance of plants. In exceptional circumstances, the trunks and branches of host plants become coated in sheets of webbing. Eggs, which are laid along the midrib, are pale yellowish and somewhat flattened, with a dorsal stipe;

the overwintering eggs are reddish-orange and are usually found between bud scales or in other sheltered positions on the bark. The mites are small (0.2–0.4 mm long), greenish or reddish, with blackish markings, and relatively flat-bodied.

Stigmaeopsis celarius Banks
syn. *Schizotetranychus celarius* (Banks)
Bamboo mite

A widely distributed oriental pest of bamboo (e.g. *Phyllostachys*). Now present in many parts of the world, including Australasia, Europe (e.g. Belgium, Britain, France, Italy and the Netherlands) and the USA. Bamboo mite is considered a complex of several social species, including *Stigmaeopsis nanjingensis* which has now become well established in northern Italy.

DESCRIPTION
Adult: 0.4–0.6 mm long; whitish to translucent. **Egg:** 0.1 mm across; spherical.

LIFE HISTORY
Overwintered adult females commence egg laying in the early spring, each initiating a small, discrete colony beneath a densely woven silken web (often described as a 'nest') on the underside of a young leaf. Mites exhibit social behaviour, and typically defaecate at one and the same point at the periphery of the communal web; adults also demonstrate parental care, defending their nests against predators. Breeding continues throughout the summer and autumn (although interrupted in hot conditions by a period of aestivation), with a succession of overlapping generations.

DAMAGE
Infested leaves develop distinct whitish or yellowish-white blotches on the upper surface, each coinciding with the presence below of a mite colony. Attacked leaves may also become pallid, and flecked with silvery white where the contents of individual plant cells have been removed. Photosynthetic activity is impaired, and leaves may be shed prematurely.

Tetranychus urticae Koch (1135–1136)
Two-spotted spider mite

An often abundant pest of greenhouse and outdoor plants, including ornamentals such as azalea (*Rhododendron*), buddleia (*Buddleja*), broom (*Cytisus*), busy lizzie (*Impatiens*), calla lily (*Zantedeschia aethiopica*), *Ceanothus*, *Chrysanthemum*, *Dahlia*, diviner's sage (*Salvia divinorum*), *Freesia*, *Fuchsia*, *Hippeastrum*, *Hydrangea*, Mexican orange (*Choisya ternata*), morning glory (*Ipomoea*), mulberry (*Morus*), passion flower (*Passiflora*), *Phormium tenax*, poinsettia (*Euphorbia pulcherrima*), primrose (*Primula vulgaris*) and rose (*Rosa*). Cosmopolitan. Present throughout Europe.

DESCRIPTION
Adult female: 0.5–0.6 mm long; pale yellowish or greenish, with two dark patches on the body (overwintering form orange); body oval, with moderately long dorsal setae; striae on the hysterosoma form a diamond-shaped pattern. **Adult male:** similar to female but body smaller, narrower and more pointed. **Egg:** 0.13 mm across; globular and translucent. **Larva:** light green, with darker markings; 6-legged. **Nymph:** light green, with darker markings; 8-legged.

LIFE HISTORY
Female mites overwinter amongst debris, in dry crevices in the soil and in other suitable shelter, often clustering in cracks in greenhouse structures, stakes and

1135 Two-spotted spider mite (*Tetranychus urticae*) damage to leaves of *Choisya*.

poles. They become active in March or April, and invade host plants to feed and eventually deposit eggs. There are several overlapping generations of summer forms annually, mites passing through egg, larval, protonymphal and deutonymphal stages before maturing; males often omit the deutonymphal stage. Development is completed in less than two weeks at temperatures above 20°C, but is greatly protracted below 12°C, extending over almost two months at 10°C; outdoors, therefore, there tend to be fewer generations than in protected situations. Colonies occur mainly on the underside of the expanded leaves, the various mite stages being sheltered by fine webs of silk. When populations are large, these webs often become extensive; they may then cover complete leaves and parts of shoots and stems. During September, in response to short days (daylight of less than 14 hours), orange-coloured winter-female forms are produced; after mating, these seek shelter for the winter. As plant vigour also declines, breeding ceases and males and summer females all die.

DAMAGE

Infested leaves are speckled with yellow and often become extensively chlorotic. This affects both the vigour and appearance of plants; hosts may also be disfigured by webbing. Heavy infestations, which are most likely to occur in hot, dry conditions, cause considerable stunting and may result in the eventual death of plants.

Tetranychus cinnabarinus (Boisduval) (**1137**)
Carmine spider mite

This subtropical species occurs throughout Europe, but in temperate areas is confined mainly to greenhouses. Infestations develop on various herbaceous plants, including cacti, the foliage often becoming coated in masses of webbing; affected plant tissue also becomes discoloured. Typically, infested leaves of carnation (*Dianthus caryophyllus*) and pink (*D. plumarius*) tend to curl downwards. Eggs are laid singly under leaves or on the webbing, the pattern of development following that of *Tetranychus urticae*. There are several overlapping generations annually but there is no winter diapause; the rate of development is also less rapid than that of the previous species, carmine spider mite being adapted to higher temperatures. Adult females are mainly red, with dark internal markings visible on either side of the body; the nymphal stages are green and the eggs whitish to pink (cf. those of *Tetranychus urticae*). This species crosses regularly with *Tetranychus urticae*, particularly in greenhouses, resulting in hybrid populations.

1136 Two-spotted spider mite (*Tetranychus urticae*) damage to leaves of *Impatiens*.

1137 Web of carmine spider mite (*Tetranychus cinnabarinus*) on *Aporocactus*.

Family **TENUIPALPIDAE** (false spider mites)

Distinguished by the absence of a thumb-claw on the palps. Also, the mites do not produce webbing.

Brevipalpus obovatus Donnadieu
syn. *Tenuipalpus inornatus* Banks

A tropical or subtropical species, accidentally introduced into Europe where it is a minor pest of greenhouse ornamentals such as azalea (*Rhododendron*), bell flower (*Campanula*), *Cissus*, *Gardenia* and ivy (*Hedera*). In Europe found in various countries, including Austria, England, Germany and the Netherlands.

DESCRIPTION

Adult female: 0.25–0.30 mm long; red; body flat and egg-shaped, with a reticulate pattern on the idiosoma; five pairs of short dorsolateral setae on the hysterosoma; palps 4-segmented; legs relatively short. **Egg:** 0.1×0.07 mm; bright red and elliptical.

LIFE HISTORY

Mites occur on both sides of young leaves but are most numerous on the underside of older foliage where they often congregate around the margins. Eggs are deposited close to the midrib, often in clusters of several hundred. They hatch in 2–3 weeks. The juvenile stages feed for 3–4 weeks before attaining maturity. Breeding continues so long as conditions remain favourable, and reproduction is typically parthenogenetic.

DAMAGE

The mites cause brown, necrotic areas to develop on host plants, the discoloration commonly occurring along either side of the midrib or appearing as a fine rusty or bronze-like speckling over the entire leaf. Heavy infestations check the growth of plants and may result in premature senescence and defoliation. On ivy, damaged leaves are often 'cupped' and reduced in size, and growth from infested buds is typically weak and pallid.

Brevipalpus phoenicis (Geijskes)

This mite is very similar to *Brevipalpus obovatus*. Although of most importance as a pest of *Citrus* and tea plant (*Camellia sinensis*), it also damages a wide range of other plants. It is reported from various parts of the world, including the Netherlands.

Brevipalpus russulus (Boisduval)

A Central American species, introduced into Europe many years ago; sometimes reported as a pest of cacti (e.g. *Mammillaria*) and succulents, most frequently in Belgium, France, Germany and the Netherlands. Infested plants often develop a reddish-grey discoloration, and growth is checked. The mites are similar in appearance to *Brevipalpus obovatus* but possess an additional pair of dorsolateral setae on the hysterosoma.

Family **SITEROPTIDAE**

Siteroptes avenae Müller
syn. *S. graminum* (Reuter)
Grass & cereal mite

This mite is reported occasionally in England and mainland Europe on carnation (*Dianthus caryophyllus*). The mites occur in association with the fungus *Fusarium poae*, causing a brown necrosis of petals and, sometimes, death of opening buds. This symptom is commonly known as 'bud rot'. Bud rot on carnation is less frequent than the condition described as 'silver top', which occurs following the same mite/fungus association on cereals and grasses. The mites breed parthenogenetically on cereals and grasses throughout the summer, passing through egg, larval and nymphal stages, all of which develop within the much swollen, sac-like body of gravid maternal females. Such females measure up to 2 mm across, and are glistening, hyaline-whitish in appearance. Infestations tend to spread onto greenhouse-grown carnations in autumns that follow a spell of hot, dry weather.

Family **ACARIDAE**

Rhizoglyphus callae Oudemans (**1138**)
A bulb mite

An often common, usually secondary pest of bulbs, corms and tubers, including *Freesia*, *Gladiolus*, hyacinth (*Hyacinthus orientalis*), lily (*Lilium*), *Narcissus* and tulip (*Tulipa*). Present throughout Europe; also found in North America.

DESCRIPTION

Adult: 0.7 mm long; body oval, smooth, whitish and very shiny, with discrete internal brownish patches and several long hairs projecting back beyond the tip of the hysterosoma; legs reddish brown and stout. **Egg:** 0.2 × 0.1 mm; hyaline to whitish, smooth and shiny. **Larva:** whitish and shiny; 6-legged. Nymph: similar to adult but smaller. **Hypopus:** 0.3 mm long; dark brown; legs short, very stout and with large claws.

LIFE HISTORY

Mites occur in large numbers within diseased and otherwise unhealthy bulbs, corms or tubers. They breed continuously under suitable conditions, passing through egg, larval and two nymphal stages. Completion of the lifecycle takes 1–4 weeks depending upon temperature. A phoretic hypopal stage appears in some populations; these hypopi attach themselves to passing insects, such as small narcissus flies (*Eumerus* spp.), and are then transported to other hosts, thereby spreading infestations.

DAMAGE

General: infestations of this secondary pest may lead to the complete breakdown and destruction of bulbs, corms or tubers, the internal tissue turning blackish and powdery. **Freesia and gladiolus:** attacked roots develop dark brown streaks, and are often mined internally. Also, if healthy corms are planted into heavily infested soil, growing points and leaves are affected; these become distorted and the leaves develop ragged or saw-toothed edges.

Rhizoglyphus robini Claparède
A bulb mite

This species also infests bulbs and tubers but appears to find diseased hosts less favourable than relatively healthy ones. The mites are very similar in appearance to *Rhizoglyphus callae*. However, *R. robini* is slightly larger, and has noticeably shorter body hairs and very short dorsal idiosomal setae. Over the years, both species have been cited under the name *R. echinopus*.

1138 Bulb mite (*Rhizoglyphus* sp.) damage to bulb of *Tulipa*.

Chapter 4
Miscellaneous Pests

WOODLICE

Family ARMADILLIDIIDAE

Woodlice with flagellar segment of the antennae divided into two sections; body strongly arched, and individuals capable of rolling into a ball.

Armadillidium nasatum Budde-Lunde (**1139**)
Blunt snout pillbug
Widespread and often common in heated greenhouses, occasionally causing damage to ornamental plants.

DESCRIPTION
Adult: 21 mm long; grey, with a pale median line and pale lateral patches; snout drawn into a narrow, square-sided projection.

LIFE HISTORY
Adults occur throughout the year, usually hibernating in piles of soil. Breeding occurs from April to October, there being two or more generations each year. Females deposit about 50 eggs, and these are retained in a special pouch on the underside of the thorax. The eggs hatch in about 1–2 months. Young, white, 12-legged woodlice (each about 1 mm long) then escape from the maternal pouch to begin feeding. Individuals acquire the typical adult coloration after about two weeks, and a seventh pair of legs appears after two moults. Further moults occur, even after the adult stage is attained, but individuals do not reach sexual maturity until the following year. Mortality of young woodlice is high and few individuals survive for more than a couple of months; adults, however, may survive for up to four years.

DAMAGE
Woodlice produce irregular holes in leaves, mainly in those close to the ground; growing points of young plants may also be destroyed. Stems of potted plants and transplants are often grazed but those of older plants are rarely attacked.

Armadillidium vulgare (Latreille) (**1140**)
Common pillbug
This species often occurs in unheated greenhouses and garden frames, causing damage to plant roots; seedlings are frequently destroyed and the growth of older plants may also be affected if soil infestations are heavy. Adults are 18 mm long, dark slate-grey to reddish, with several pale patches along the body; the snout is slightly raised.

1139 Blunt snout pillbug (*Armadillidium nasatum*).

1140 Common pillbug (*Armadillidium vulgare*).

Family **ONISCIDAE**

Woodlice with flagellar segment of the antennae divided into three sections.

Oniscus asellus Linnaeus (**1141**)
Grey garden woodlouse
Widespread and common, often sheltering in considerable numbers beneath bricks, stones and pieces of rotting wood. Damage is often caused in greenhouses to the young, tender growth of ornamentals, especially carnation (*Dianthus caryophyllus*) and sweet pea (*Lathyrus odoratus*); aerial roots of orchids are also attacked. Adults are 16 mm long, shiny grey, with irregular pale patches; the body is relatively broad.

Family **PORCELLIONIDAE**

Woodlice with flagellar segment of the antennae divided into two sections; body distinctly flattened.

Porcellio dilatatus Brand
This species occurs in abundance beneath seed boxes and flower pots in cold-frames and in unheated greenhouses, often attacking the roots of cultivated plants. Adults are 15 mm long, and greyish brown with distinctive lateral stripes; the body is noticeably roughened, and the telson, which reaches just beyond the endopodites, usually has a rounded tip.

Porcellio laevis Latreille
A locally common species in manure heaps and, if introduced into greenhouses, damaging to the roots of ornamental plants, including various ferns. Adults are 18 mm long, brown, smooth and glossy; the telson has a pointed tip and the uropods are very long.

Porcellio scaber Latreille (**1142**)
Garden woodlouse
This widespread and common species sometimes damages plants in cold-frames and seed boxes, but is not usually a problem in greenhouses. Adults are 17 mm long, dark slate-grey, with a roughened body and a pointed telson; the basal antennal segment is usually orange. Juveniles are often yellow or orange.

1141 Grey garden woodlouse (*Oniscus asellus*).

1142 Garden woodlouse (*Porcellio scaber*).

MILLEPEDES

Millepedes are vegetarian creatures that feed mainly in the soil on soft, decaying plant tissue and debris. Although possessing weak mouthparts, millepedes cause direct damage to plant roots, seeds and germinating seedlings. Attacked tissue is often then invaded by rot-inducing bacteria and fungi. Millepedes also feed on bulbs, corms and other parts of plants, usually enlarging wounds initiated by pests such as slugs and wireworms. Millepedes are light-shy; they tend to be most numerous amongst leaf mould and vegetable rubbish, and in soil with a high organic content. Although favoured by damp conditions, millepedes often cause damage during spells of dry weather by tunnelling into plant tissue in search of moisture.

Family **BLANIULIDAE** (spotted snake millepedes)

Blaniulus guttulatus (Bosc) (**1143**)
Spotted snake millepede
A generally abundant millepede, often causing damage to outdoor seeds, seedlings and tulip (*Tulipa*) bulbs. Attacks also occur in greenhouses, particularly on carnation (*Dianthus caryophyllus*) roots, *Chrysanthemum* stools and *Cyclamen* corms. Adults are 8–18 mm long, snake-like, mainly white and translucent, with a series of reddish or purplish spots along each side; the body is composed of about 50 segments, most of which bear two pairs of short legs. Eggs are laid in groups in the soil during the spring and summer. They hatch into first-stage juveniles that possess few body segments and just three pairs of legs. Maturity is reached in about a year, development to adulthood occurring after several moults at which additional body segments and pairs of legs are added. This species lack eyes but the skin is photo-sensitive, individuals soon becoming active if exposed to light.

Family **POLYDESMIDAE** (flat millepedes)

Polydesmus angustus Latzel (**1144**)
syn. *P. complanatus* Linnaeus
Flat millepede
An often common millepede, particularly amongst leaf litter; frequently associated with damage to cultivated plants in the British Isles and in various other parts of western Europe. Adults are 25 mm long, flat-bodied and mainly reddish brown, with 20 rough-textured body segments.

1143 Spotted snake millepede (*Blaniulus guttulatus*).

1144 Flat millepede (*Polydesmus angustus*).

Family **IULIDAE** (snake millepedes)

Cylindroiulus londinensis (Leach)
syn. *C. teutonicus* (Pocock)
A black millepede
This widely distributed and often common millepede sometimes causes damage to cultivated plants. Adults (up to 50 mm long) are brownish black to black, the body being of similar diameter throughout its length.

Tachypodoiulus niger (Leach) (**1145**)
syn. *T. albipes* (Koch)
A black millepede
A common, black-bodied millepede that sometimes causes damage to nursery and garden plants. Unlike that of the previous species, the body is distinctly narrowed at the hind end.

1145 A black millepede (*Tachypodoiulus niger*).

SYMPHYLIDS

Family **SCUTIGERELLIDAE**

Scutigerella immaculata (Newport)
Glasshouse symphylid
A generally common soil-inhabiting pest of various greenhouse and outdoor plants, including seedlings and transplants. Widely distributed in Europe; also present in North America and Hawaii.

DESCRIPTION
Adult: 5–9 mm long; body white, with 12 pairs of legs and a pair of pointed anal cerci; antennae long. **Egg:** white, rounded and sculptured. **Nymph:** white, with from six to ten pairs of legs (additional pairs being added at each moult).

LIFE HISTORY
Breeding occurs throughout the year, eggs being laid near the soil surface in batches of up to 20. Eggs hatch in 1–3 weeks, and nymphs reach maturity several months later. Nymphs and adults feed on plant roots when the soil is warm and moist. In dry conditions, however, they migrate to deeper levels. Individuals are capable of surviving for several years.

DAMAGE
Fine roots are grazed away, weakening and sometimes leading to the death of infested seedlings and young plants.

NEMATODES

Nematodes are important pests of ornamental plants, and some species are of significance as virus vectors. Brief details of the main groups associated with ornamentals are given below.

Cyst nematodes

Cyst nematodes are associated mainly with the roots of herbaceous plants. Affected plants are weakened, and often become stunted and discoloured; they may also wilt in strong sunlight. The root system is typically attenuated, plants often developing a mass of new rootlets ('hunger roots') close to the soil surface in an attempt to compensate for the reduction or loss of normal root activity. Cyst nematodes invade the young roots of host plants as minute second-stage juveniles, the first-stage having moulted whilst still within the egg. Having gained entry to the host, each nematode settles down to feed in the centre of a young root. Feeding induces the development of giant cells which interrupt normal vascular activity within the root. Some of the nematodes develop into small worm-like males which eventually escape into the soil. Females, however, mature into whitish or yellowish, lemon-shaped (typical of the genus *Heterodera*) or rounded bodies, *c.* 0.5–1.0 mm long, which burst through the root surface but remain attached at the head end. After mating, the females die and their bodies darken into hard-walled protective cysts packed with minute, oval eggs. These cysts eventually break away from the roots and drop into the soil. Hatching of the eggs is often dependent upon the presence in the soil of chemicals exuded from the young roots of host plants. Cysts commonly remain viable in the soil for many years before eventually releasing the infective second-stage juveniles. Species of *Heterodera* most often noted on ornamental plants include: *H. cacti* on poinsettia (*Euphorbia pulcherrima*) and various cacti; *H. cruciferae* on wallflower (*Cheiranthus cheiri*) and other brassicaceous plants; *H. fici* on *Ficus*, including rubber plant (*F. elastica*) and weeping fig (*F. benjamina*); and *H. trifolii* on carnation (*Dianthus caryophyllus*).

Leaf nematodes (1146)

Leaf nematodes (*Aphelenchoides* spp.) affect a wide variety of hosts, occurring as ectoparasites or endoparasites of the buds, leaves and growing points; they sometimes also feed in the epidermal layers of green stems. The pests often attack greenhouse plants and propagation material, and they thrive in warm, moist conditions. Chrysanthemum nematode (*Aphelenchoides ritzemabosi*) is a major pest of *Chrysanthemum* and one of the most important species to attack ornamentals. Adults are about 1 mm long, and usually overwinter in the dormant buds and shoot tips of chrysanthemum stools. They are capable of surviving in dried leaves for several years, but survival in the soil is poor. The nematodes slither over host plants in films of water and usually enter their hosts through stomata. They reproduce sexually, and development from egg to adult takes about two weeks. Affected tissue becomes yellowish and then brown or blackish, these symptoms often appearing as distinctive wedge-shaped areas between the major veins. Seriously affected leaves eventually wither and die. In moist conditions symptoms spread rapidly up the plant, and new leaves emerging from infested shoots are often distorted and thickened; heavy infestations lead to stunting of growth and the production of small, malformed flowers. Other plants attacked by this species include African violet (*Saintpaulia hybrida*), buddleia (*Buddleja*), *Ceratostigma*, *Cineraria*, *Crassula*, *Dahlia*, *Delphinium*, *Doronicum*, lavender (*Lavendula*), Michaelmas daisy (*Aster*), peony (*Paeonia*), pyrethrum (*Chrysanthemum coccineum*), wallflower (*Cheiranthus cheiri*) and *Weigelia*. Various other species of *Aphelenchoides* are also pests of ornamentals. These include: *A. blastophthorus*, primarily a pest of scabious (*Knautia* and *Scabiosa*) but also associated with plants such as *Anchusa*, *Anemone*, *Begonia*, bulbous *Iris*, *Cephalaria*, globe flower (*Trollius europaeus*), lily-of-the-valley (*Convallaria majalis*), *Narcissus* and sweet violet (*Viola odorata*); *A. fragariae*, which affects mainly Liliaceae, Primulaceae, Ranunculaceae and many kinds of fern; and *A. subtenuis* on *Phlox* and various bulbous hosts, including *Allium*, *Colchicum*, *Crocus*, *Narcissus*, squill (*Scilla*) and tulip (*Tulipa*).

1146 Chrysanthemum nematode (*Aphelenchoides ritzemabosi*) damage to leaves of *Weigelia*.

1147 Galls of northern root-knot nematode (*Meloidogyne hapla*).

Migratory nematodes

Migratory nematodes usually feed externally on the roots of plants, infestations most often occurring on rosaceous hosts growing in light, sandy soils. Attacked roots are gnarled and distorted, affected plants lacking vigour and becoming distinctly stunted. Migratory nematodes are also capable of transmitting important plant viruses. Several groups of migratory nematodes cause damage to ornamental plants. These include dagger nematodes (*Xiphinema* spp.) and needle nematodes (*Longidorus* spp.), which are often a problem on rosaceous plants, and stubby-root nematodes (*Trichodorus* spp.), which commonly affect not only woody plants but also bulbs.

Root-lesion nematodes (*Pratylenchus* spp.) are also classified as migratory species, although unlike the other groups the adults and juveniles enter the roots of host plants to feed internally; the nematodes also breed within the host. Roots of infested plants develop short, elongate lesions which afford ideal sites for the entry of pathogenic bacteria and fungi; affected plants lack vigour and wilt under stress. Roots often break off at the point of damage, particularly once bacterial or fungal rots have gained a hold. When conditions within the roots become unfavourable, the nematodes disperse to seek more suitable hosts. Root-lesion nematodes affect various herbaceous ornamentals, including *Anemone*, *Begonia*, Christmas rose (*Helleborus*), *Delphinium*, hyacinth (*Hyacinthus orientalis*), lily (*Lilium*) and lily-of-the-valley (*Convallaria majalis*). More specifically, the nematode *Pratylenchus bolivianus* infests Peruvian lily (*Alstroemeria*), notably in the Netherlands but also in England, and *Pratylenchus penetrans* is implicated in the rotting of *Narcissus* bulbs, particularly in the

Netherlands and in the Scilly Isles. Other root-lesion nematodes (e.g. *Pratylenchus fallax* and *P. thornei*) are of significance in mainland Europe on field-grown trees and shrubs, especially Rosaceae; finally, *Pratylenchus vulnus*, although a pest of greenhouse-grown roses (*Rosa*) in northern Europe, cannot survive outdoors except in warmer climates.

Root-knot nematodes (1147)

Root-knot nematodes (*Meloidogyne* spp.) attack the roots of various trees, shrubs and herbaceous plants. Infested roots become distorted and develop rounded or irregular galls. These galls measure anything from 1 to 20 mm across and often coalesce, causing considerable distortion. The nematodes also exacerbate the deleterious effects of pathogenic bacteria and fungi. Root-knot nematodes are associated mainly with light soils but most damage is caused under glass, particularly in hot conditions where certain tropical and subtropical species, e.g. the Javanese root-knot nematode (*Meloidogyne javanica*), have become established. Pot plants such as *Begonia*, *Coleus*, *Cyclamen*, *Gloxinia* and various cacti may suffer considerable damage, severely affected plants appearing discoloured, lacking vigour and wilting under stress. Northern root-knot nematode (*Meloidogyne hapla*) is a widely distributed, polyphagous pest in northern Europe; it attacks many different kinds of plant, including various ornamentals. Root-knot nematodes invade host plants as second-stage juveniles; these settle down to feed in the young roots and usually reach maturity about 1–2 months later. Adult females are translucent-whitish, pear-shaped and about 0.5–1.0 mm long. They may be found within the galled tissue, often attached to a gelatinous sac that contains

1148 Stem nematode (*Ditylenchus dipsaci*) damage to bulb of *Narcissus.*

masses of eggs. In some cases development of the pest is parthenogenetic; in others, minute worm-like males mate with the females before eggs are laid. First-stage juveniles develop within the eggs, second-stage individuals eventually breaking free and either migrating inside the root or escaping into the soil to commence feeding elsewhere. These infective nematodes are capable of surviving in moist soil for about three months. In dry conditions they persist for no more than a few weeks.

Stem nematodes and tuber nematodes (1148)

Stem nematode (*Ditylenchus dipsaci*) is a major pest of herbaceous and bulbous plants, including many ornamentals. The pest exists as several distinct races or strains – the hyacinth, narcissus, onion, phlox and tulip races are examples. Some races affect a wide variety of host plants, including cultivated plants and weeds, whereas others are more specific. Plants are invaded by adults and final-stage (fourth-stage) juveniles, which move through the soil in moisture films. The nematodes gain entry to host plants via wounds and lenticels in the basal parts of stems or through stomata in leaves in contact with the ground; roots are not attacked. Adults are minute (*c.* 1.2 mm long) and thread-like. They feed and breed continuously in suitable hosts, development from egg to adult taking about three weeks at 15°C. In the absence of host plants, stem nematodes are able to survive in moist soil for up to a year; however, they are more resistant to desiccation, fourth-stage juveniles often congregating in their thousands and drying out to form yellowish, woolly masses ('nematode wool'). These desiccated nematodes are able to survive unfavourable conditions, remaining viable and potentially infective for several years. Damage

symptoms vary from host to host but commonly include stunting, crinkling, twisting, swelling or malformation of leaves, leaf stalks, flowers and stems; infested tissue may also split. *Aubretia*, bell flower (*Campanula*), evening primrose (*Oenothera*), golden-rod (*Solidago virgaurea*), *Gypsophila*, *Heuchera*, sneezewort (*Helenium*), *Hydrangea, Phlox*, primrose (*Primula vulgaris*) and sweet william (*Dianthus barbatus*) are often affected. Bulbous plants such as daffodil (*Narcissus*), hyacinth (*Hyacinthus orientalis*), snowdrop (*Galanthus nivalis*) and tulip (*Tulipa*) are also important hosts. Leaves arising from infested daffodil bulbs often develop small, yellowish swellings ('spickels'); also, when infested bulbs are sliced open, brown rings of damaged tissue may be visible where the scales have been destroyed. In tulips, flower stalks from heavily infested bulbs are bent and develop lesions which eventually split open; flowers also fail to colour properly, the petals remaining partly green. Bulbs damaged by nematodes may eventually rot, and they are often invaded by secondary organisms such as small narcissus flies (*Eumerus* spp.) (p. 190) and bulb mites (*Rhizoglyphus* spp.) (p. 433).

Potato tuber nematode (*Ditylenchus destructor*) is similar to stem nematode but lacks a resistant stage capable of surviving periods of desiccation. Also, unlike stem nematode, it is restricted mainly to the subterranean parts of plants, producing dry, brownish or blackish lesions on bulbs, corms, roots and tubers. Various ornamentals are affected, including bulbous *Iris, Colchicum, Dahlia, Gladiolus* and tulip (*Tulipa*); leaves developing from damaged bulbs are weakened and often have yellow tips. Damaged tissue is frequently invaded by secondary organisms, including bacteria, fungi and mites.

SLUGS AND SNAILS

Slugs and snails are often important pests in gardens and nurseries, attacking seedlings and herbaceous plants such as *Anemone*, bell flower (*Campanula*), *Doronicum*, *Gladiolus*, *Hosta*, hyacinth (*Hyacinthus orientalis*), *Iris*, *Narcissus*, *Petunia*, primrose (*Primula vulgaris*), *Rudbeckia*, strawflower (*Helichrysum*), sweet pea (*Lathyrus odoratus*), *Tagetes*, tulip (*Tulipa*), violet (*Viola*) and various lilies. Slugs and snails feed at night, and often then cause severe damage to seedlings, young shoots, foliage and flowers. Such damage is sometimes confused with that inflicted by caterpillars and various other pests, but slime trails (if not the pests themselves) on or in the immediate vicinity of attacked plants readily betray the identity of the true culprits.

Slugs (1149–1150)

Unlike snails, which usually hibernate during the winter months, slugs breed throughout the year and remain active in all but the coldest and driest of conditions. Their translucent, often pearl-like eggs are deposited in groups in the soil or amongst surface vegetation. The eggs usually hatch within a few weeks but those deposited during the winter might not complete their development until the following spring. Juvenile slugs are similar in appearance to adults but smaller and usually paler. They take anything from five months to two years to reach maturity, the rate of development varying from species to species and according to conditions.

Several species are responsible for damaging ornamentals; these include the garden slug (*Arion hortensis*), the field slug (*Deroceras reticulatum*) and various keeled slugs. The garden slug is a relatively small (25–30 mm long), tough-skinned species, with a rounded tail; the adults are mainly black above and yellow or orange below. The field slug is larger (30–40 mm long), soft-bodied and mainly yellowish brown, with a distinctly pointed tail and a short dorsal keel at the hind end. Both of these species readily attack the aerial parts of plants. Keeled slugs, such as *Tandonia* (= *Milax*) *budapestensis*, are characterized by the presence of a distinct dorsal ridge that extends from the mantle to the tail. Unlike other slugs, they feed mainly below ground level, often boring into bulbs, corms, rhizomes and tubers of ornamental plants; they also damage plant roots.

1149 Slug damage to flower of *Narcissus.*

1150 Field slug (*Deroceras reticulatum*).

1151 Garden snail (*Cornu aspersum*).

Snails (1151)

Particularly on calcareous soils, snails are troublesome pests of ornamentals, attacking seedlings as well as the young shoots and foliage of established plants. Species most likely to cause damage in gardens and nurseries are banded snails (*Cepaea* spp.), the garden snail (*Cornu aspersum*) (syn. *Helix aspersa*) and the strawberry snail (*Trichia striolata*). Strawberry snails often cause extensive damage to seedlings being raised in cold-frames.

EARTHWORMS

When burrowing in the soil, some species of earthworm regularly deposit excreted soil (worm casts) on the surface, their activity being greatest in the spring and autumn. Worm casts are a particular nuisance on fine lawns, and are especially unwelcome on bowling greens, golf greens and tennis courts. The species most often causing a problem are *Allolobophora longa* and *A. nocturna*. On balance, however, earthworms are beneficial and important components of the soil fauna.

BIRDS

In the winter and spring, birds frequently attack dormant or opening buds of trees and shrubs, affected shoots often developing with any remaining blossoms restricted to the extreme tips. Bud stripping is sometimes of considerable importance on almond (*Prunus dulcis*), crab-apple (*Malus*), flowering cherry (*Prunus*), *Forsythia*, lilac (*Syringa*) and *Viburnum carlesii*, and is usually caused by bullfinches. Other birds, including blue tits, chaffinches and greenfinches, also damage dormant buds but are usually of only minor importance.

House sparrows are notorious for the damage they cause to *Crocus*, primrose (*Primula vulgaris*), sweet pea (*Lathyrus odoratus*), violet (*Viola*) and various other flowers, attacked flowers being torn to pieces and the stripped petals left lying on the ground, a typical sign of bird damage. House sparrows also invade greenhouses and plastic tunnels. They also steal seeds from newly sown lawns, and cause disturbance in dry seedbeds by taking dust baths.

In autumn and winter, various birds feed on the berries or fruits of trees and shrubs, the colourful displays on ornamentals such as barberry (*Berberis*), *Cotoneaster*, firethorn (*Pyracantha*), holly (*Ilex*) and *Sorbus* often being depleted. Birds responsible for such damage include blackbirds, fieldfares, jays, redwings and woodpigeons.

The aesthetic value of birds usually far outweighs the damage they cause, and many are of direct benefit to the gardener or nurseryman. During the spring and summer, for example, insectivorous species such as tits, warblers and wrens destroy vast numbers of insect pests, including many of those attacking the foliage or flowers of ornamental trees and shrubs. Some birds, notably blue tits and great tits also feed on overwintering insects and mites hiding beneath bud scales or within bark crevices. Others, including rooks, seek out soil pests such as leatherjackets, swift moth larvae, vine weevil larvae and wireworms; large numbers of slugs and snails are also killed by birds.

MAMMALS

Various mammals cause damage to cultivated plants, particularly in rural or semi-rural areas. The following are examples of those most frequently reported affecting ornamentals in gardens and nurseries.

Badgers

Although rarely a significant problem, badgers sometimes enter private gardens. They then cause damage to lawns and cultivated plants, usually by scrabbling into the ground in search of food. During the winter they often unearth bulbs and corms, leaving behind tell-tale scrapings and piles of soil.

Cats and dogs

Domestic cats are often troublesome in gardens when they dig into seedbeds, the fine soil making attractive toilet areas. They also cause damage by scratching the bark of trees and shrubs. Dogs are also renowned for digging in gardens. However, they cause most trouble by urinating on low-lying plants and killing the foliage, affected tissue turning brown. Such damage is often severe on dwarf conifers growing alongside paths, and sometimes results in the death of plants. Urinating dogs, especially bitches, also cause brown patches on lawns.

Deer

In some areas ornamental trees and shrubs, especially rose (*Rosa*) bushes, may be damaged by deer. Attacks usually take place from March to May, when the animals browse on the young leaves and new shoots. Deer cannot bite cleanly through plant material, as they possess teeth in the lower jaw only; damaged shoots and stems, therefore, are left with a distinctly ragged edge on one side where the partly severed tissue has been torn away. Damage to ornamentals is most often caused by fallow deer, Reeves' muntjac and roe deer.

Hares and rabbits

Hares and rabbits attack the foliage and stems of various ornamental plants, often causing considerable damage in inadequately protected rural gardens and nurseries. Also, particularly during hard winters when food supplies are scarce, they often gnaw the bark of young trees. In severe cases, stems or trunks may be completely ringed and the plants killed. Fencing off gardens or plantations and the use of protective wire or plastic sleeves around the bases of vulnerable trees will reduce the likelihood of damage. However, in deep snowfall the animals may gain access to normally secure areas and feed on bark above the level of any netting or other guards. Rabbits scrabbling on the ground are also capable of causing considerable damage to lawns and other grassland areas.

Mice, rats and voles

Various small mammals may prove troublesome in gardens and nurseries, some species commonly entering outbuildings during the winter to feed on stored bulbs, corms and seeds. Small mammals may also eat bulbs and corms already in the ground, such damage being caused most often by long-tailed field mice. Voles, especially short-tailed voles, frequently gnaw the bark of young trees and shrubs; they also attack herbaceous ornamentals such as *Chrysanthemum*. Growth of affected plants is often checked and, if the tissue is completely ringed, branches or whole plants may be killed. Voles also ascend young ornamental trees and shrubs to feed on the buds, young shoots and berries.

Moles

Moles can be a problem in lawns, parks and sports grounds as they burrow through the soil in search of food (earthworms, slugs and other soil invertebrates). Infested areas become disfigured by the presence on the surface of numerous mole-hills. In addition, surface soil sometimes collapses into the subterranean workings, and turf over shallow tunnels may be forced upwards in ridges. Moles are also a nuisance in flower borders and seedbeds, as they accidentally disturb plant roots and cause established plants and seedlings to wilt and die.

Squirrels

Grey squirrels are notorious pests of woodland trees and shrubs, and are also of some importance on ornamentals in gardens and nurseries. During the winter they strip the bark from the shoots and branches of trees such as ash (*Fraxinus excelsior*), sycamore (*Acer pseudoplatanus*) and spruce (*Picea*), and damage caused is often extensive. Grey squirrels also dig up bulbs of plants such as *Crocus* and tulip (*Tulipa*). In spring, young shoots, buds and flowers of trees and shrubs are attacked, and the animals show a particular liking for ornamentals such as flowering currant (*Ribes sanguineum*) and saucer magnolia (*Magnolia* × *soulangeana*).

Glossary

abdomen The hindmost of the three main divisions of the body of an insect.

abdominal Pertaining to the abdomen.

adventitious bud A plant bud that does not arise from a leaf axil.

adventitious root A plant root that does not arise from the primary root system.

aestivation Summer diapause.

alate A winged form, as applied to an aphid.

anal Pertaining to the anus or to the hindmost body segment.

anal claspers The pair of prolegs on the last abdominal segment of certain lepidopterous larvae.

anal comb A pronged, comb-like structure at the tip of the abdomen of certain lepidopterous larvae, used to expel frass from the anus.

anal plate A dorsal, sclerotized plate on the hindmost segment of a lepidopterous larva.

anchor process *see* **Sternal spatula**

antenna (*pl.* **antennae**) One of a pair of sensory 'feelers' on the head of an insect and certain other arthropods.

antennal Pertaining to an antenna.

apical At or towards the tip (apex) of a structure (cf. basal).

apodous Without legs.

appendix As in the venation of the leafhopper genus *Alebra*, a separately demarcated marginal area around the wing tip.

aptera (*pl.* **apterae**) A wingless form, as applied to an aphid.

apterous Wingless.

arista A bristle-like outgrowth on the antenna of certain flies.

axil In a plant, the angle formed by the upper surface of a leaf and the stem.

axillary In the axil of a plant.

basal At or towards the base of a structure (cf. apical).

base plate The reduced (disc-like) stem at the base of a plant bulb.

basitarsus (*pl.* **basitarsi**) On the leg of an insect, the basal (and often the largest) subsection of the tarsus.

bifid Two-pronged.

big bud A swollen (galled) bud, as induced by eriophyid mites.

bilobed With two lobes.

biordinal Of two sizes.

bipectinate Feather-like, as applied to the antennae of certain insects.

bisexual Of two sexes.

blotch A broad, non-linear leaf mine. Also, an area in a patterned wing.

brassicaceous Pertaining to the plant family Brassicaceae (formerly Cruciferae).

breast-bone *see* **Sternal spatula**

bulb A usually underground plant organ, consisting of a short disc-like stem that bears fleshy scale leaves and one or more buds.

callosity A hardened lump.

calyx The outermost non-reproductive part of a flower, surrounding the corolla (q.v.).

cambium Plant tissue that is the source of cells for secondary growth, producing either new cork (from cork cambium) or new vascular tissue (from vascular cambium).

capitate With a distinct head.

carapace A dorsal body shell.

case A protective, often tubular, habitation.

caterpillar A colloquial name for a leaf-feeding larva of certain butterflies, moths and sawflies.

cauda A tail, including the tail-like structure at the hind end of an aphid.

caudal Pertaining to the cauda.

cell On an insect wing, an area of the membrane partly (an open cell) or entirely (a closed cell) surrounded by veins.

cephalopharyngeal skeleton In larvae of certain flies, the structure formed from the reduced, retracted head and the mouthparts.

cercus (*pl.* **cerci**) One of a pair of segmented appendages arising from the eleventh abdominal segment of certain insects.

chelicerate Bearing chelicerae.

chelicera (*pl.* **chelicerae**) One of a pair of much modified structures at the tip of the pedipalp of a mite.

chitin A polysaccharide which is the main chemical component of the cuticle.

chitinized Containing chitin.

chlorosis An unhealthy condition, owing to a lack of chlorophyll, that results in the green parts of a plant becoming yellowish.

chlorotic Plant tissue that has become yellow or white, following the loss of chlorophyll.

cilium (*pl.* **cilia**) A fine, hair-like projection arising from a cell. Also, as cilia, fine hairs fringing insect wings.

clavus The posterior part of the fore wing of a heteropteran bug.

clypeus The lowest part of the face of an insect, immediately above the labrum.

compound eye One of a pair of multi-faceted eyes on the head of an insect (cf. ocellus).

corm A swollen, bulb-like but solid underground stem.

corolla The innermost non-reproductive part of a flower, formed by petals and surrounded by the calyx (q.v.).

costa The anterior-most vein of an insect wing, running along the costal margin.

costal Pertaining to the costa.

cotyledon One of the first leaves arising from the embryo of a seedling plant.

coxa (*pl.* **coxae**) The first (basal) segment on the leg of an insect or a mite.

cremaster A small cluster of hooks or spines at the tip of a pupa, often borne on a distinctive outgrowth.

crotchet One of several chitinized, hook-like structures on the abdominal proleg of a lepidopterous larva.

cultivar A variety of a cultivated plant, produced by breeding or selection.

cv. Cultivar.

cuticle The three-layered, non-cellular 'skin' of an insect or mite.

deuterogenous Pertaining to deuterogeny.

deuterogeny Having distinct summer and winter forms.

deutogyne The winter female form of a deuterogenous mite (cf. protogyne).

deutonymph The first-stage nymph of a mite.

diapause A period of rest. *See* **Aestivation** *and* **Hibernation**.

digitate Finger-like.

dimorphism The existence of two distinct forms.

distal That part of an appendage furthest from the body.

dorsal Pertaining to the upper side of an animal (cf. ventral).

dorsolateral Pertaining to sides of the dorsal surface of an organism.

dorsoventral Pertaining to the upper and lower parts of an organism.

ecdysis The process of moulting from one growth stage to the next, when the 'old' skin is cast and replaced by another.

ectoparasite A parasite that develops outside its host (cf. endoparasite).

elytral Pertaining to the elytra.

elytral suture The suture running down the back of a beetle, between the elytra.

elytron (*pl.* **elytra**) One of a pair of hardened, horny fore wings (as in a beetle and certain other insects), that often form a protective cover over the abdomen.

emarginate With a distinct indentation or notch.

endocuticle The inner layer of the cuticle (cf. epicuticle and exocuticle).

endoparasite A parasite that develops inside its host (cf. ectoparasite).

endopodite The inner part of the locomotive appendage of a crustacean.

epicuticle The middle layer of the cuticle (cf. endocuticle and exocuticle).

epidermal Pertaining to the epidermis.

epidermis In a plant, the protective outermost layer of cells covering leaves, stems etc.

erineum (*pl.* **erinea**) A hair-lined, blister-like, mite-induced gall.

Eurasiatic The zoogeographical region represented by Europe and Asia.

exocuticle The outer layer of the cuticle (cf. endocuticle and epicuticle).

exoskeleton The hardened external skeleton of an invertebrate, to which the muscles are attached.

exuvia (*pl.* **exuviae**) The cast skin (cuticle) of an insect or other arthropod.

eye bridge In certain flies, that part of the compound eyes that meets at the top of the head.

eye-cap In certain moths, a hood-like structure formed from the base of the antenna, that partly covers the eye.

family A major taxonomic group of genera with common technical characters.

feather-claw A feather-like outgrowth at the tip of the leg of an eriophyid mite.

femur (*pl.* **femora**) In an insect, the third and frequently the largest (but not necessarily the longest) segment of the leg, located between the trochanter and the tibia. Also, in a mite, the third segment of the leg, located between the trochanter and the genu.

flagellar Pertaining to the flagellum.

flagellum The third (distal) part of a geniculate antenna beyond the pedicellus. Subdivisions of the flagellum are not strictly segments and are more correctly termed 'flagellomeres'.

frass Solid excreta produced by an insect, especially a caterpillar.

frons The upper part of the face of an insect, lying between and below the antennae and usually including the central ocellus.

fundatrix (*pl.* **fundatrices**) A parthenogenetic, viviparous, usually wingless female aphid that emerges from an overwintered egg, and typically establishes a colony in the spring.

gall Abnormal growth of plant tissue, induced by certain insects, mites, fungi and other organisms.

gallicola (*pl.* **gallicolae**) A winged female adelgid that emerges from a gall whose development it helped (as a nymph) to induce.

galligenous tissue Tissue produced by a plant in response to an external stimulus.

gaster In certain Hymenoptera, that part of the abdomen located behind the petiolus.

geniculate Abruptly bent, elbowed.

genu The fourth segment of the leg of a mite, located between the femur and the tibia.

genus The main taxonomic category between family and species, including one or more species considered to be of a single evolutionary origin.

glabrous Without hairs, hairless.

gnathosoma The first (anterior-most) of the two main sections of the body of a mite, comprising the mouthparts and bearing the paired chelicerae and pedipalps (q.v.).

gravid Pregnant.

grub A colloquial name for an insect larva, such as a 'chafer grub'.

haemolymph The blood-like fluid filling the body cavity of an insect.

haltere One of a pair of club-shaped or drumstick-like balancing organs on the metathorax of a true fly, formed from the modified hind wings.

hemelytron (*pl.* **hemelytra**) An elytron that is mainly hardened but is membranous distally.

hemimetabolous insect An insect with incomplete metamorphosis, i.e. no pupal stage in the lifecycle (cf. holometabolous insect).

heteroecious Species whose 'summer' and 'winter' hosts are different, as in certain aphids.

hibernaculum (*pl.* **hibernacula**) The structure or shelter (e.g. a cocoon) in which hibernation takes place.

hibernation Winter diapause.

Holarctic The area encompassing the Nearctic and Palaearctic zoographical regions.

holometabolous insect An insect with complete metamorphosis, i.e. with a pupal stage in the lifecycle within which transformation from the larval stage to adulthood takes place (cf. hemimetabolous insect).

honeydew A sugary fluid excreted through the anus of certain polyphagous bugs, including aphids, mealybugs, psyllids, scale insects and whiteflies.

horn As in hawk moth caterpillars, a term of convenience to describe an often spinose outgrowth arising from the eighth abdominal segment.

hyaline Transparent and translucent.

hypha (*pl.* **hyphae**) The long, filamentous structure that forms the mycelium of a fungus.

hypocotyl That part of a seedling below the cotyledons which gives rise to the roots.

hypodermis In a plant, the layer of cells immediately below the epidermis.

hypopal Pertaining to a hypopus.

hypopus (*pl.* **hypopi**) The morphologically distinct, often phoretic, dispersal stage of a mite.

hysterosoma The hind section of the idiosoma of a mite, bearing the third and fourth pairs of legs.

idiosoma The main body section of a mite, divided into the propodosoma and the hysterosoma.

idiosomal Pertaining to the idiosoma.

inflorescence A flower cluster.

inquiline An animal that lives in the habitation (e.g. a gall or a nest) of another and shares its food.

instar One of usually several larval or nymphal growth stages in the development of an insect.

internode The part of a plant stem between adjacent nodes.

intersegmental Between segments.

interveinal Between the veins, as in a plant leaf.

keel In certain slugs, a distinct dorsal ridge running down the body.

knee In certain flies (e.g. family Agromyzidae), a term of convenience to describe the differentially coloured section at the base of the tibia which demarcates the point of articulation with the femur.

labial Pertaining to the labium

labium The 'lower lip' of the mouthparts of an insect.

labrum The 'upper lip' of the mouthparts of an insect.

lamella (*pl.* **lamellae**) A thin, leaf-like flap or plate.

lamellate Composed of several lamellae.

lanceolate Spear-shaped.

larva (*pl.* **larvae**) The immature growth stage of a holometabolous insect, which is very different in appearance from the adult into which it metamorphoses during a quiescent, non-feeding pupal stage; compound eyes absent (cf. nymph). Also, the first immature, six-legged growth stage of a mite.

larval Pertaining to a larva.

lateral At or close to the sides.

legume A plant in the family Fabaceae (formerly Leguminosae).

lenticel A pore in a stem that allows gases to pass between the inside of a plant and the outside air.

macropterous Fully winged.

mandible One of a pair of jaws.

mandibulate With mandibles suitable for biting or chewing.

mantle In a mollusc, a protective fold of skin covering part or all of the body; often enclosing a mantle cavity.

marginal At the margin.

maxilla (*pl.* **maxillae**) One of a pair of appendages forming the mouthparts of an insect, lying just behind the mandibles.

maxillary Pertaining to a maxilla.

melanic Exhibiting melanism, a darkness of colour that results from the presence of black pigment.

mesothorax The second (middle) segment of the thorax, bearing the first pair of wings (fore wings) and the second pair of legs.

metamorphosis Change of form from one state to another.

metathorax The third (hindmost) segment of the thorax, bearing the second pair of wings (hind wings) and the third pair of legs.

micropyle A tiny pore in an insect egg.

midrib The central or main vein of a leaf.

moniliform Bead-like.

mouth-hooks The rasping, hook-like part of the cephalopharyngeal skeleton of a fly larva, formed from the mandibles.

Nearctic The zoogeographical region encompassing the arctic and temperate parts of North America.

notifiable pest A non-native pest, suspected outbreaks of which must be reported immediately to Plant Health authorities.

nymph The immature growth stage of a hemimetabolous or semi-hemimetabolous insect, usually similar in general appearance to the adult and often sharing similar feeding habits; wing buds (when present) develop externally and increase in size at each successive moult; compound eyes often present (cf. larva). Also, the usually eight-legged growth stage of a mite between the six-legged 'larva' and the adult.

nymphal Pertaining to a nymph.

ocellus (*pl.* **ocelli**) A simple, non-focussing, light-sensitive, usually oval or round eye; in an adult insect, three ocelli often form a triangle at the top of the head, between the compound eyes (q.v.).

oligophagous Feeding on a limited range of related plants.

omnivorous Feeding on all kinds of material, both animal and vegetable.

ootheca (*pl.* **oothecae**) The purse-like, egg-containing structure produced by cockroaches and other Dictyoptera.

operculum (*pl.* **opercula**) A lid-like structure.

ovariole One of several egg-producing tubules into which an ovary is divided.

ovipara (*pl.* **oviparae**) A female aphid that lays eggs.

oviposition The act of egg laying.

ovipositor The egg-laying apparatus of an insect.

ovisac An often silken, cocoon-like sac which contains eggs.

Palaearctic The zoogeographical region encompassing Europe, Africa north of the Sahara and most of Asia north of the Himalayas.

palp A segmented, sensory structure arising from the maxilla or the labium of an insect. *See also* **pedipalp**.

papilla (*pl.* **papillae**) A small, often finger-like, projection.

parasite An organism which lives in or on another (the host), obtaining nourishment from it but giving nothing in return.

parasitoid A parasite in which only the immature stages are parasitic, the adult being free-living. Unlike a parasite, a parasitoid usually kills its host.

parthenogenesis Reproduction that does not involve sex.

parthenogenetic Reproducing asexually.

pathogen An organism that causes disease.

pathogenic Disease-causing.

pedicel The second segment of an insect antenna, between the scape and the flagellum.

pedipalp One of a pair of sensory, leg-like structures arising from the gnathosoma of a mite.

petiolus In certain Hymenoptera, the narrow 'waist' or stalk between the gaster and the propodeum.

phoresy The involuntary transportation of one organism by another, without involving parasitism.

phoretic Undertaking phoresy.

phytophagous Plant feeding.

pinaculum (*pl.* **pinacula**) A small, chitinized, pimple-like plate on the body of an insect larva.

pinna (*pl.* **pinnae**) A leaflet or primary division of the compound leaf of a plant.

pleuron The lateral section of a body segment.

plumose With numerous feathery branches; feathery.

polyphagous Feeding on a wide range of hosts.

prepupa (*pl.* **prepupae**) The non-feeding stage of development of a holometabolous insect between the final-instar larva and the pupa.

prepupal Pertaining to the prepupa.

primordium (*pl.* **primordia**) Tissue at an early stage of development that will eventually give rise to a specific structure or organ.

prodorsal shield The more or less triangular shield-like part of an eriophyid mite, which covers the propodosoma and lies in front of the hysterosoma.

proleg A false, unjointed leg, as on the abdomen of a caterpillar.

pronotal Pertaining to the pronotum.

pronotum The often enlarged, shield-like dorsal surface of the prothorax.

propodeum In certain Hymenoptera, the first part of the abdomen located in front of the petiolus and appearing to be part of the thorax, to which it is fused.

propodosoma The anterior part of the idiosoma of a mite, bearing the first and second pairs of legs.

propupa (*pl.* **propupae**) The non-feeding stage in the development of a thrips, between the final-instar nymph and the pupa.

prothoracic Pertaining to the prothorax.

prothorax The first (anterior) segment of the thorax, bearing the first pair of legs.

protogyne The summer female form of a deuterogenous mite (cf. deutogyne).

protonymph The second-stage nymph of a mite.

pseudo-fundatrix (*pl.* **pseudo-fundatrices**) A similar form to the fundatrix of certain aphids, but produced parthenogenetically rather than arising from an overwintered egg.

pseudo-gall A distortion, such as leaf curling, induced by a phytophagous insect or other such organism.

pseudo-pupa A pupa-like stage in the development of certain hemimetabolous insects, as in whiteflies.

pseudo-pupal Pertaining to a pseudo-pupa.

pterostigma A small pigmented area present near the tip of an otherwise 'clear' wing of certain insects; also known as a stigma.

pubescence A coating of fine, soft hairs.

pupa (*pl.* **pupae**) The pre-adult stage of a holometabolous insect within which metamorphosis to the adult form takes place.

pupal Pertaining to a pupa.

puparium (*pl.* **puparia**) In certain Diptera, a hardened, barrel-shaped structure (formed from the cast skin of the final-instar larva) within which the pupa is formed.

pupation The act of becoming a pupa.

pygidium The dorsal surface of the last visible abdominal segment.

radula The rasping 'tongue' of a mollusc.

reniform stigma A kidney-shaped marking on the fore wing of certain moths (family Noctuidae).

reticulate Net-like.

rhizome A root-like stem, lying horizontally on or beneath the soil surface and bearing buds, shoots and roots.

rosaceous Pertaining to the plant family Rosaceae.

rostrum The beak or snout of an insect, as in the piercing mouthparts of a true bug or the trunk-like projection or snout at the front of the head of a weevil.

scape The basal segment of an insect antenna.

sclerite One of the hardened (sclerotized) plates which form the body wall of an insect.

sclerotization The process whereby the cuticle is hardened and tanned by the formation of the protein sclerotin.

sclerotized Hardened and tanned by the protein sclerotin.

scutellum The third (hindmost) of the main dorsal components of each of the three sections of the thorax, but usually obvious only on the mesothorax.

seta (*pl.* **setae**) A small bristle.

siphuncular Pertaining to a siphunculus.

siphunculus (*pl.* **siphunculi**) One of a pair of cone-like or tube-like structures (also known as cornicles) on the abdomen of an aphid, through which alarm pheromones and other defensive compounds are discharged.

sooty mould A black fungal growth that develops on honeydew.

sp. (*pl.* **spp.**) Species.

species The basic unit of classification, being a group of closely related organisms that can interbreed and produce fertile offspring.

sternal spatula In many midge larvae, the so-called 'anchor process' or 'breast bone', a longish sclerite on the underside of the first thoracic segment.

spatulate Spade-like.

spickel A gusset-like injury in the surface of tissue.

spinose Spiny.

spiracle One of several breathing pores or openings to the tracheal (respiratory) system of an insect.

spiracular Pertaining to a spiracle or the spiracles.

ssp. Subspecies.

stellate Star-shaped.

sternal Pertaining to the sternum.

sternite A ventral sclerite on the abdomen of an insect. Also, in an eriophyid mite, a ventral annulation on the hysterosoma.

sternum The ventral section of a body segment.

stigma (*pl.* **stigmata**) A breathing pore in a mite. *See also* **pterostigma** *and* **reniform stigma**.

stipe On an egg, a dorsal spine-like projection.

stolon A lateral stem that grows out horizontally at ground level, rooting and budding at the nodes to produce new plants.

stoma (*pl.* **stomata**) A small pore, as in the epidermis of a leaf, that allows gases to pass between the inside of a plant and the outside air.

stool The stump of a plant from which new shoots arise.

stria (*pl.* **striae**) A groove running across or along the body cuticle, as on the elytra of beetles. Also, as in the pattern on a moth wing, a streak, line or dash.

strobe A furrow-like depression on the side of the rostrum of a weevil.

subanal Close to the anus.

subcostal Close to the costa.

subdorsal Close to the middle of the back.

subhyaline Almost transparent.

subjugal furrow The division on the idiosoma of a mite, between the propodosoma and the hysterosoma.

submarginal Close to the margin.

subspecies Taxonomically, a subdivision of a species.

subspiracular Close to but below the level of the spiracles.

subterminal Close to the apex.

suctorial Sucking.

suture A division on the body surface, dividing one sclerite from another.

syn. Synonym.

synonym An alternative name for an organism.

tarsal Pertaining to the tarsus.

tarsus (*pl.* **tarsi**) The distal part ('foot') of the leg of an insect; strictly, subdivisions of the tarsus are not segments and are usually termed 'tarsomeres'. Also, the last segment of the leg of a mite.

tegmen (*pl.* **tegmina**) The hardened, leathery fore wing of certain insects, such as a cockroach.

tegula (*pl.* **tegulae**) A small lobe covering the base of the fore wing.

telson The hindmost segment-like division of the body of a crustacean.

tergite A dorsal sclerite on the abdomen of an insect. Also, in an eriophyid mite, a dorsal annulation on the hysterosoma.

tergum The dorsal section of a body segment.

terminal bud A bud arising at the apex or tip of a plant stem.

terminal process On an aphid antenna, a narrow extension at the tip of the apical 'segment'.

test In a scale insect, the protective covering formed from wax and cast nymphal skins.

thorax The second of the three main divisions of the body of an insect, from which arise the legs and wings.

thoracic Pertaining to the thorax.

thumb-claw An enlarged spine-like seta arising from the tibia of an eriophyid mite.

tibia (*pl.* **tibiae**) The fourth segment of the leg of an insect. Also, the fifth segment of the leg of a mite.

trachea (*pl.* **tracheae**) One of the breathing tubes that permeate the body of an insect, opening to the outside via the spiracles.

tritonymph The third-stage nymph of a mite.

trochanter The small and often inconspicuous second segment of the leg of an insect. Also, the second segment of the leg of a mite.

tuber A swollen underground root or stem, containing food reserves.

tubercle A small rounded projection.

umbelliferous Pertaining to the plant family Apiaceae (formerly Umbelliferae).

unicolorous Of one colour throughout.

unilocular Single-chambered.

uniordinal Of one size.

uropod One of a pair of appendages arising from the last abdominal segment of a crustacean.

urticating Stinging or irritating.

var. Variety

variety As in a plant, a taxonomic rank below that of species.

vascular Pertaining to the vessels that transport water and nutrients within a plant.

vascularized Supplied with vessels that circulate fluids.

vector An organism that carries and transfers a virus or other organism from one plant or animal to another.

venation The arrangement of veins in the wing of an insect.

ventral Pertaining to the underside of an animal (cf. dorsal).

verruca (*pl.* **verrucae**) A wart-like outgrowth, often bearing one or more bristles or setae.

viviparous Giving birth to live young, rather than laying eggs.

viceral mass In a mollusc, the region of the body that contains most of the digestive, excretory and nervous systems.

Selected Bibliography

Balachowsky, A. S. (ed.) (1962–72). Entomologie appliquée à l'agriculture (4 vols). Masson: Paris.

Barnes, H. F. (1948). *Gall midges of economic importance. Vol. IV: Gall midges of ornamental plants*. Crosby Lockwood: London

Barnes, H. F. (1951). *Gall midges of economic importance. Vol. V: Gall midges of trees*. Crosby Lockwood: London.

Becker, P. (1974). *Pests of Ornamental Plants*. MAFF Bulletin 97. HMSO: London.

Bevan, D. (1987). *Forest Insects. A guide to insects feeding on trees in Britain*. Forestry Commission Handbook 1. HMSO: London.

Blackman, R. L. & Eastop, V. F. (2000). *Aphids on the World's Crops. An Identification and Information Guide*. 2nd edition. Wiley: Chichester.

Blackman, R. L. & Eastop, V. F. (2000). *Aphids on the World's Trees: An Identification and Information Guide*. Wiley: Chichester.

Bradley, J. D., Tremewan, W. G. & Smith, A. (1975–79). *British tortricoid moths. Cochylidae and Tortricidae (2 vols)*. Ray Society: London.

Buhr, H. (1964–65). *Bestimmungstabellen der Gallen (Zool- und Phytocecidien) an Pflanzen Mittel- und Nordeuropas (2 vols)*. Fischer: Jena.

Cameron, P. (1882–89). *A monograph of the British phytophagous Hymenoptera (3 vols)*. Ray Society: London.

Carter, C. I. (1971). *Conifer Woolly Aphids (Adelgidae) in Britain*. Forestry Commission Bulletin 42. HMSO: London.

Carter, C. I. & Maslin, N. R. (1982). *Conifer Lachnids*. Forestry Commission Bulletin 58. HMSO: London.

Carter, D. J. & Hargreaves, B. (1986). *A Field Guide to Caterpillars of Butterflies & Moths in Britain and Europe*. Collins: London.

Chinery, M. (1993). *A Field Guide to the Insects of Britain and Northern Europe. 3rd edition*. Collins: London.

Darlington, A. (1968). *The Pocket Encyclopaedia of Plant Galls in Colour*. Blandford Press: London.

Emden, H. F. van & Harrington, R. (eds) (2007). *Aphids as Crop Pests*. CAB International: Oxford.

Heath, J. (ed.) (1976). *The Moths and Butterflies of Great Britain and Ireland. Vol. 1*. Blackwell: Oxford.

Heath, J. & Emmett, A. M. (eds) (1979–85). *The Moths and Butterflies of Great Britain and Ireland (3 vols)*. Harley Books: Colchester.

Emmett, A. M. (ed.) (1996). *The Moths and Butterflies of Great Britain and Ireland. Vol. 3*. Harley Books: Colchester.

Hering, E. M. (1951). *Biology of the Leaf Miners*. Junk: The Hague.

Hussey, N. W., Read, W. H. & Hesling, J. J. (1969). *The Pests of Protected Cultivation*. Edward Arnold: London.

Jeppson, L. R., Keiffer, H. H. & Baker, E. W. (1975). *Mites Injurious to Economic Plants*. University of California Press: Berkeley.

Kosztarab, M. & Kozár, F. (1988). *Scale Insects of Central Europe*. Dr W. Junk: Dordrecht.

Lane, A. (1984). *Bulb Pests*. MAFF Reference Book 51. HMSO: London.

Lorenz, H. & Kraus, M. (1957). *Die Larvalsystematik der Blattwespen*. Akademie Verlag: Berlin

Mound, L. A. & Halsey, S. H. (1978). *Whitefly of the World. A Systematic Catalogue of the Aleyrodidae (Homoptera) with Host Plant and Natural Enemy Data*. Wiley: New York.

Nalepa, A. (1929). Neuer Katalog der bisher beschriebenen Gallmilben ihrer Gallen und Wirtspflanzen. *Marcellia* **25:** 67–183.

Newstead, R. (1901–3). *Monograph of the Coccidae of the British Isles (2 vols)*. Ray Society: London.

Pape, H. (1964). *Krankheiten und Schädlinge der Zierpflanzen und ihre Bekämpfung*. Paul Parey: Berlin.

Pirone, P. P. (1978). *Diseases and Pests of Ornamental Plants*. Wiley: New York.

Pittaway, A. R. (1993). *The Hawkmoths of the Western Palaearctic*. Harley Books: Colchester.

Redfern, M. (2011). *Plant Galls*. Collins: London.

Reitter, E.(1908–16). *Fauna Germanica. Die Käfer des Deutschen Reiches (5 vols)*. Lutz: Stuttgart.

Schwenke, W. (ed.) (1972–82). *Die Forstschädlinge Europas (4 vols)*. Parey: Hamburg.

Seymour, P. R. (1989). *Invertebrates of economic importance in Britain. Common and scientific names*. MAFF Reference Book 406. HMSO: London

Spencer, K. A. (1973). *Agromyzidae (Diptera) of Economic Importance*. Junk: The Hague.

Strouts, R. G. & Winter, T. G. (1994). *Diagnosis of ill-health in trees*. Research for Amenity Trees No. 2. HMSO: London.

Wheeler, Jr. A. G. (2001). *Biology of the Plant Bugs (Hemiptera: Miridae)*. Cornell University Press: New York

Host Plant Index

General Index